A Concise Introduction to
Engineering Graphics

Including Worksheet Series A

Fifth Edition

Timothy J. Sexton, Professor
Department of Industrial Technology
Ohio University

ISBN: 978-1-63057-129-0

PLUS – Bonus Feature:

Technical Graphics

Meyers • Croft • Miller • Demel • Enders
Ohio State University

Publications

SDC Publications
P.O. Box 1334
Mission, KS 66222
913-262-2664
www.SDCpublications.com

Publisher: Stephen Schroff

A CONCISE INTRODUCTION TO ENGINEERING GRAPHICS

TECHNICAL GRAPHICS

Examination Copies:

Electronic Files:

Printed in the United States of America.

Table of Contents:

Table of Contents

Table of Problems:

Chapter 3
MULTIVIEW:

Chapter 4
MULTIVIEW / ISOMETRIC:

Chapter 6
AUXILIARY VIEWS:

Chapter 7
SECTIONING:

Chapter 8
MEASUREMENT and SCALE:

Chapter 9
DIMENSIONING:

GEOMETRIC CONSTRUCTION:

WORKING DRAWING:

PARAMETRIC MODEL:

Bonus Feature:

Technical Graphics

Meyers • Croft • Miller • Demel • Enders
Ohio State University

Table of Contents

CHAPTER 1 Getting Started

CHAPTER 2 Technical Sketching

CHAPTER 3 Orthographic Projection

CHAPTER 4 Pictorial Drawings

CHAPTER 5 Sections and Conventions

CHAPTER 6 Dimensions and Tolerances

CHAPTER 7 Dimensioning for Production

CHAPTER 8 Fastening, Joining, and Standard Parts

CHAPTER 9 Production Drawings

CHAPTER 10 Three-Dimensional Geometry Concepts

CHAPTER 11 3-D Geometry Applications

CHAPTER 12 Graphical Presentation of Data

CHAPTER 13 Design Process

APPENDIX Tables

Chapter 1

Sketching Techniques & Materials

Equipment:

Sketching is the most efficient way to illustrate an object or explain a process. Sketching can be done anywhere, anytime and needs no special equipment. All that is required is a 0.5mm and 0.7mm mechanical pencil with HB lead and a white vinyl eraser as illustrated in Figure STK 1. However, it is much easier to keep your sketches to scale by using blue lined grid paper. The most common blue lined grid paper has .25" squares as illustrated in Figure STK 2.

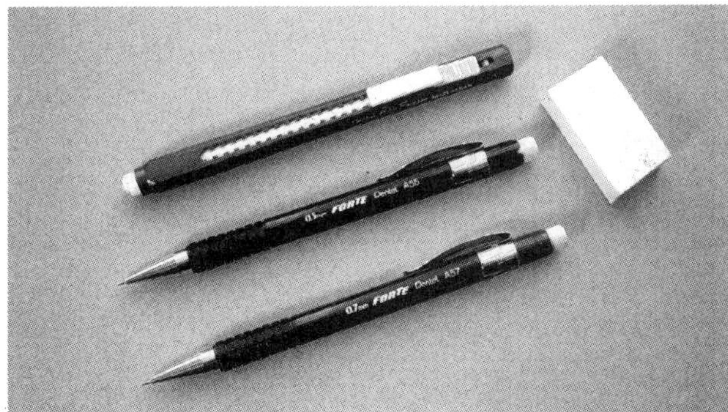

Figure SKT 1: 0.5mm and 0.7mm mechanical pencils with a block and pencil style vinyl erasers used for sketching.

Figure SKT 2: .25″ square blue lined grid paper used for sketching (lines do not appear blue in this figure).

Straight Lines:

The easiest way to draw straight lines is to use the "dot-to-dot" method. With this method you always look at your destination point because you hand will always follow your eye. Figure SKT 3 illustrates the four steps used in the dot-to-dot method:

1. mark the beginning and ending points,
2. place your pencil on the starting point,
3. look at the ending point, and
4. sketch the line while keeping your eyes on the end point.

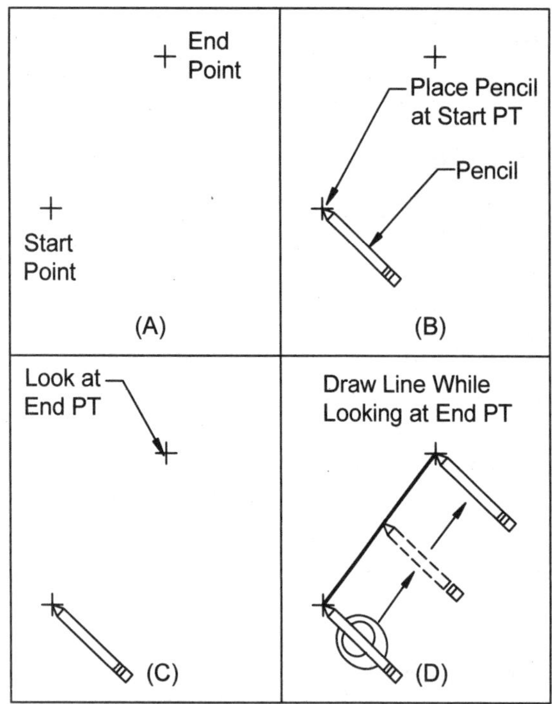

Figure SKT 3: Sketching a straight line using the "dot-to-dot" method.

When drawing straight lines it is easier to tilt the paper at different angles depending on the angle of the line being drawn. Figure STK 4 illustrates the technique used to draw horizontal, vertical, 30°, 45°, and 60° lines. In each illustration A-H you start at point 1 and follow the arrow to point 2. Note that Figure STK 4 is for right handed people. Left handed people need to rotate the paper in the opposite direction and change the direction of the pencil stroke.

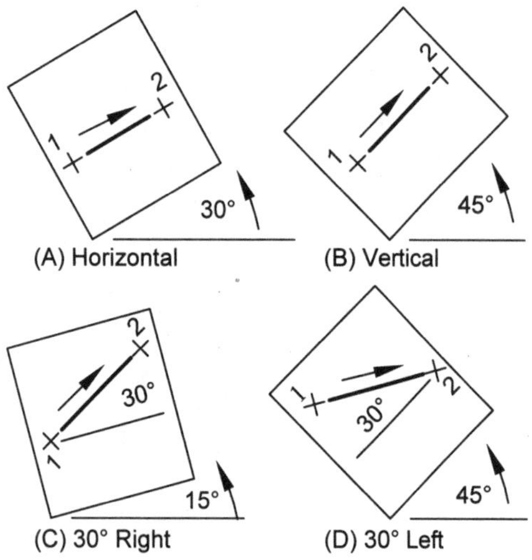

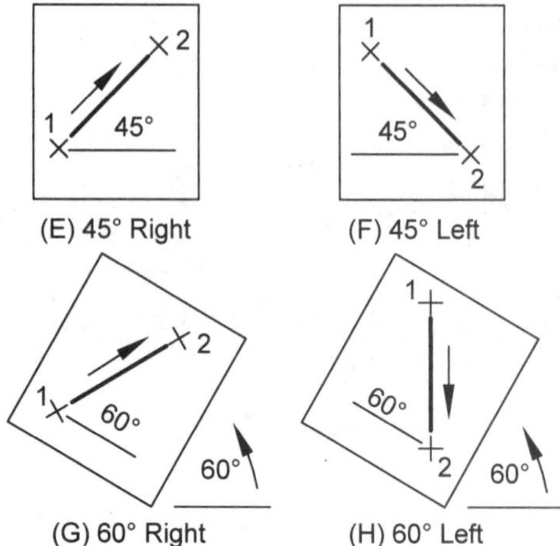

Figure SKT 4: Horizontal, vertical, and angled lines for right-handers. (Left-handers: rotate the paper in the opposite direction and reverse pencil stroke from 1 to 2.)

Sketching Arcs and Circles:

When sketching small arcs up to about 40mm (1.5in) your hand can stay stationary while it rotates along with the fingers forming the arc. Figure SKT 5 illustrates the steps to sketch an arc:

A. Layout the 90° corner of the arc and mark the points of tangency.

B. Sketch the arc from point of tangency to point of tangency.

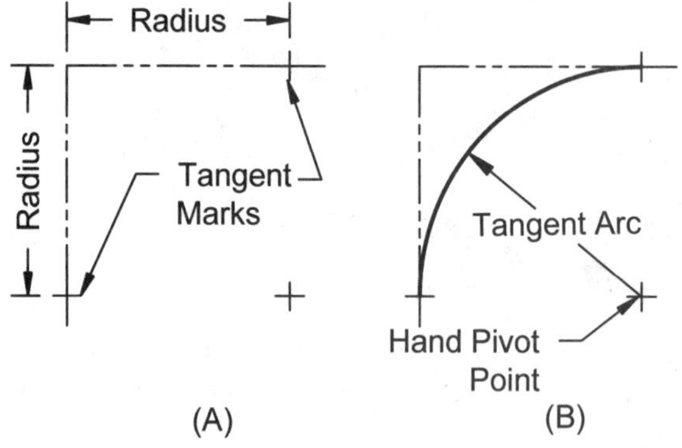

Figure SKT 5: Sketching small arcs A) 90° corner and tangent marks, B) sketched arc.

Figure SKT 6 illustrates the steps to sketch a circle:

A. Layout a square where the length of each side is equal to the circle's diameter. Mark the center point of the square and the points of tangency on each side of the square as illustrated in Figure SKT 6A.

B. Sketch an arc representing ¼ of the circle as illustrated in Figure SKT 6B.

C. Sketch the arc opposite the first arc as illustrated in Figure SKT 6C.

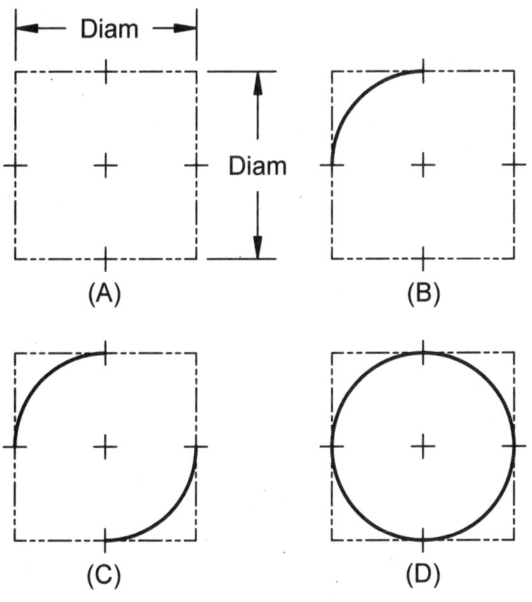

Figure SKT 6: Four steps to sketch a circle: A) square and tangent marks; B) ¼ circle; C) ¼ circle opposite first arc; D) fill in remaining two arcs.

Sketching to Scale:

When you sketch an object using grid paper, you can assign each small square a specific measurement. This way you can fit large objects on a small piece of paper or sketch small objects larger in order to better study its details. For example, if you want to lay out a floor plan, you need to shrink the size of the actual floor plan in order to fit it on a sheet of paper. You could assign each .25" square on the grid paper to represent 1 foot. Your sketch will be drawn at a scale of .25" = 1'-0". This is a common architectural scale used to draw floor plans. Sketching an object proportionally smaller or larger than the actual object is called sketching to scale. Scale is simply a ratio of how much larger or smaller the drawing is with respect to the actual object. Our example used the scale of .25" = 1'-0". This scale has the ratio of 1 = 48 (there are 48 quarter inches in one foot). If you decide to let the .25" square represent 1", the scale would be .25" = 1" or a ratio of 1 = 4. Similarly, if a .25" square represented .12", the scale would be .25" = .12" or a ratio of 2 = 1.

Drafting Materials:

This text concentrates on engineering graphics emphasizing sketching and computer graphics. However, your sketches may lead to renderings using shading or you may decide that your sketches need more clarity by using very thin or very thick lines. To render sketches you need different mechanical pencils and different types of lead. Mechanical pencils hold four different diameter leads 0.3mm, 0.5mm, 0.7mm, and 0.9mm. Lead Diameters 0.5 and 0.7 work for most sketching. The 0.3mm can help clarify small complex details or lessen the importance of a line by being so thin. The 0.9mm makes a bold thick line and helps with shading a rendering.

Paper:

It is easiest to sketch on blue lined graph paper that is available with the inch being divided into 4, 5, 6, 8, and 10 divisions per inch. Blue grid paper is also available in an isometric grid (isometric drawing is discussed in a later chapter).

Manual drafting has been traditionally done on either 100% cotton vellum (often mistakenly referred to as tracing paper) or plastic drafting film. Table ST-1 lists the available paper sizes with the first two columns listing inch based sizes and the third column metric based sizes.

Table ST-1
Paper Sizes

	Standard Inches	(Architect) Inches		Metric
A	8.5″ x 11″	9″ x 12″	**A4**	210 x 297
B	11″ x 17″	12″ x 18″	**A3**	297 x 420
C	17″ x 22″	18″ x 24″	**A2**	420 x 594
D	22″ X 34″	24″ X 36″	**A1**	594 x 841
E	34″ x 44″	36″ x 48″	A0	841 x 1188

Pencil lead:

Pencil lead is available in eighteen different degrees of hardness from soft to hard (7B, 6B, 5B, 4B, 3B, 2B, B, HB, F, H, 2H, 3H, 4H, 5H, 6H, 7H, 8H and 9H):

7B, 6B, 5B, 4B, 3B, 2B, B are very soft with 7B being the softest and are good for shading and artistic renderings.

HB is the most popular general purpose lead and it is best for sketching. It is equivalent to a #2 pencil.

The midrange HB, F, H, 2H, and 3H is the degrees of hardness used for manual drafting.

4H, 5H, 6H, 7H, 8H and 9H are used for drafting when extremely thin lines are required.

Chapter 2

Hand Lettering

Lines on a drawing describe the shape of an object. But to specify the size of the object and clarify special details, dimensions and notes are required. Clarity when lettering numbers and text is vital to the clear interpretation of your drawing or sketch. Traditionally only uppercase letters are used. Some computer generated drawings are beginning to use both upper and lower case letters. But standards call for all uppercase lettering.

This text illustrates only the most common style of lettering called single stroke gothic. In this style each stroke of your pencil is equal in width, i.e., “single stroke” and the letters are sans serif, meaning without extensions or tails, i.e., “Gothic.” Notes and dimensions are normally 3mm (.12") in height. Be sure to keep all lettering uniform in height by using guidelines which are lines drawn with a straight edge the height of the lettering. Guide lines are similar to the blue lines on notebook paper. Titles and page numbers can be up to 6mm (.25") in height. The width of letters varies depending on the individual letter, for example the letter “I” is narrower than the almost square letter “M.” In general the width of lettering should be a factor of .75 to 1 times the height of the lettering. CAD programs refer to this height to width ratio as a width factor or an aspect ratio. Just remember that wide or “fat” lettering is easier to read than tall narrow lettering. Never crowd your letters leaving about two letters width between words. If your note has multiple rows, double space so the space between the rows is equal to the height of the lettering. The space between notes should be triple spaced if possible to improve the clarity of each note. Your lettering and the spacing of your lettering will improve as you practice. So practice, practice, practice!

Figure LET 1 illustrates how letters and numbers are formed by showing the order and direction of the pencil strokes. The grid background provides a guide for determining letter and number heights and widths. Note that the fractions go slightly above and below the guidelines. Architectural drawings often use stylized lettering unique to the individual draftsman. Figure LET 2 illustrates two examples of unique lettering styles. Figure LET 3 illustrates lettering done with a wide tip pen.

Figure LET 1: Directions on how to letter text and numbers. The numbers and arrows indicate the direction and order of the strokes.

ABCDEFGH
IJKLMNOP
QRSTUVW
XYZ
1234567890

ABCDEFGHIJK
LMNOPQRSTU
VWXYZ
1234567890

Figure LET 2: Two examples of unique stylized lettering.

ABCDEFGHI
JKLMNOPQ
RSTUVWXY
Z
12345678
910

Figure LET 3: Lettering done with a wide tipped pen.

Chapter 3

Multiview Drawing

The Engine of Industry:

Multiview drawings are the most common type of technical drawings for both the machine and architectural industries. As the name implies, a *multiview drawing* is "multiple" views of a single object. For example, Figure MVD 1 is a multiview sketch of a table lamp showing the lamp's front, top, and right side views. A single view of the lamp would not describe its exact shape. You need at least these three views to envision its shape.

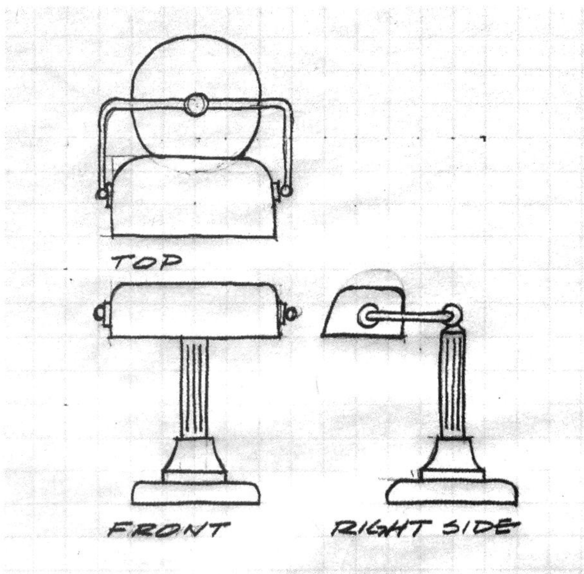

Figure MVD 1: Multiview sketch showing the front, top, and right views of a table lamp.

Normally when people refer to "blue prints" they are not referring to traditional drawing that have white lines on a blue background. They are referring to any technical drawing. These technical drawings are multiview drawings and they are the engine that runs industry because nothing is manufactured or constructed without first making multiview drawings.

Orthographic Projection:

Orthographic projection is the projection theory used to make multiview drawings. The easiest way to understand orthographic projection is to envision an object placed inside a rectangular box made of sheets of plastic as illustrated in Figure MVD2. The arrows illustrate viewing positions of the six principle orthographic views.

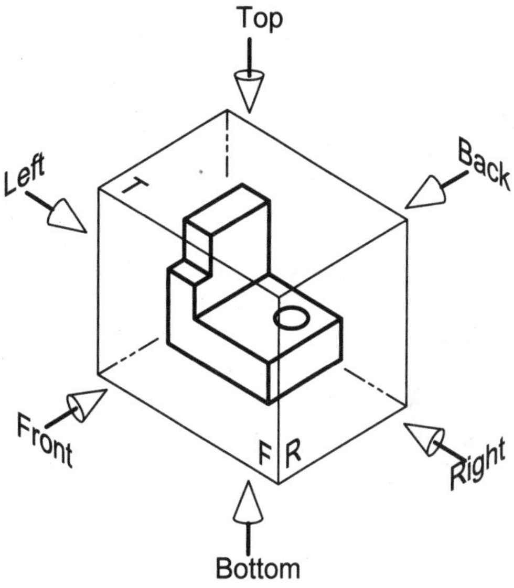

Figure MVD 2: Object encased in a clear plastic box with arrows locating the viewing position of the six principle orthographic views.

Each of its six clear plastic sides is called a *plane of projection*. Each side of the object is projected onto one of the six planes of projection. For example, in Figure MVD 3 the front view of the object is projected onto the front plane of projection with lines called *projection rays*.

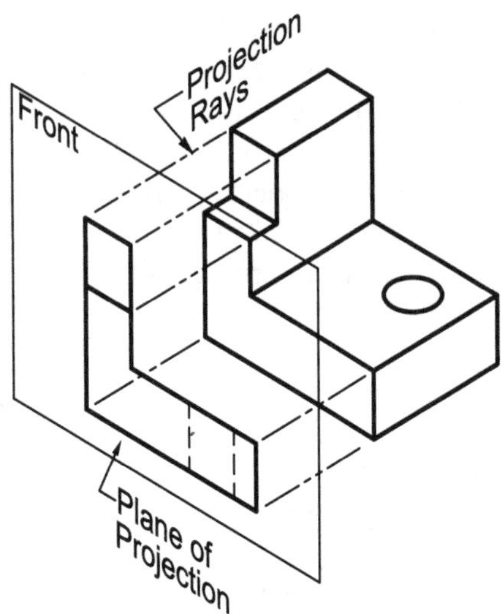

Figure MVD 3: Projection rays projecting the front view of the object onto the front plane of projection.

Projection rays are parallel to each other and perpendicular to the plane of projection. The resulting front orthographic view is illustrated in Figure MVD 4.

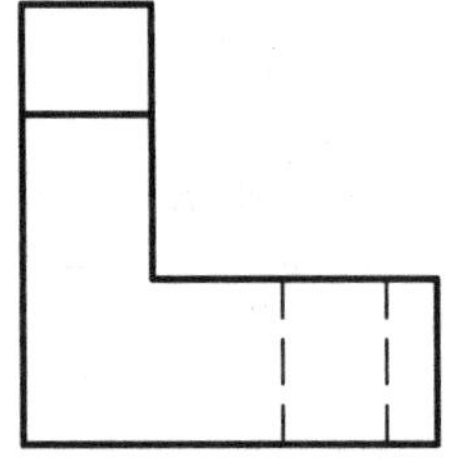

Figure MVD 4: Front orthographic view as projected onto the front plane of projection.

To obtain another view of the object, the viewer must either change their viewing position or rotate the object until the desired view is parallel to the plastic plane you are viewing the object through.

The Six Plastic Planes of Projection:

The top, front, right, left, bottom, and back make up the *six principal orthographic views* of an object. In order to describe a 3-D object on a 2-D piece of paper, the six principle orthographic views must be unfolded into a single plane. Figure MVD 5 illustrates the plastic box unfolding with the views being hinged about the front view. Figure MVD 6 illustrates the six principle projection planes completely unfolded into a single plane. The result is the correct layout of the six principle orthographic views. Note that the projectability of views is maintained. This means the top, front, and bottom views line up and the front, right, left, and back views line up. Projectability of views must ALWAYS be maintained when sketching, drafting, or laying out views in CAD.

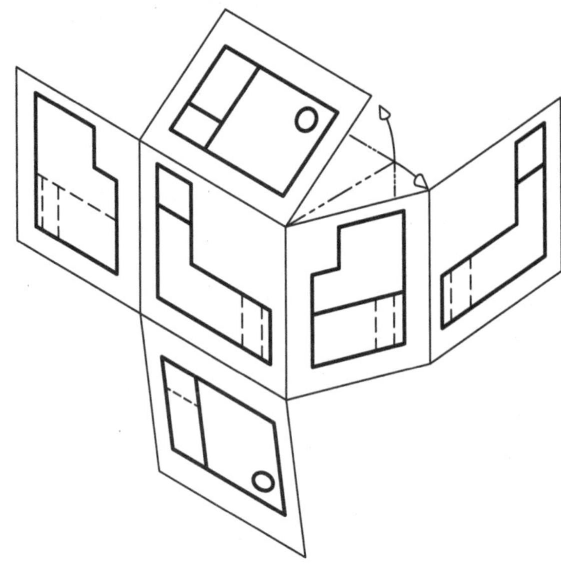

Figure MVD 5: All six planes of projection being unfolded about the front plane.

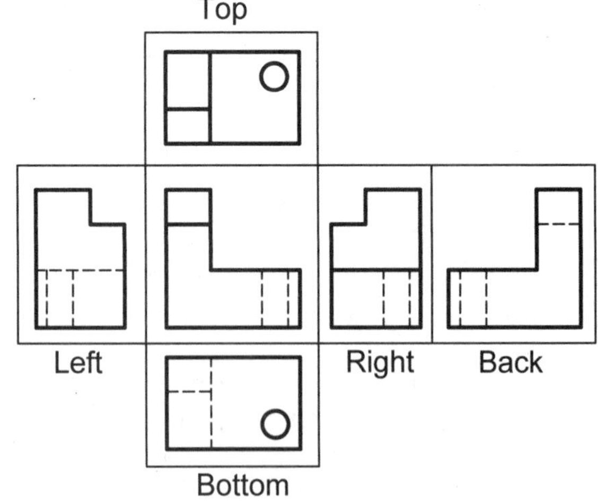

Figure MVD 6: All six principal sides of the plastic box unfolded into the 2-D plane of a piece of paper.

Figure MVD 7 illustrates the relationship of height, width, and depth between the views. Note that each individual view only has two of the three dimensions height, width, or depth (thus 2-D drawing).

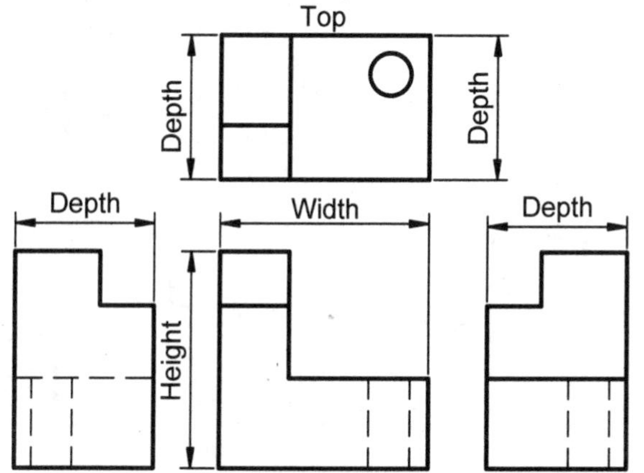

Figure MVD 7: Relationship between height, width, and depth.

Figure MVD 8 provides a look of how a 45° miter line is used to transfer depth between the top and right views. The 45° miter line starts at the intersection of lines 1 and 2. Lines 1 and 2 radiate from the edge views (EV) of the frontal planes in the top and right views. Line 3 helps transfer the small notch from the right to the top view. Line 4, 5, and 6 help transfer the hole from the top to the right view. Using the 45 ° miter line makes it both faster to construct a multiview drawing and allows you to check yourself with respect to the right/top and left/top relationships.

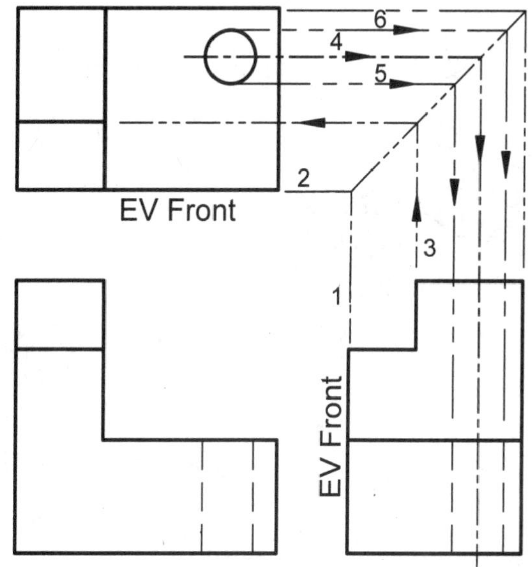

Figure MVD 8: The 45° miter line is used to transfer details between the top and right views.

Even though each orthographic view in a multiview is only 2-D, it is helpful to envision each view as if it were 3-D. Figure MVD 9 illustrates the six principle views of the 3-D letter "L". Planes in or parallel to a principle plane of projection are shaded while planes perpendicular to a principle plane of projection are labeled EV for edge view. Note that the planes labeled EV appear as lines in their respective planes of projection.

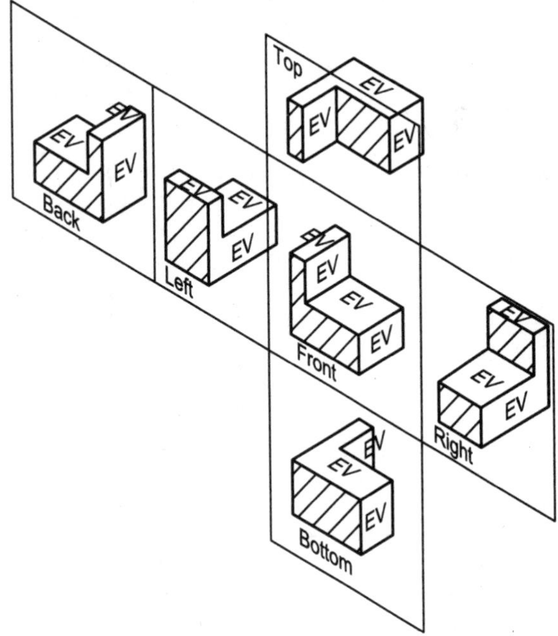

Figure MVD 9: The 3-D letter "L" with planes in or parallel to a principle plane shaded and planes perpendicular to a principle plane labeled EV.

2-D Planes:

Each orthographic view of a multiview drawing is two dimensional. For example, if you look at just the front view of Figure MVD 10, you cannot tell that plane A is the front most plane while B and C are further back. Similarly, if you look at the top view of Figure MVD 10, you cannot tell that plane D is higher than E, and if you look at the right view, you cannot tell that plane F is closer than plane G.

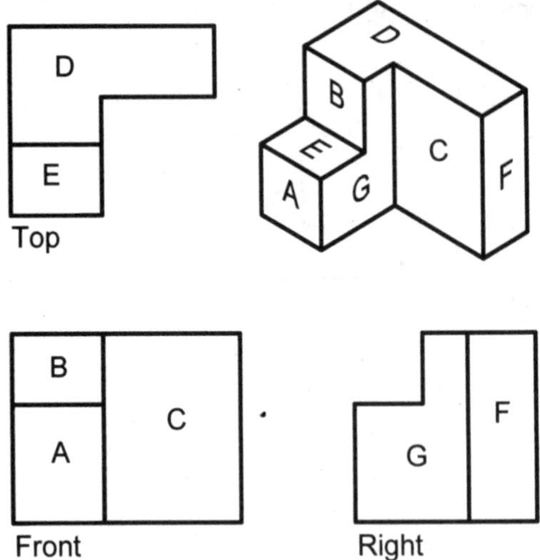

Figure MVD10: No way of telling which plane is closer in the 2-D principal views.

Types of Planes:

The examples and exercises in this text are limited to: normal, inclined, oblique, and single curved surfaces. Each of these four will be discussed and illustrated in this text. Double curved and warped surfaces are considered too advanced for this text.

Normal Planes:

A *normal plane* is parallel to two principle planes and projects as an edge view in the other four. When a plane is parallel to a principle plane, it projects its true size and shape (TSS). For example, the normal surface "A" in Figure MVD 11 projects its TSS in the front view and projects as an edge view (line) in both the top and the right views.

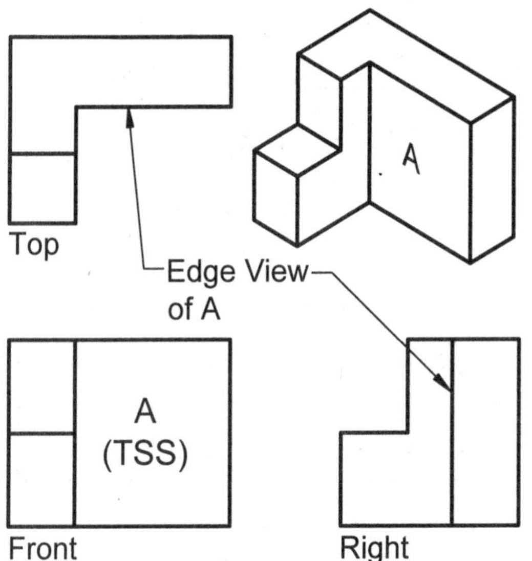

Figure MVD11: Normal plane "A" is true size and shape in the front view and is in edge view in the top and right views.

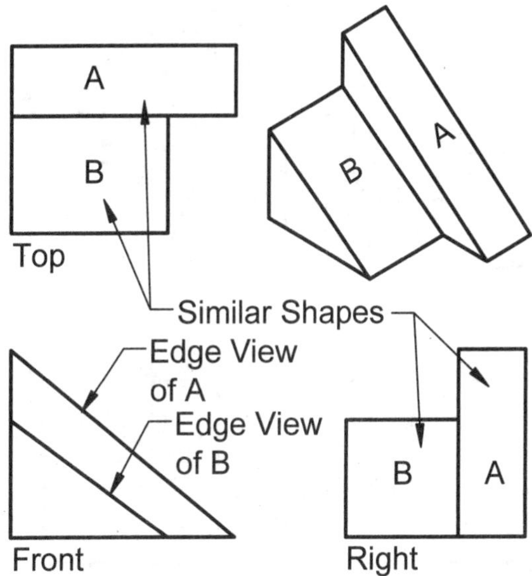

Figure MVD 12: Inclined plane "B" is in edge view in the front view and similar though foreshortened shapes in the top and right views.

Inclined Planes:

An *inclined plane* is not parallel to any of the principle planes. Therefore, it does not show its true size and shape in any of the principle views. An inclined plane will be perpendicular to two principle planes and display as a similar shape in the other four. For example, plane B in Figure MVD 12 displays as an edge view in the front view, because plane B is perpendicular to the front plane of projection and as similar foreshortened rectilinear shapes in both the top and right views.

Oblique Planes:

An *oblique plane* is not parallel nor is it perpendicular to any of the six principle planes of projection. Therefore, it does not display its true size and shape nor as an edge view in any view. An oblique plane shows the same number of sides and the same number of corners in all six principle views as plane "C" illustrates in Figure MVD 13.

Oblique planes can be difficult to draw unless the following rules are understood and applied:

1. parallel lines are ALWAYS parallel; and
2. when parallel planes are sliced at an angle by another plane, the resulting lines of intersection are parallel.

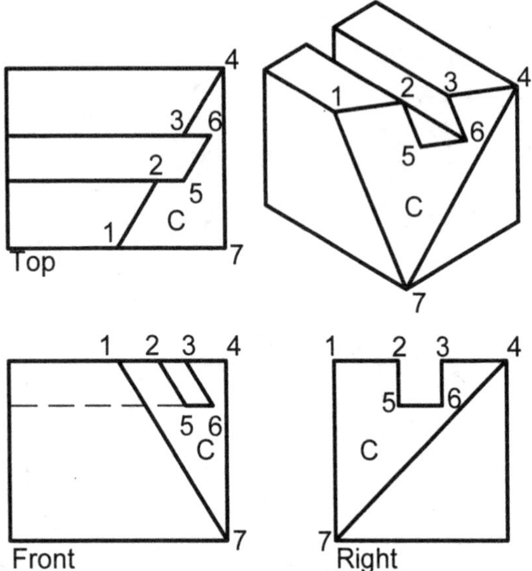

Figure MVD 13: Oblique plane "C" with the same number of sides and corners in all views.

Single Curved Surfaces:

A *single curved surface* is formed by extruding (stretching out) a circle, an arc, or an irregular curve or by having a line element rotate about a central axis following a circular or arc path. Figure MVD 14 illustrates an extruded circle and two extruded arcs. An extruded circle forms either a hole, or a solid cylinder like a coffee can. An extruded arc forms a smooth curved surface such as the rolled front edge of many kitchen counters. An extruded irregular curve forms a rolling curved surface like a child's slide found at a playground. The front view in Figure MVD 14 displays the EV of two arcs and one circle. The top view displays the edge lines of the cylindrical hole and the surface view of arc A. The right view displays the edge line of the two arcs A and B, the surface of arc A, and the edges of a cylindrical hole.

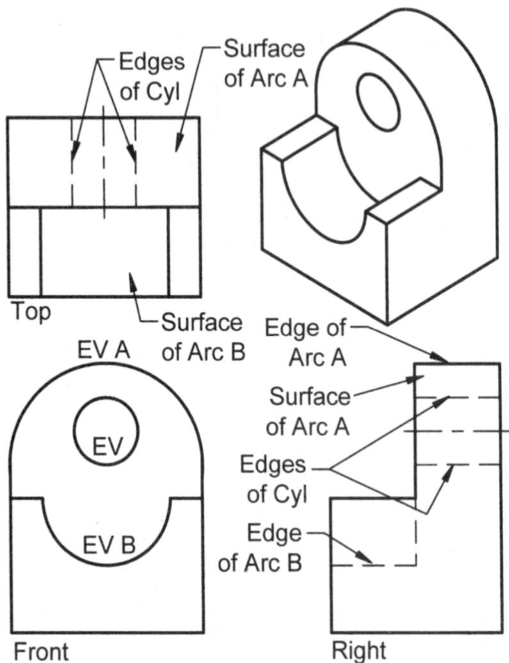

Figure MVD 14: Single curved surfaces display as edge views in the front view and as lines and surfaces in the top and right views.

Line and Surfaces:

This text concentrates on flat and single curved surfaces. It is easier to visualize an object from multiview drawings if you know how lines, arcs and circles are formed. Refer to Figure MVD 15 when visualizing the ways geometry is formed.

A straight line is formed when:

1. viewing the EV of a flat plane
2. two flat planes intersect
3. viewing the edge lines of a cylinder, cone, torus (doughnut), or extruded arc

An arc or circle is formed when:

1. viewing the edge view of a curved surface
2. looking down into a hole
3. viewing the end of a cylinder

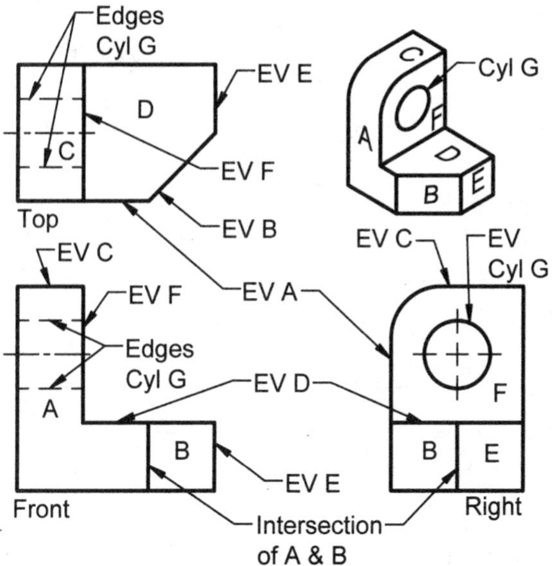

Figure MVD 15: Lines are formed by the intersection of planes, EV of planes, and EV of arcs or cylinders.

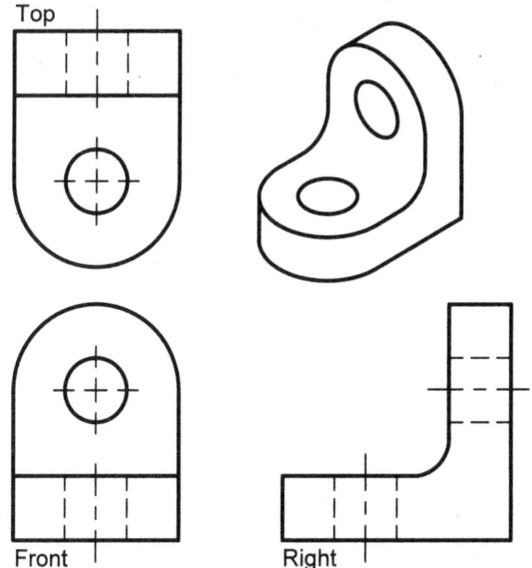

Figure MVD 16: Circles and arcs in normal planes.

Circles vs. Ellipses:

Holes and solid cylinders appear as circles when they are on normal surfaces as illustrated in Figure MVD 16. But when a hole passes through an inclined plane it appears as ellipse as illustrated in Figure MVD 17.

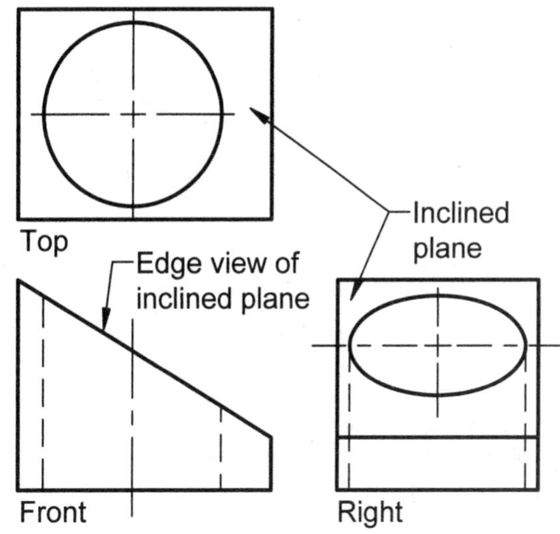

Figure MVD 17: A hole appears as an ellipse when in an inclined plane.

When a circle is rotated about an axis it changes into an ellipse. The shape of the ellipse depends on how much the ellipse is rotated. Figure MVD 18 shows a circle being rotated about a horizontal axis. When the circle is in the 0° position it appears as an edge view (line) in the front view. A full 90° rotation produces a true circle in the front view. A 30° or 60° rotation of the circle produces an ellipse. The flatness of the ellipse depends on how much it is rotated. A 30° rotation produces an ellipse that is flatter than the ellipse produced by rotating the circle 60°. Note that the length of the ellipses major axis AB in the front view of Figure MVD 18 does not change. It remains equal to the circle's original diameter. But the minor axis CD of the 30° rotated circle is shorter than the minor axis EF of the 60° rotated circle.

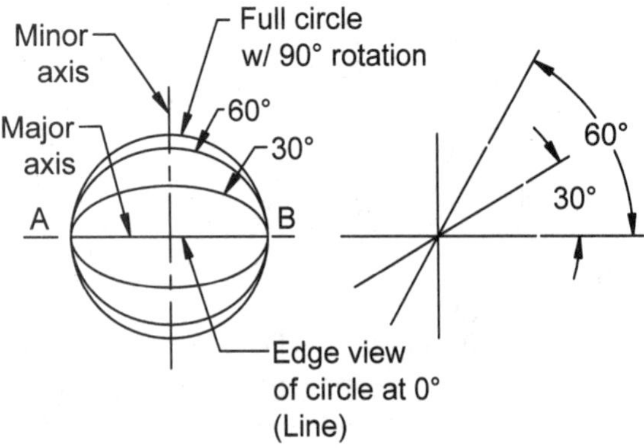

Figure MVD 18: The rotating circle starts as a line then forms two different ellipses at 30° and 60° and finally a full circle.

Fillets and Rounds:

A fillet is an internal rounding between intersecting surfaces. A round is an external rounding between intersecting surfaces as illustrated in Figure MVD 19. Fillets can also be formed at the intersection of planar and cylindrical surfaces as illustrated in Figure MVD 20.

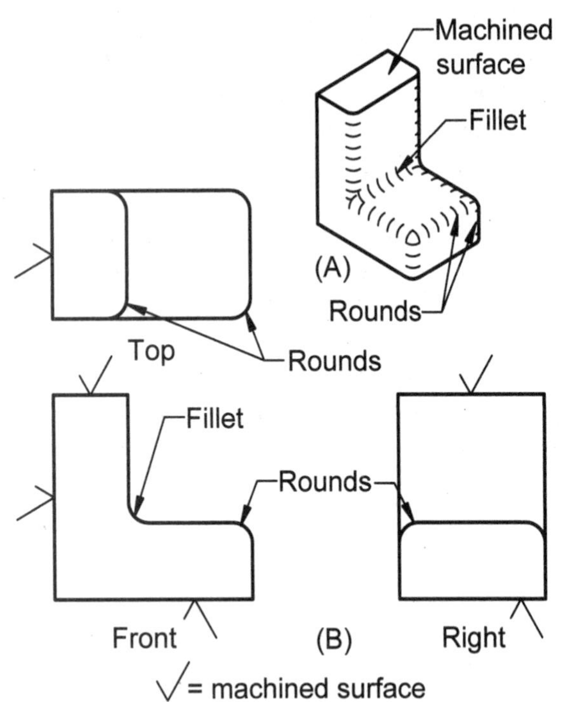

Figure MVD 19: A fillet is an internal rounding between intersecting surfaces while a round is an external rounding between intersecting surfaces.

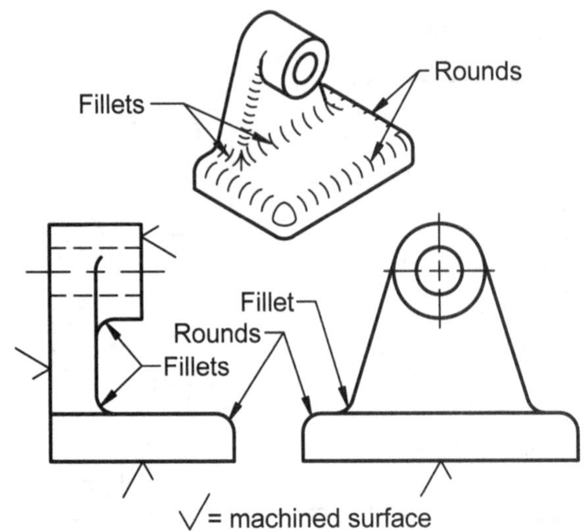

Figure MVD 20: Fillets can be formed between planar and cylindrical surfaces.

An example familiar to almost everyone is the fillets and rounds on a kitchen counter as illustrated in Figure MVD 21. A sheet of plastic laminate is heated slightly to make it pliable. Then it is glued to a wooden form which already has the fillets and rounds formed in it. The front edge of the counter and the top of the backsplash form rounds while the intersection of the main counter and the vertical backsplash forms a fillet.

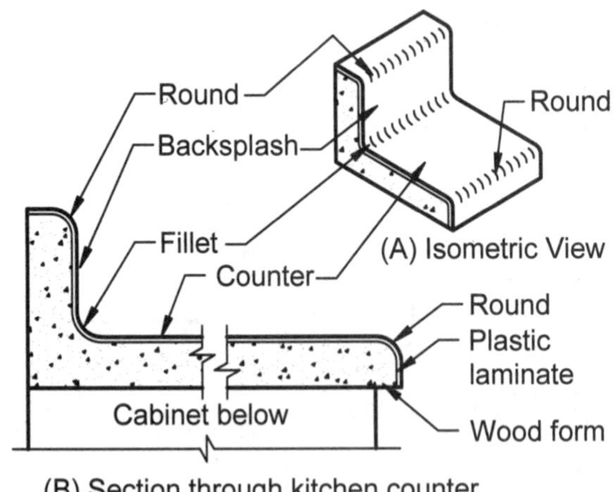

Figure MVD 21: A kitchen counter is a familiar example of the use of fillets and rounds.

Fillets and Rounds in Castings:

Fillets and rounds are formed using a variety of manufacturing processes. Sand casting is one of the most common processes, e.g., automobile engine blocks are made using the sand casting method. Molten metal is poured into a cavity shaped by a pattern. The molten metal assumes the shape of the cavity, less slight shrinkage, and the casting is removed from the sand once it is solidified. The pattern is typically made up of two halves split down the center along its central plane called the parting line as illustrated in Figure MVD 22. To make it easier to remove, the pattern is slightly tapered (e.g., 7°) getting smaller as it moves away from the parting line into the sand as illustrated in Figure MVD 22. This tapering is called draft or the draft angle. Fillets and rounds are used on patterns for the purpose of: appearance, softening edges, increasing strength at interior intersecting planes, and because it is difficult to obtain sharp corners in sand. Since the pattern is surrounded by sand, the product's final surface will be rough. Only a machining operation will smooth the surface and form sharp edges.

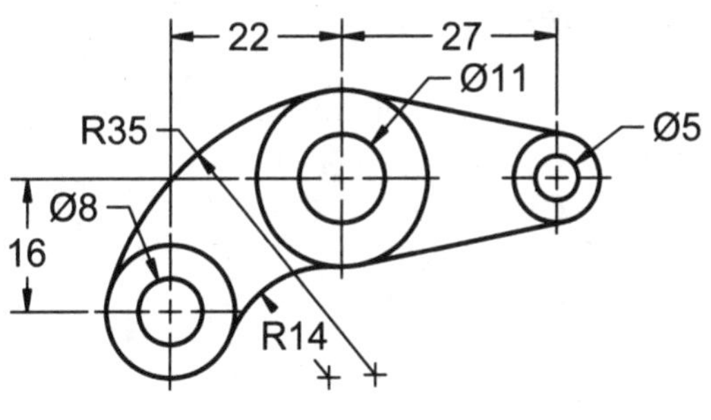

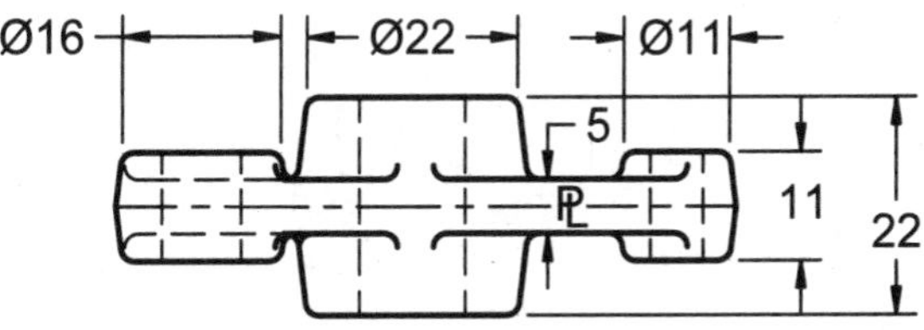

ALL FILLETS & ROUNDS R1

Figure MVD 22: Drawing of a part to be cast showing the parting line, draft angle, and fillets/rounds.

The sand casting process follows these steps and is illustrated by Figure MVD 23:

1. A match plate is placed on a surface supported by its ends leaving the cope split pattern face hanging down and the drag split pattern facing up. Besides the object's pattern, the drag side of the split pattern has additional forms which make gate channels in the sand. Gate channels carry the molten metal from the sprue to the cavity and a second gate channel carries the molten metal from the cavity to the riser.
2. A drag is placed around the pattern with its alignment pins going through the holes in the match plate then sand is compacted around the pattern.
3. A molding board is placed on top of the drag and the entire assembly is turned over.
4. A cope is positioned around the upward facing pattern using its alignment holes and the drag's alignment pins. Hollow tubes (slightly funnel shaped to help with removal) are strategically positioned so the pattern's cavity will fill properly with molten metal. One tube called the sprue receives the molten metal and the other tube called the riser receives the excess molten metal. The cope is then filled with compacted sand.
5. The cope and drag are separated from the match plate leaving cavities in both the cope and drag and then the match plate is set aside. The cope is then repositioned on top of the drag using alignment pins to ensure the cavities align.
6. Molten metal is poured into the sprue filling the pattern cavity by way of the gate channel and the excess metal flows through the opposite gate channel and up into the riser.
7. Once solidified the casting is removed from the sand.

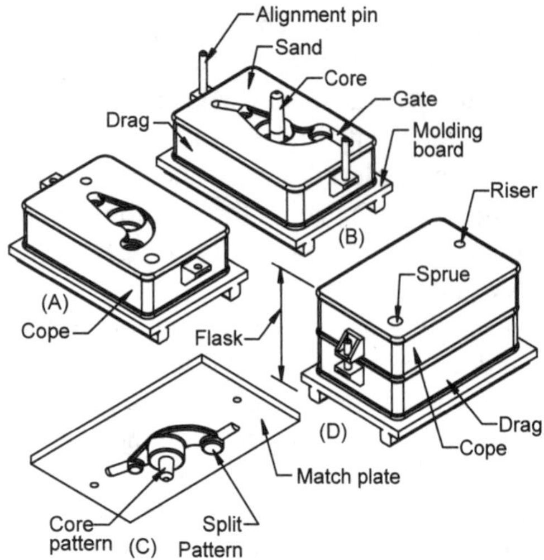

Figure MVD 23: A) The cope, compacted sand, and split pattern's cavity; B) the molding board, drag, compacted sand, and split pattern's cavity; C) match plate with a split pattern on both sides; D) illustrates all the components assembled and the holes for the sprue and riser showing. The flask is made up of the cope, drag, and the sand cavity in each (does not include the match plate).

Forming Fillets & Round Using Bended Metal:

The bending of sheet metal also forms fillets and rounds. For example, when a piece of sheet metal is bent at a 90° angle, the exterior of the bend, which is under tension, forms a round while the interior of the bend, which is in compression, forms a fillet. The radial axis where the metal changes from compression to tension is called the neutral axis as illustrated in Figure MVD 24. The neutral axis is normally assumed to be 0.44 of the thickness from the inside of the metal. To calculate the sheet metal's "stretch-out" or length of the metal before it is bent, add up the distances a+b+c where c is the length along the neutral axis as illustrated in Figure MVD 24. The distance along the neutral axis is called bend allowance. It is calculated using tables based on the thickness of the metal and the angle of the bend. These tables can be found in resources such as the *Machinery's Handbook*. When dimensioning objects with bends the dimension is measured to the inside mold line (IML) or outside mold line (OML) as illustrated in Figure MVD 24. A mold line is developed by extending the surfaces adjacent to the bend.

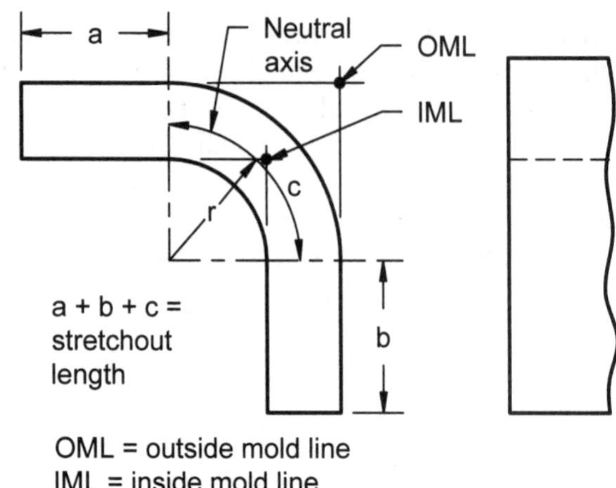

Figure MVD 24: Bent metal forms both a fillet (in tension) and a round (in compression). The total length of a piece of metal stock before bending is called its stretch-out length and is equal to a + b + c where c is determined using tables based on the metals thickness and degree of bend.

3-D Parametric Models:

When creating a 3-D parametric model, the basic geometry, without fillets and rounds, is created first as illustrated in Figure MVD 25A. Then fillets and rounds are generated using a separate operation. Figure MVD 25B illustrates the model after the filleting operation has been applied.

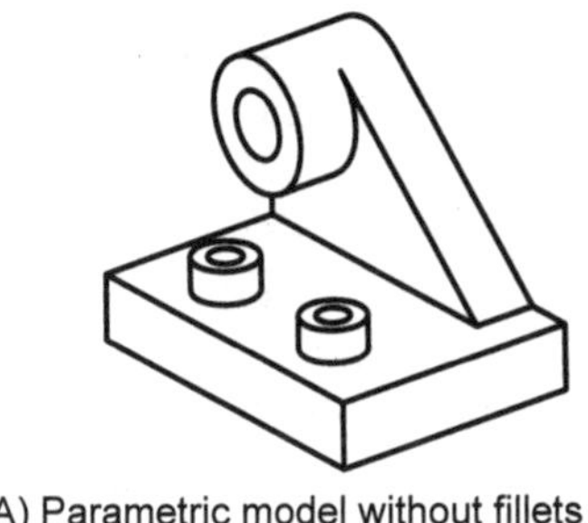

(A) Parametric model without fillets & rounds

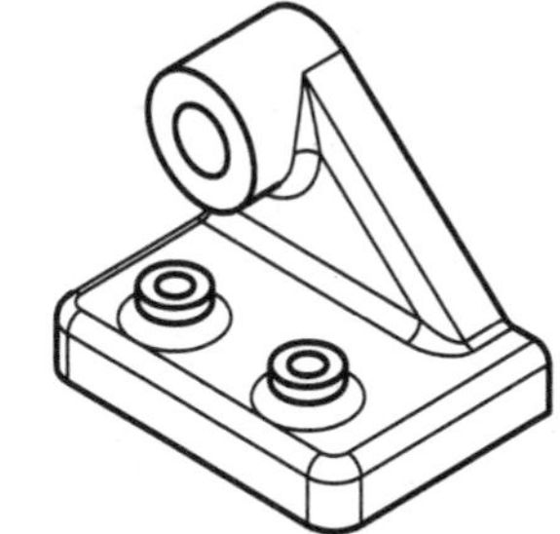

(B) Parametric model with fillets & rounds

Figure MVD 25: A) Illustrates a basic 3-D parametric model; B) illustrates the model after the filleting operation.

Intersecting Surfaces Using Fillets, Rounds, and Runouts:

A runout is a method for representing fillets and/or rounds at the intersection of planar surfaces or at the intersection of a cylindrical or spherical shaped surface and planar surface. Figure MVD 26 illustrates examples of four cylinders and rectangular prisms intersecting and the resulting fillets, rounds, and runouts. Figures MVD 26A, 26B, 26E, and 26F illustrate intersecting cylinders and planar surfaces without fillets and rounds. Figures MVD 26C, 26D, 26G and 26H illustrate intersecting cylinders and planar surfaces with fillets and rounds. Figure MVD 26C illustrates a runout when the planar surface is tangent to the cylinder; Figure MVD 26D illustrates a runout when the planar surface intersects the cylinder and is not tangent; Figure MVD 26G illustrates a runout when the planar surface is tangent beyond the cylinder's center line; and Figure MVD 26H illustrates a runout when the planar surface is tangent before the cylinder's center line. The location of the end of a runout is located by finding its point of tangency in the top view and projecting it into the front view as illustrated in Figure MVD 26C, 26G, and 26H. The point of tangency is located by drawing a line through the center of the circle and perpendicular to the line representing the edge of the planar surfaces. The radius of the runout is the same as the fillet's radius and its length is one eighth of the radius' circle.

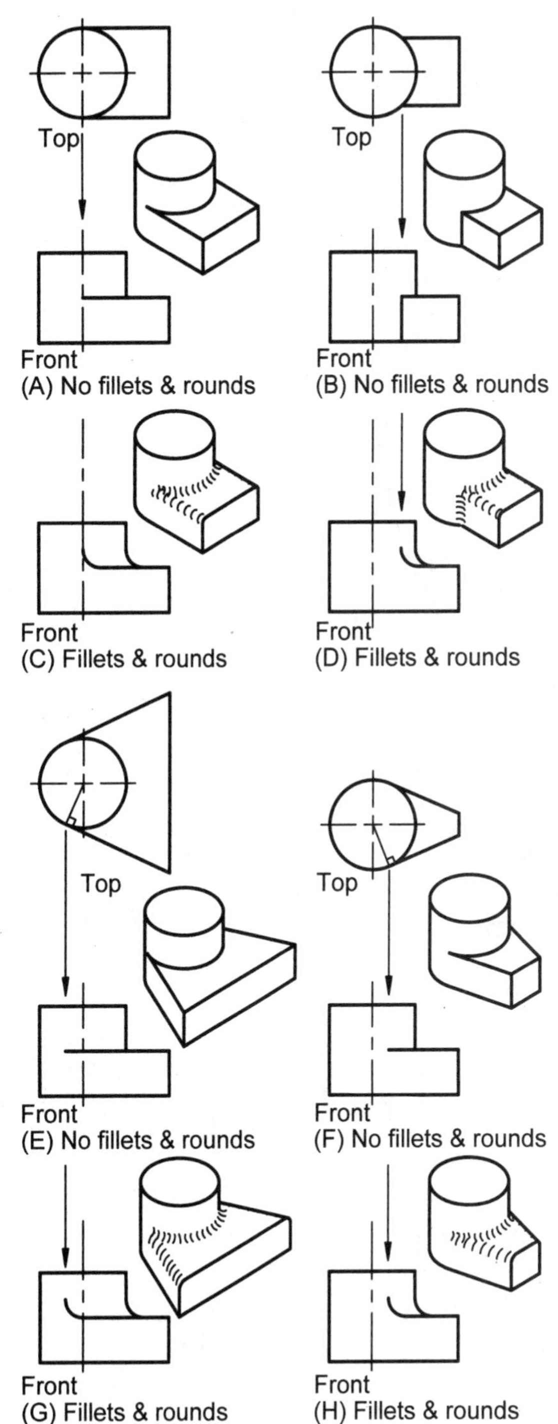

Figure MVD 26: Examples of intersecting cylinders and rectangular prisms with and without fillets and rounds and an illustration of how to locate the planar surface's point of tangency which locates the end of the runout.

Rounded Shapes Intersecting Cylinders:

Figure MVD 27 illustrates runouts when four different rounded shapes intersect a cylinder. Figure MVD 27A illustrates a rectangular cross section with rounded edges intersecting a cylinder; Figure MVD 27B illustrates a smaller cylinder intersecting a larger cylinder; Figure MVD 27C illustrates an elliptical cross section intersecting a cylinder; and Figure MVD 27D illustrates a rectangular prism with rounded ends and intersecting at a location that is tangent beyond the cylinder's center line. The location of the end of a runout is located by finding its point of tangency in the top view and projecting it into the front view as illustrated in Figure MVD 27A, 27B, and 27D. The point of tangency is located by drawing a line through the center of the circle and perpendicular to the line representing the edge of the planar or cylindrical surface.

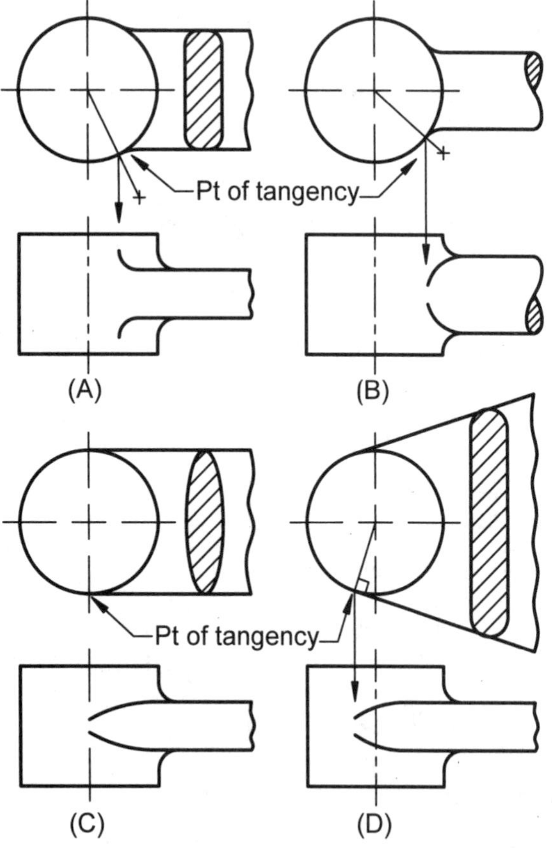

Figure MVD 27: Examples of runouts when rounded shapes intersect cylindrical shapes.

Intersecting Planar Surfaces:

Figure MVD 28 illustrates a variety of situations where filleted and rounded surfaces intersect. In Figure MVD 28A two small radius fillets intersect. Their intersection is shown using a 45° line the length of the fillet's radius. Figure MVD 28B illustrates a similar object when the fillets are larger. Figure MVD 28C and MVD 28D provide additional examples.

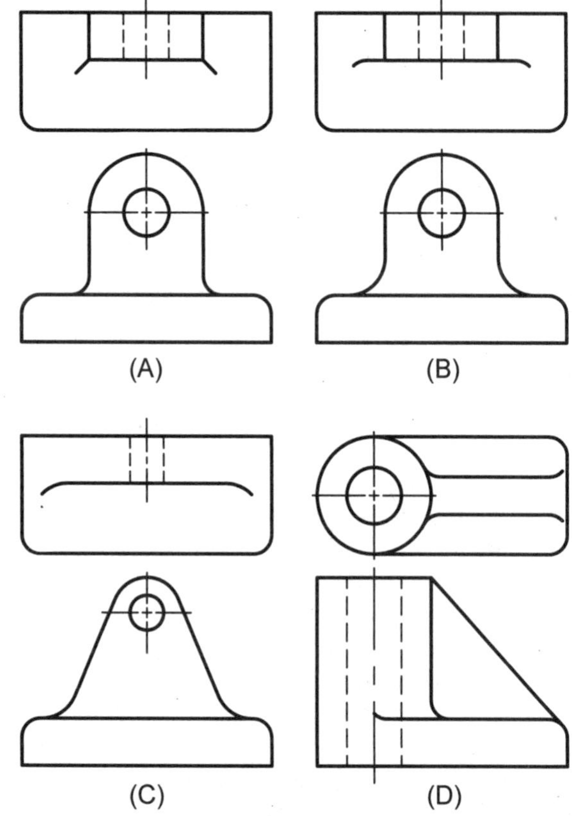

Figure MVD 28: Additional example of runouts.

Angle Blocks:

An angle block is used when a vertical surface perpendicular to a machining table is required. Figure MVD 29A illustrates how fillets and rounds appear when the angled support has a rectangular shape with small rounded edges. Figure MVD 29B illustrates how fillets and rounds appear when the angled support is rounded.

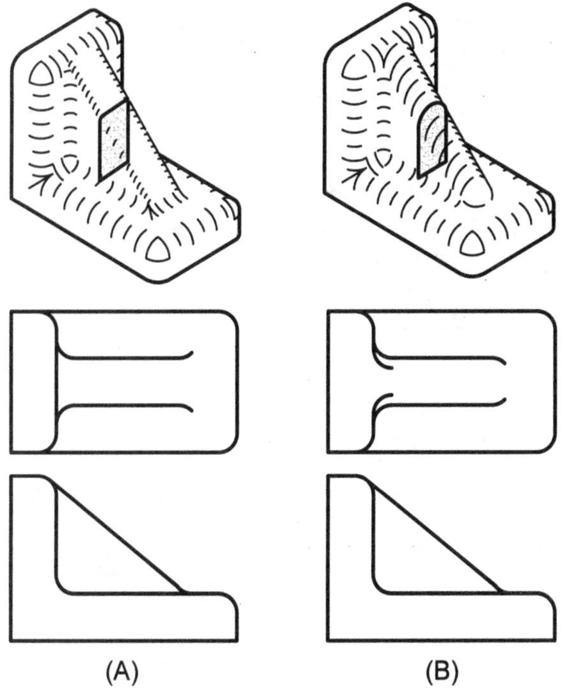

Figure MVD 29: Illustration of runouts (A) when the angled rib is square with rounded edges and (B) when it is fully rounded.

Conventional Edges:

One way a line is formed is when two surfaces intersect. But when fillets and rounds are applied, the edges are softened and no sharp edge line appears. In Figure MVD 30A and MVD 30B the top most view provides a strict interpretation of drawing rules. But these views are misleading. To better describe the shape of either object, conventional practices are applied. The center views of Figure MVD 30A and MVD 30B apply conventional practice by adding phantom lines to help delineated shape. The phantom lines are located where regular object lines would be if the fillets and rounds were removed. Figure MVD 30C illustrates the top, front and right side of a plastic scoop with phantom lines in the top view to better delineate its shape.

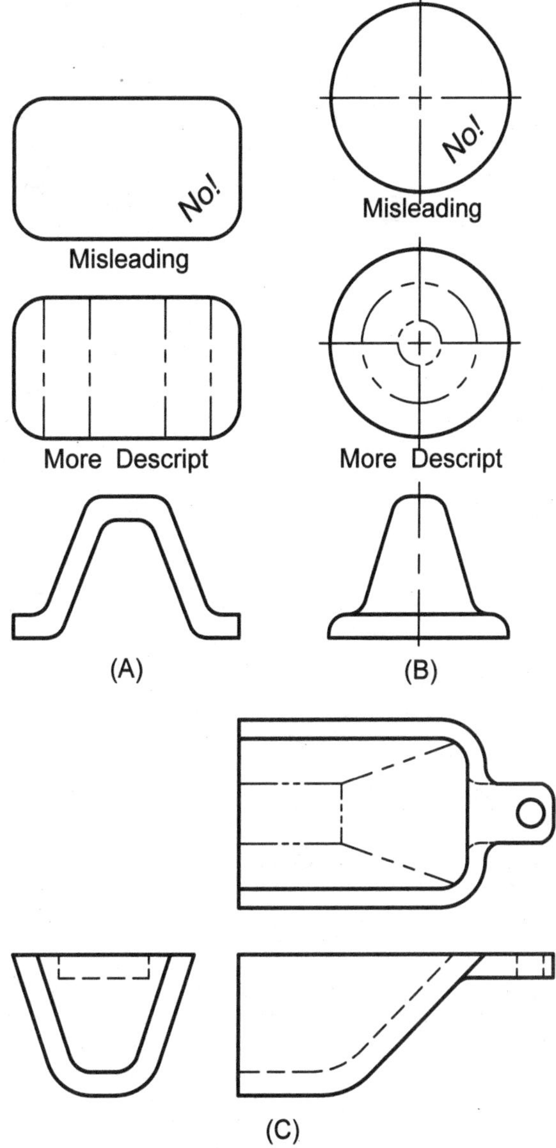

Figure MVD 30: Fillets and rounds soften intersecting surfaces and thus remove the line of intersection. Conventional practices use phantom lines where object lines would be if no fillets and rounds were used to help describe a view's shape more clearly.

3rd Angle vs. 1st Angle Projection:

To this point, and for the remainder of this text, multiview drawing has been discussed and illustrated using 3rd angle projection. This is the projection system used in the United States. However, most countries use 1st angle projection to create multiview drawings. In 1st angle projection the object is positioned in front of the plastic planes as illustrated in Figure MVD 31. Views are still hinged about the front view and result in the view placement illustrated in Figure MVD 32. Note how the positions of the views differ from 3rd angle projection. To ensure that those reading the drawing know whether it is 3rd or 1st angle projection, place one of the two symbols illustrated in Figure MVD 33 in or near the title block.

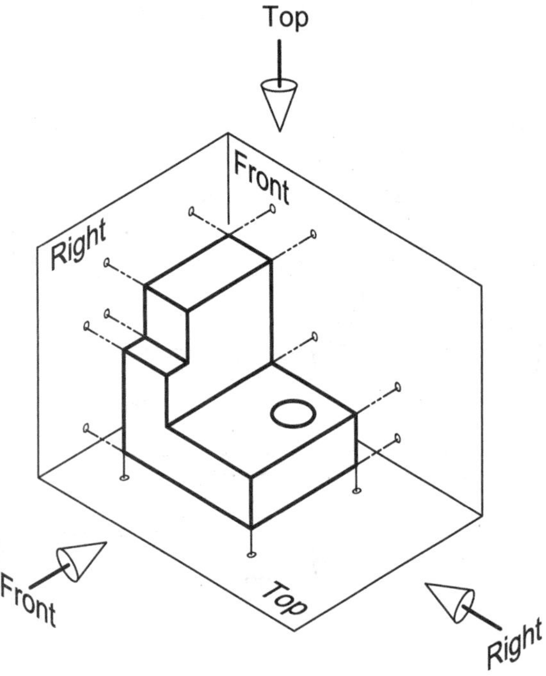

Figure MVD 31: The object is placed in front of the plastic planes in 1st angle projection.

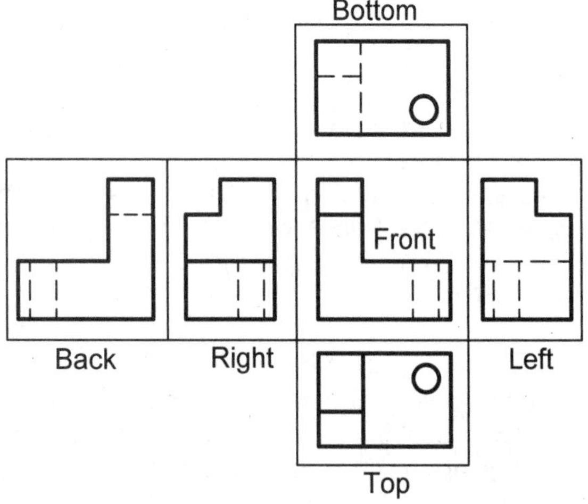

Figure MVD 32: Positioning of the six principle views in 1st angle projection.

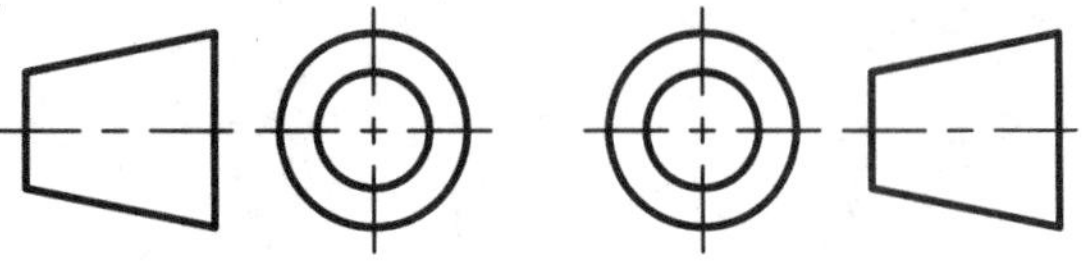

1st Angle
Projection Symbol
(SI)

3rd Angle
Projection Symbol
(ANSI)

Figure MVD 33: 1st and 3rd angle projection symbols placed in or near the title block.

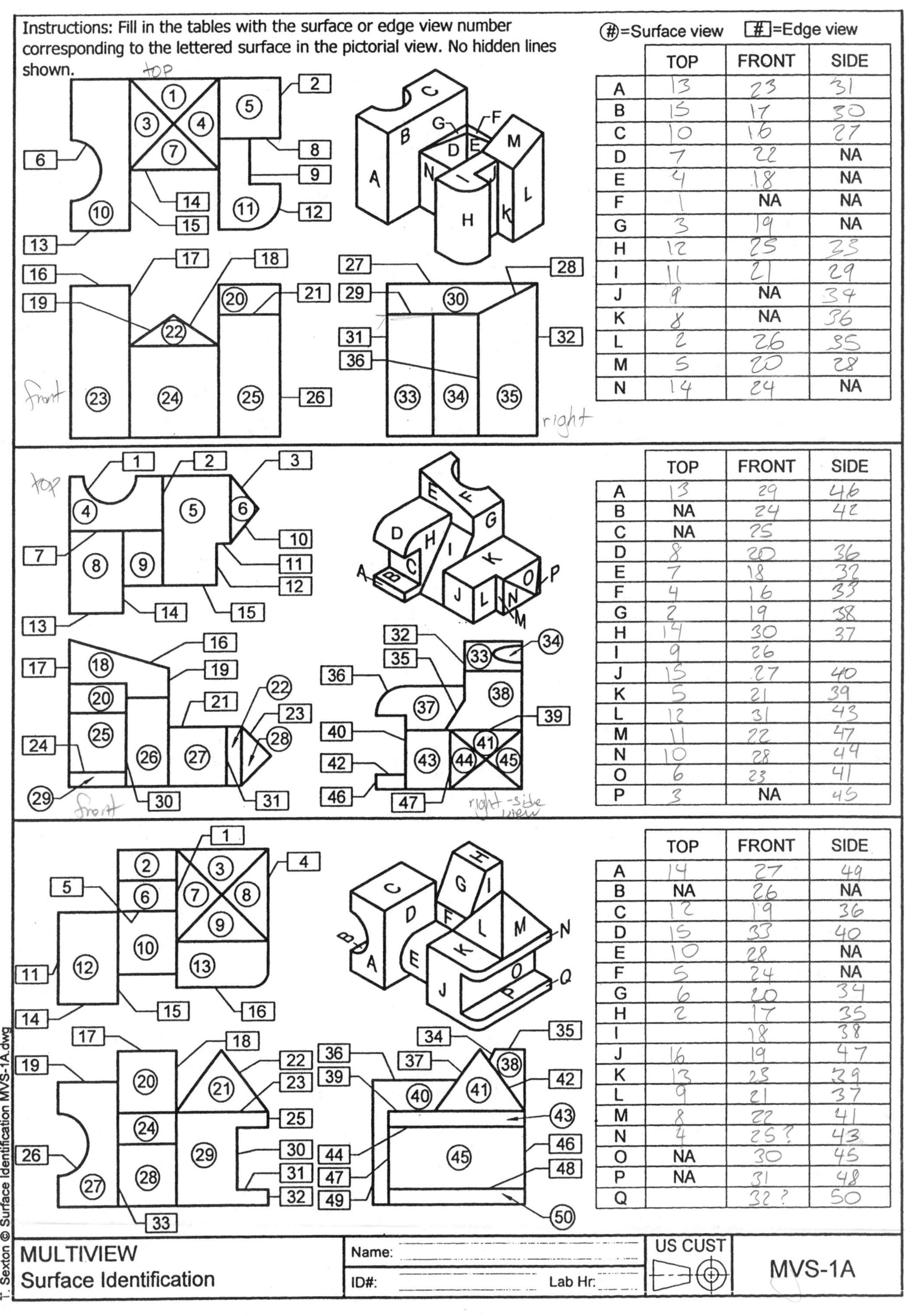

Instructions: Fill in the tables with the surface or edge view number corresponding to the lettered surface in the pictorial view. No hidden lines shown.

(#)=Surface view [#]=Edge view

	TOP	FRONT	SIDE
A	13	23	31
B	15	17	30
C	10	16	27
D	7	22	NA
E	4	18	NA
F	1	NA	NA
G	3	19	NA
H	12	25	33
I	11	21	29
J	9	NA	34
K	8	NA	36
L	2	26	35
M	5	20	28
N	14	24	NA

	TOP	FRONT	SIDE
A	13	29	46
B	NA	24	42
C	NA	25	
D	8	20	36
E	7	18	32
F	4	16	33
G	2	19	38
H	14	30	37
I	9	26	
J	15	27	40
K	5	21	39
L	12	31	43
M	11	22	47
N	10	28	44
O	6	23	41
P	3	NA	45

	TOP	FRONT	SIDE
A	14	27	49
B	NA	26	NA
C	12	19	36
D	15	33	40
E	10	28	NA
F	5	24	NA
G	6	20	34
H	2	17	35
I		18	38
J	16	19	47
K	13	23	39
L	9	21	37
M	8	22	41
N	4	25?	43
O	NA	30	45
P	NA	31	48
Q		32?	50

MULTIVIEW
Surface Identification

Name:
ID#: Lab Hr:

US CUST

MVS-1A

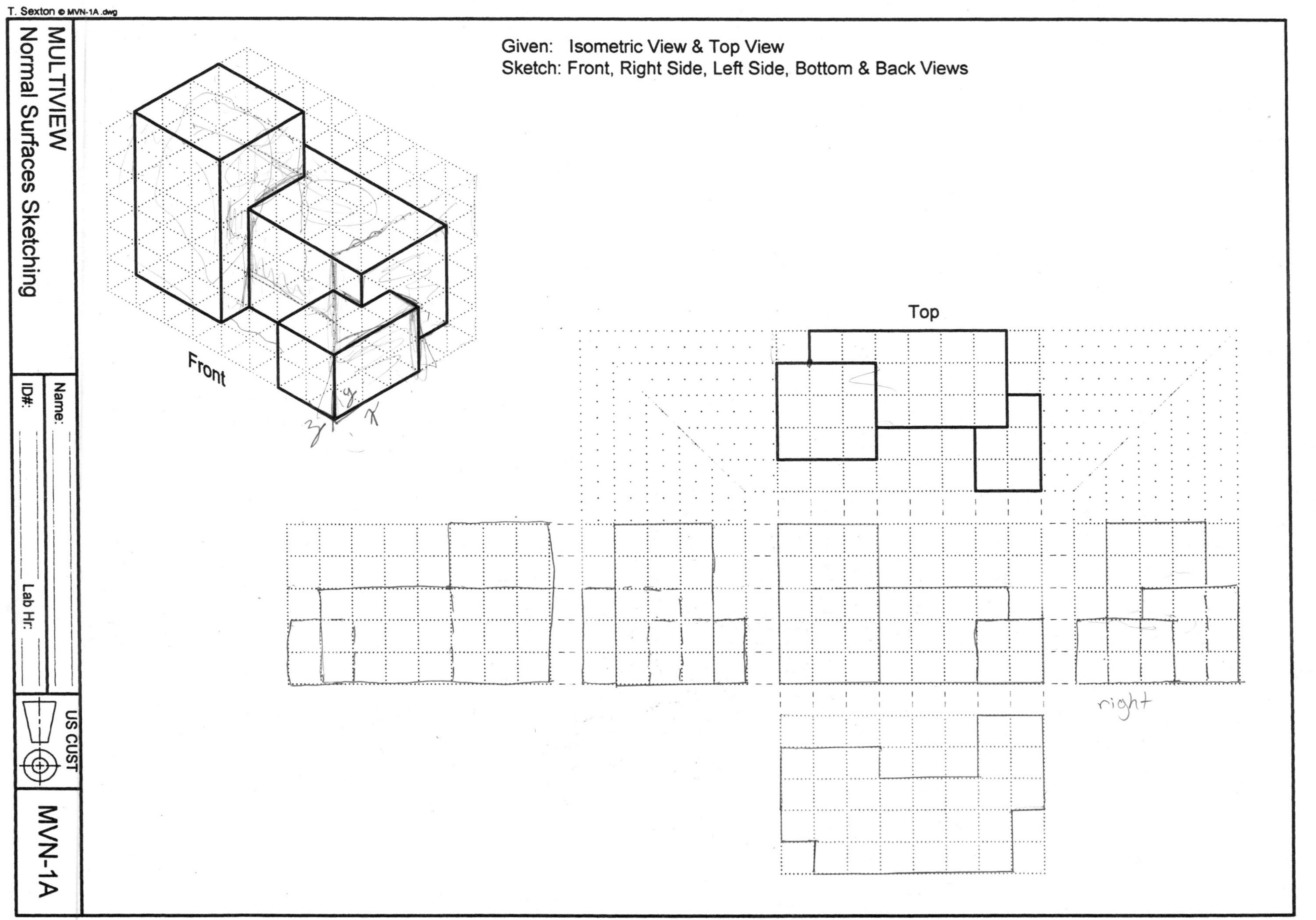

Given: Isometric View & Top View
Sketch: Front, Right Side, Left Side, Bottom & Back Views
Top
Front
right
MULTIVIEW
Normal Surfaces Sketching
Name:
ID#:
Lab Hr:
US CUST
MVN-1A
T. Sexton © MVN-1A.dwg

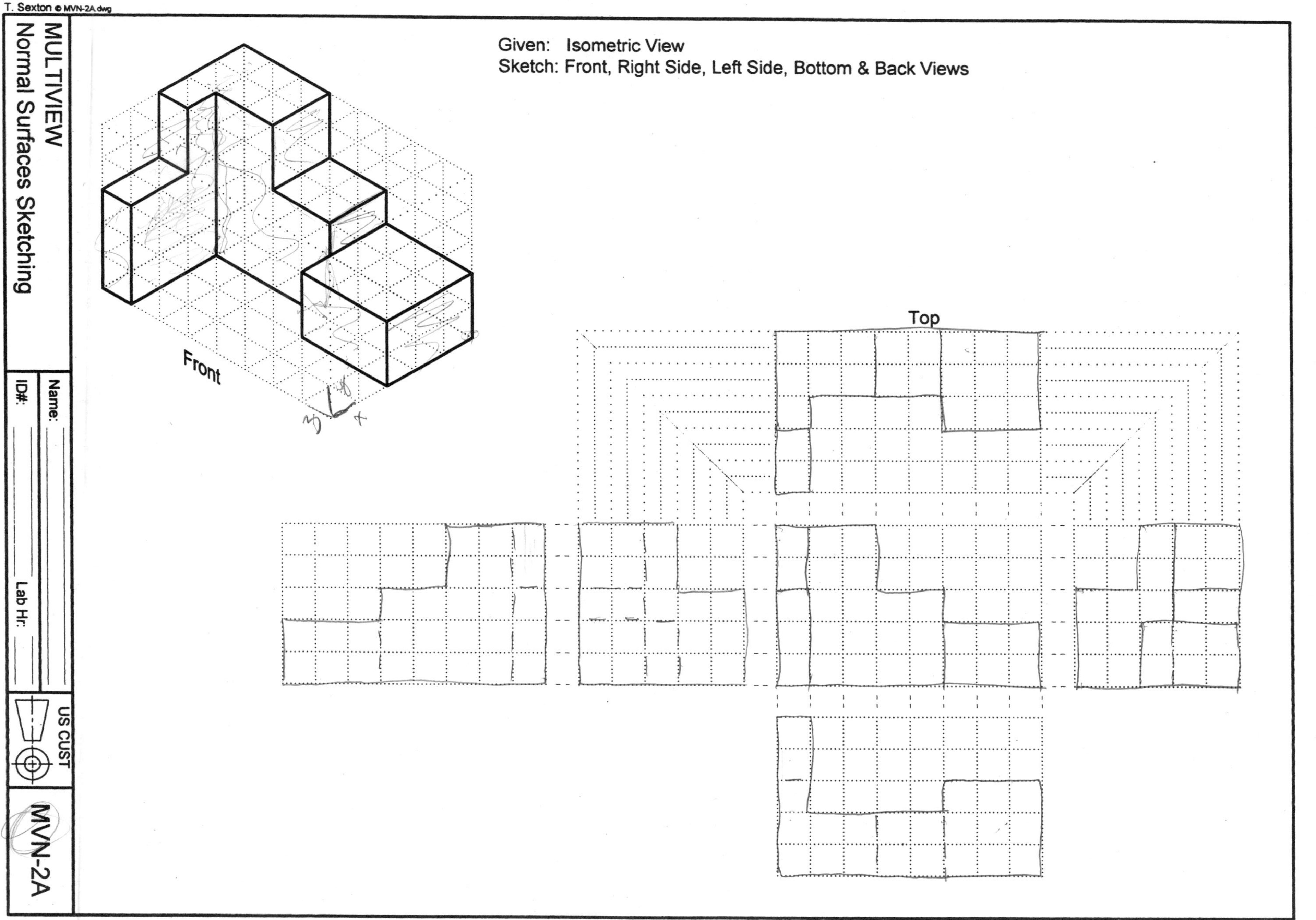
Given: Isometric View
Sketch: Front, Right Side, Left Side, Bottom & Back Views
Front
Top
T. Sexton © MVN-2A.dwg
MULTIVIEW
Normal Surfaces Sketching
Name:
ID#:
Lab Hr:
US CUST
MVN-2A

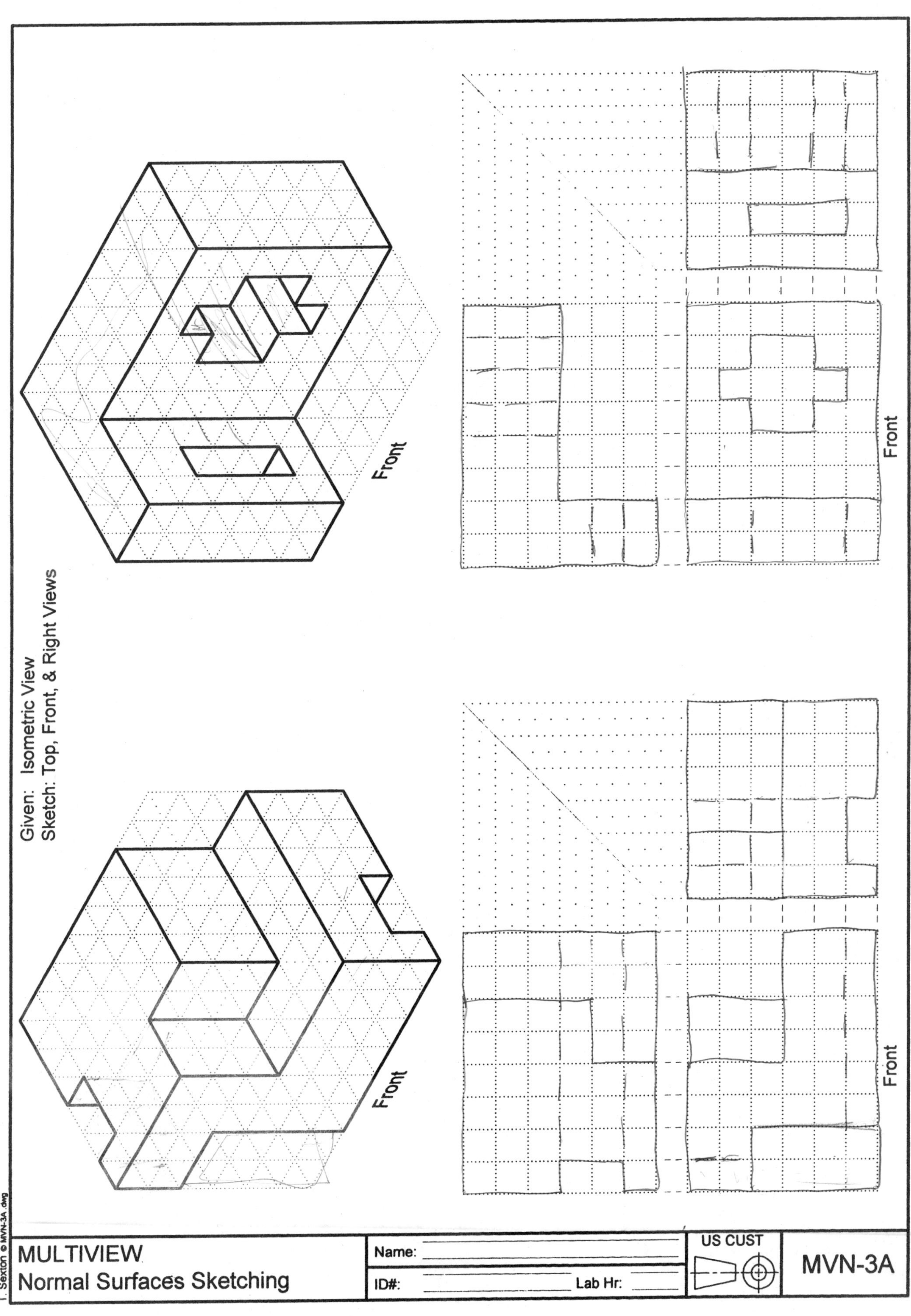
Given: Isometric View
Sketch: Top, Front, & Right Views
Front
Front
Front
Front
MULTIVIEW
Normal Surfaces Sketching
Name:
ID#:
Lab Hr:
US CUST
MVN-3A
T. Sexton © MVN-3A .dwg

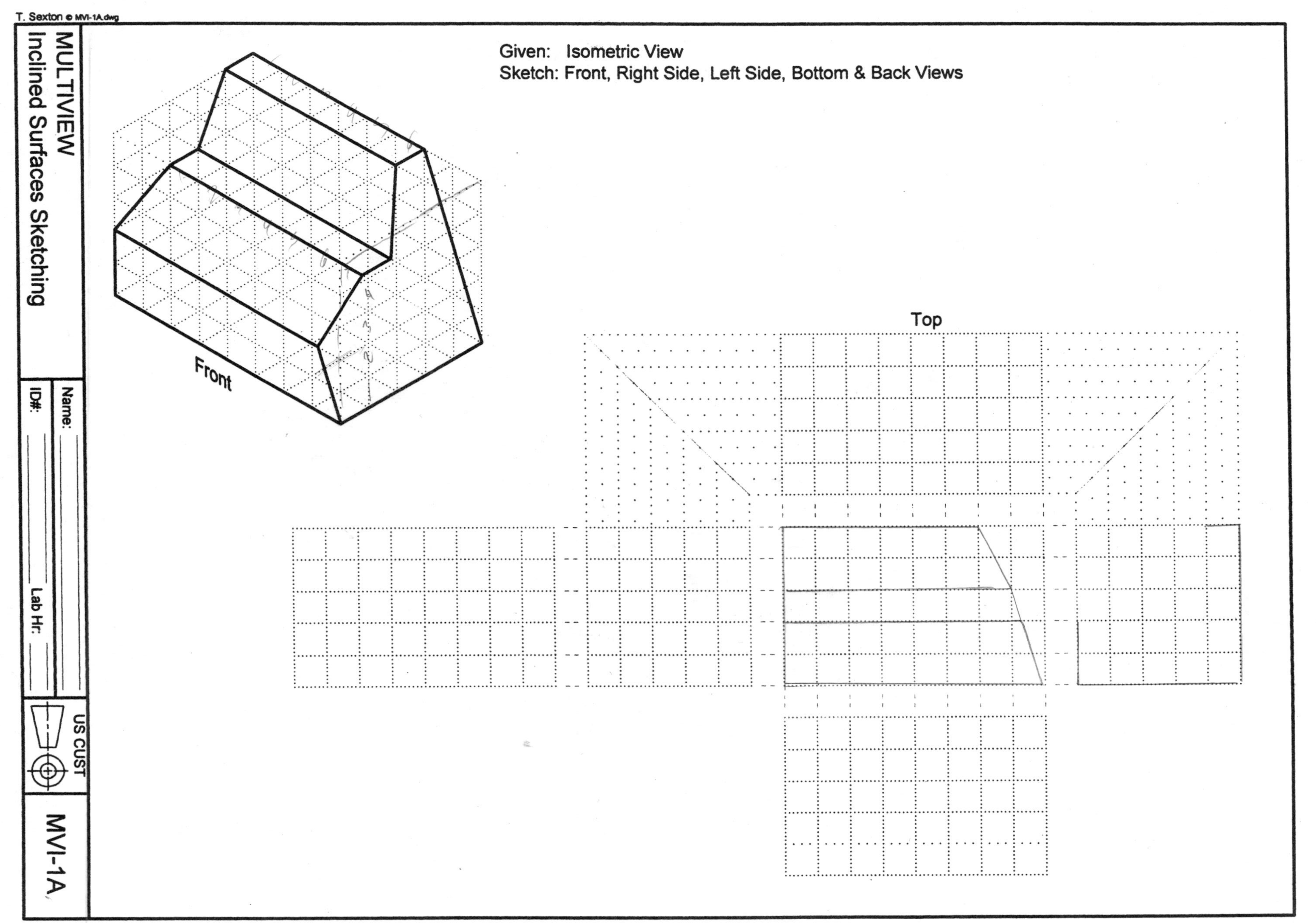

Given: Isometric View
Sketch: Front, Right Side, Left Side, Bottom & Back Views
Front
Top
T. Sexton © MVI-1A.dwg
MULTIVIEW
Inclined Surfaces Sketching
Name:
ID#:
Lab Hr:
US CUST
MVI-1A

MULTIVIEW
Inclined Surfaces Sketching

Name:
ID#: Lab Hr:
US CUST
MVI-2A

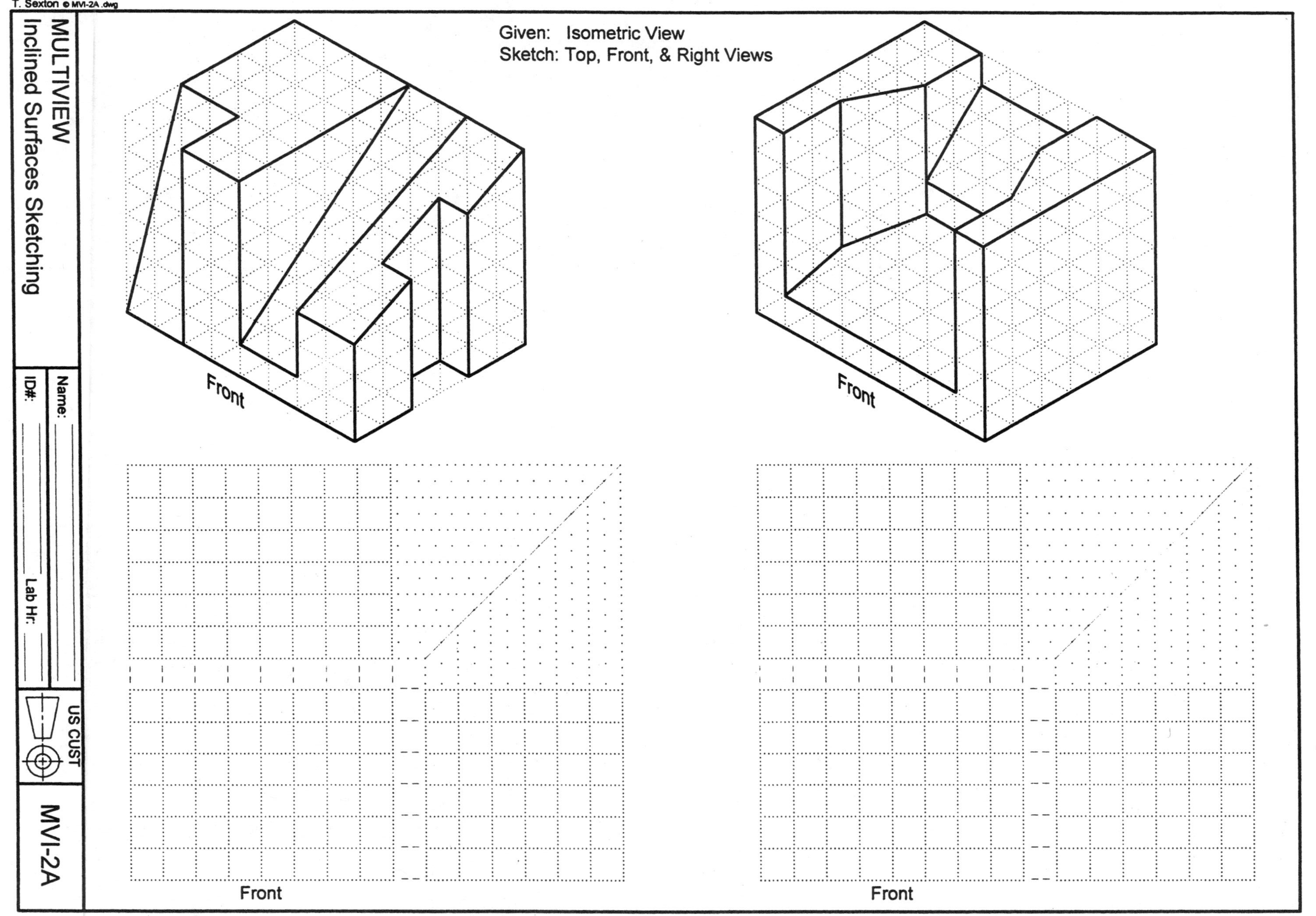

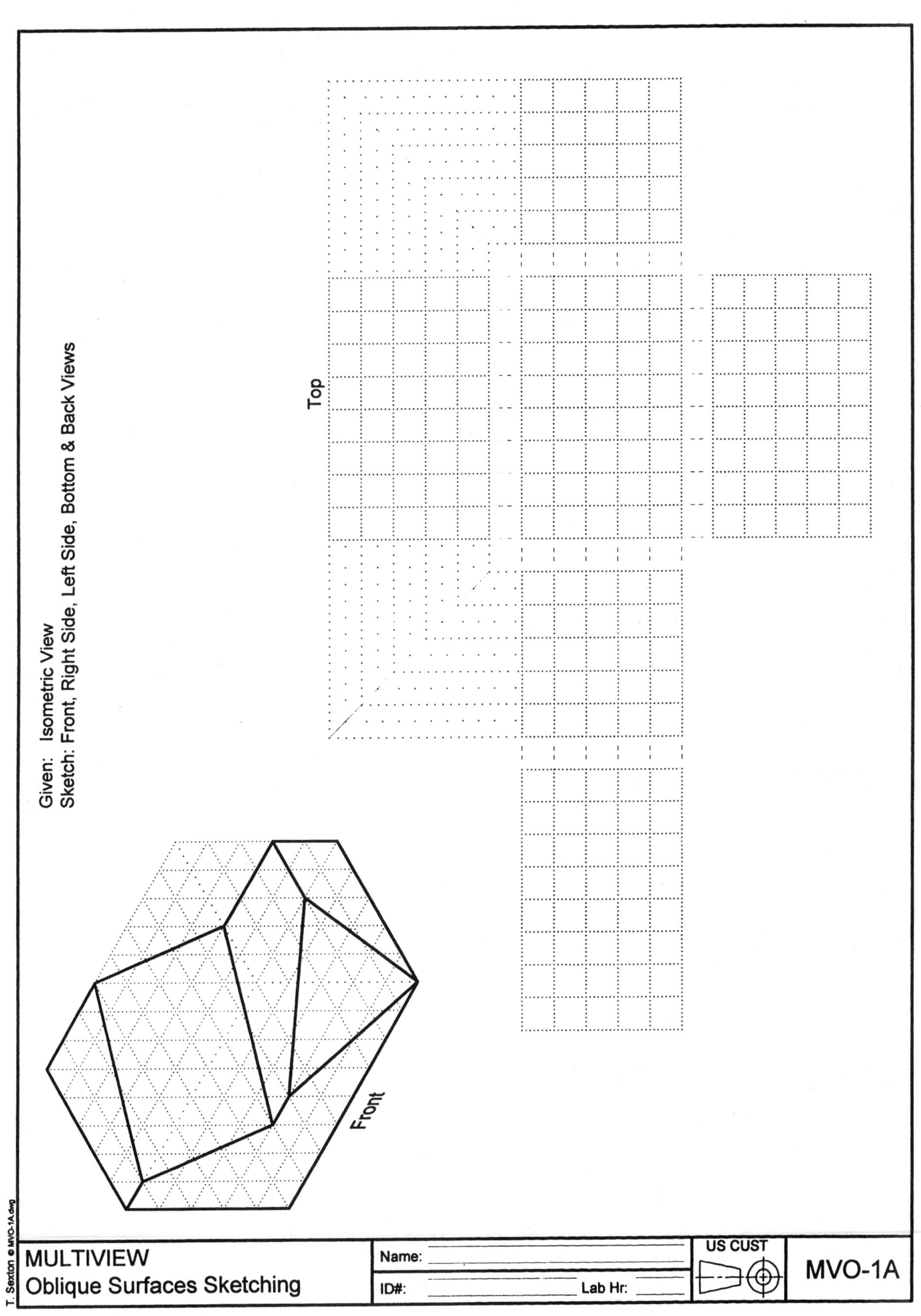
Given: Isometric View
Sketch: Front, Right Side, Left Side, Bottom & Back Views
Top
Front
T. Sexton © MVO-1A.dwg
MULTIVIEW
Oblique Surfaces Sketching
Name:
ID#:
Lab Hr:
US CUST
MVO-1A

Chapter 4

Isometric Drawing

Why Isometric Drawing?

The multiview drawings illustrated in Figures ISO 1 and ISO 2 are indispensable when trying to describe the exact shape of an object. That is why they make up the vast majority of technical drawings or "blueprints". However, they can be difficult to interpret and visualize. To help visualize the shape of an object, pictorial drawings are used. A pictorial drawing is a 3-D "picture like" drawing that represents an object in a way that is much closer to what the eye actually sees. In Figures ISO 3 and ISO 4 isometric pictorial views have been added to the multiview drawings of Figures ISO 1 and ISO 2. Notice how much easier it is to visualize the shape of the object and interpret the multiview drawings. Pictorial drawings are especially helpful to non-technically trained people needing to read multiview drawings. There are several types of pictorial drawings: perspective, oblique, isometric, diametric, and trimetric. This text will discuss isometric drawings only. You can find a discussion of the other types of pictorial drawings in a more advanced graphic textbook. *Isometric drawings* are the most common type of pictorial used in technical drawings. Just as important, the isometric environment is used when building 3-D models in all 3-D modeling CAD software.

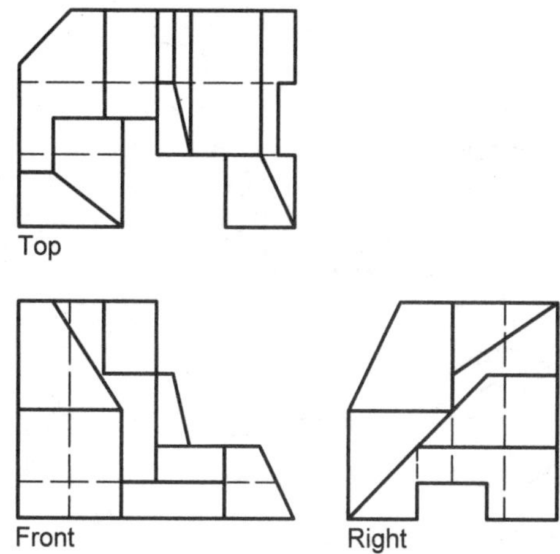

Figure ISO 1: Multiview drawing.

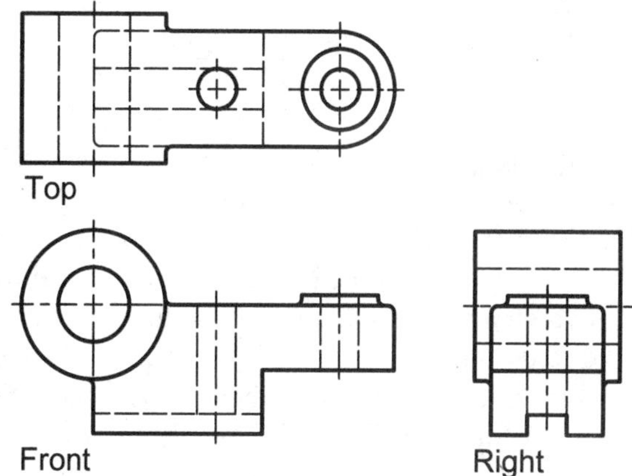

Figure ISO 2: Multiview with curved surfaces.

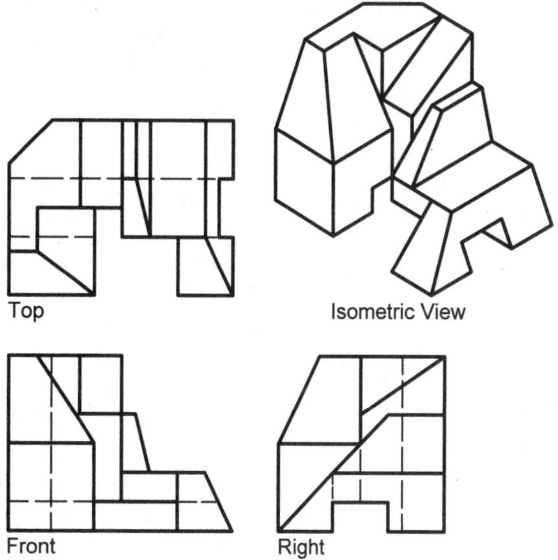

Figure ISO 3: Multiview with isometric pictorial view added

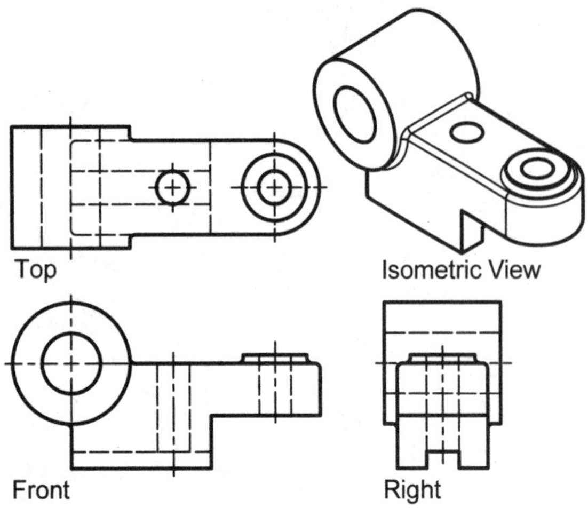

Figure ISO 4: Multiview with isometric pictorial view added.

Isometric Projection Theory:

In Figure ISO 3 the isometric pictorial combines the top, front and right views of the multiview into one isometric drawing. This is a significant aid in visualizing an object in three dimensions. An isometric view is obtained through a two step process. Using a cube to help illustrate, the steps are:

1. Rotate the cube 45° about a vertical axis as illustrated in Figure ISO 5A. In the front view, the front and right sides will project equally onto the frontal picture plane.
2. Revolve the cube 36° 15' about a horizontal axis until the cube's diagonal is horizontal and appears as point 1, 2 in the front view. The front view of Figure ISO 5B illustrates the final isometric view.

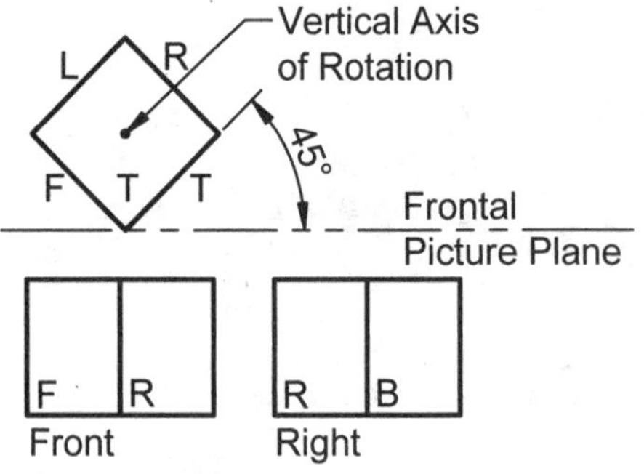

(B) Rotate Cube 45° about a vertical axis

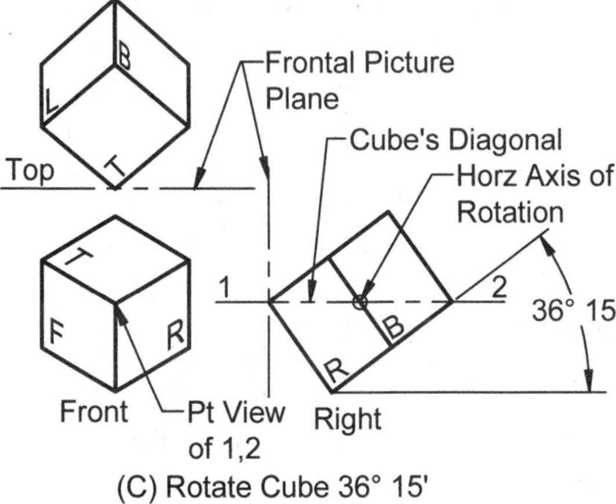

(C) Rotate Cube 36° 15'

Figure ISO 5: Two steps required to generate an isometric view.

The front view now displays the isometric view. The front, top and right views project equally onto the frontal picture plane of projection. The intersecting edges of these three views will be equal in length.

The lines of intersection between the top, front, and right views form the isometric axis as illustrated by the thick lines in Figure ISO 6A. The isometric axes are 120° apart as illustrated in Figure ISO 6A. When constructing an isometric drawing it is often easier to start at the bottom front corner of the box using two 30° lines as illustrated in Figure ISO 6B. This does not change the fact that the isometric axes are still 120° apart.

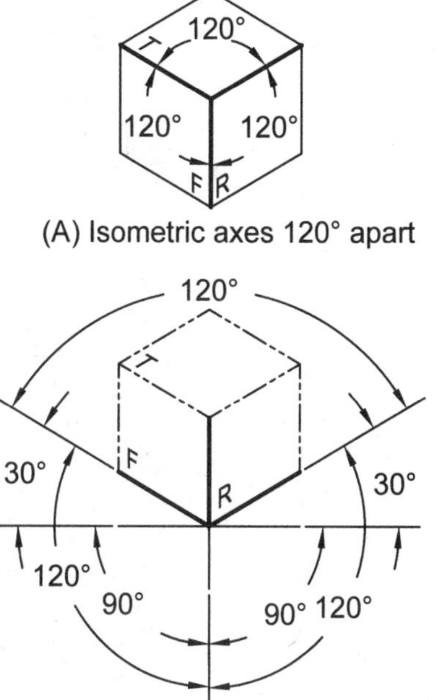

(A) Isometric axes 120° apart

(B) Starting axes in the front bottom corner with 30° angled lines

Figure ISO 6: A) isometric axes 120° apart; B) isometric axes laid out from bottom front corner using two 30° lines.

In the previous isometric figures, the top, front, and right side views were shown. This orientation is illustrated in Figure ISO 7A which shows the top/front/right "bird's eye view." But you are not limited to this orientation. Figure ISO 7B illustrates the bottom/front/right "worm's eye" view. Although less common, Figures ISO 7C and 7D illustrate two additional orientations of the isometric axes. The important criterion is that the three axes be 120° apart.

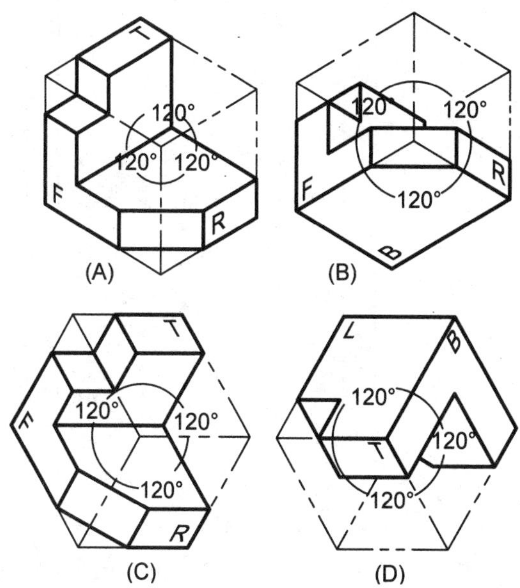

Figure ISO 7: A) bird's eye view, B) worm's eye view, C) alternate view 1, and D) alternate view 2.

Constructing an Isometric Sketch:

Before beginning to sketch an isometric drawing, Rule #1 must be emphasized:

> **Isometric Rule #1**
>
> *Measurement can only be made* ***on*** *or* ***parallel*** *to the isometric axis.*

The easiest way to construct an isometric sketch is to use isometric grid paper which has light grid lines (usually blue) running vertically and at 30° to the right and left as illustrated in Figure ISO8.

Figure ISO 8: Isometric grid with vertical and 30° lines to the right and left.

To construct an isometric sketch use the following steps:

1. Study the multiview drawing such as Figure ISO 9 or measure the actual object.
2. Lay out the isometric axes and the overall height, width, and depth to form an isometric box as illustrated in Figure ISO 10A.
3. Start cutting away at the isometric box one detail at a time. This may take several steps as illustrated in Figures ISO 10B, 10C, 10D, 10E and 10F. Remember measure on or parallel to the isometric axes!

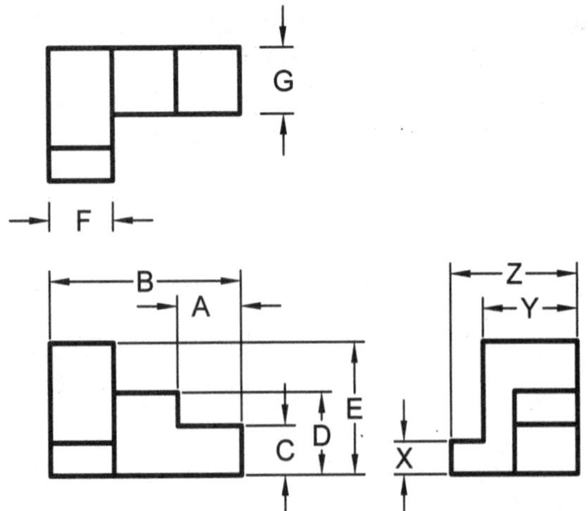

Figure ISO 9: Study the multiview drawing.

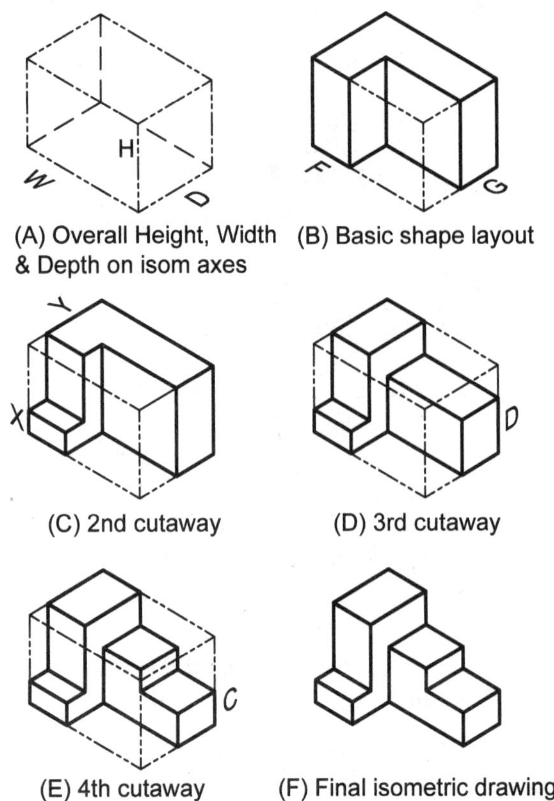

Figure ISO 10: Steps for constructing an isometric drawing.

Isometric Drawing vs. Isometric Projection:

When laying out an isometric drawing, the object's full dimensions are used. But when an object is rotated 45° and then 36° 15', to present an isometric view, the object's planes and edges will be foreshortened. The true sized projection, which takes into account foreshortening, is called an *isometric projection* and is approximately 80% the size of an isometric drawing. Figure ISO 11 compares the full length isometric drawing to an 80% foreshortened isometric projection. It is so much easier to ignore the foreshortening and use the full dimensions, that full length isometric drawings are the accepted practice.

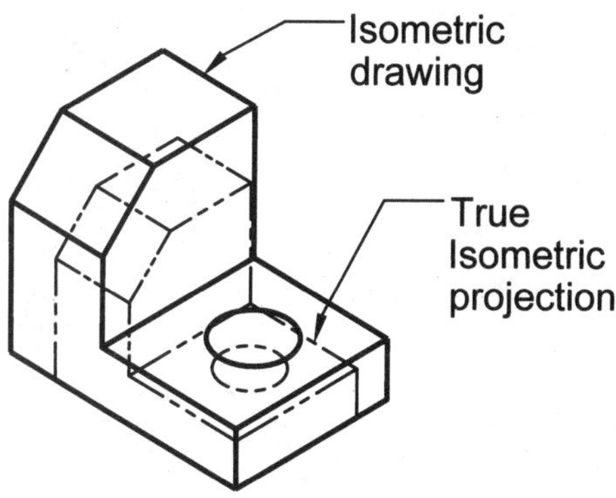

Figure ISO 11: Isometric drawing vs. true isometric projection which is 80% of the size of an isometric drawing.

Inclined and Oblique Surface:

Isometric Rule #1 states that you can measure on or parallel to the isometric axes only. Therefore you cannot measure an isometric inclined or oblique line in an isometric drawing because they are not parallel to an isometric axis. Figure ISO 12 illustrates an isometric drawing having inclined surfaces. Line AB cannot be measured because it is not parallel to an isometric axis. Figure ISO 13 illustrates an isometric drawing of an object having an oblique surface. The oblique surface in Figure ISO 13 has been rotated around both the horizontal and vertical axes. Line CD cannot be measured because it is not parallel to an isometric axis. To draw an isometric inclined or isometric oblique line the x, y and z coordinates of the line's end points must be plotted.

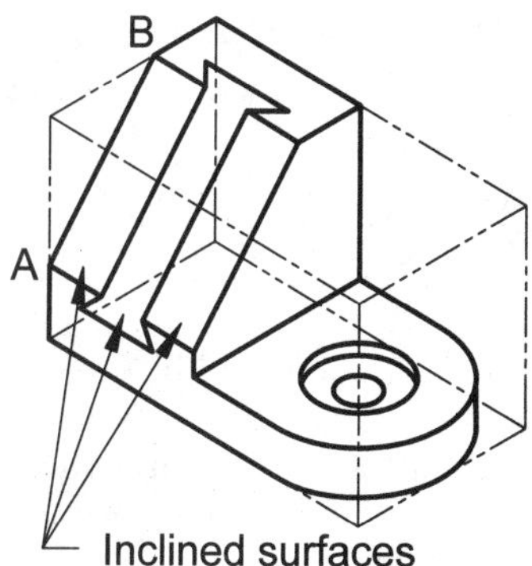

Figure ISO 12: Isometric drawing with inclined surfaces. Line AB cannot be measure because it is not parallel to an isometric axis.

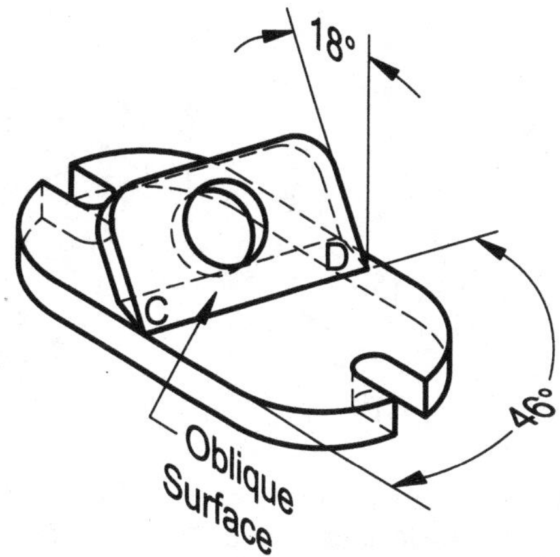

Figure ISO 13: Isometric drawing with an oblique plane. Line CD cannot be measure because it is not parallel to an isometric axis.

Figure ISO 14 shows a multiview drawing of an object containing the inclined surface ABCD. To draw the inclined plane ABCD each corner of the plane must be located using the X, Y, and/or Z coordinates of each of its corners A, B, C, and D. Figure ISO 15A illustrates how the corners of the inclined plane are located. Figure ISO 15B shows points A and B and points C and D being connected to form the inclined plane. Remember the lines AB and CD cannot be measured because they are not parallel to an isometric axis.

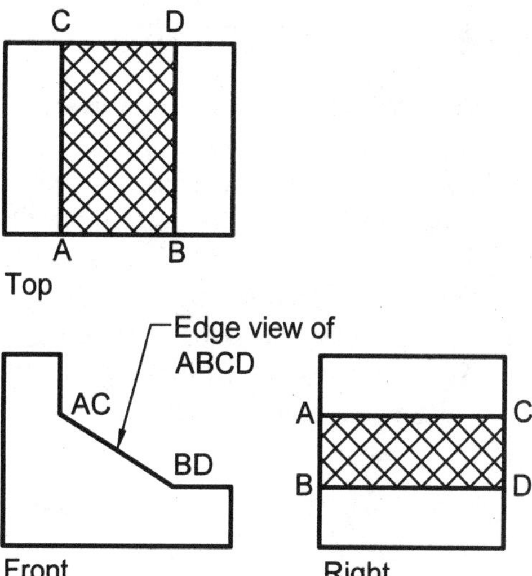

Figure ISO 14: Inclined surface ABCD.

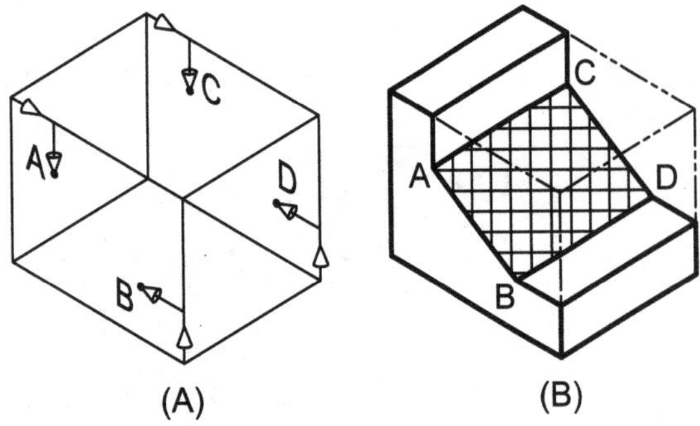

Figure ISO 15: A) The X, Y, and Z coordinates are measured out to find points A, B, C, and D; B) the inclined surface is completed by connecting points A to B and C to D.

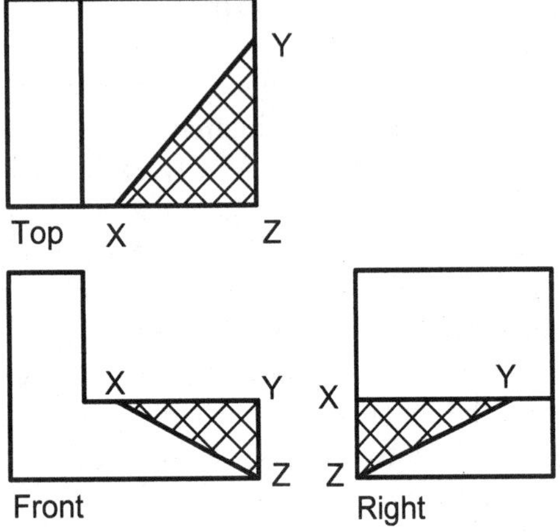

Figure ISO 16: Oblique surface XYZ.

Figure ISO 16 shows a multiview drawing containing the oblique plane A, B, C. Figure ISO 17A shows the two steps required to layout point Y from the base of the isometric axis. Figure ISO 17B shows the two steps required to layout point Z. Figure ISO 17C shows line AB, BC, and CA connected to form the isometric oblique plane XYZ. Again because of Rule #1 the sides of the oblique plane cannot be measured!

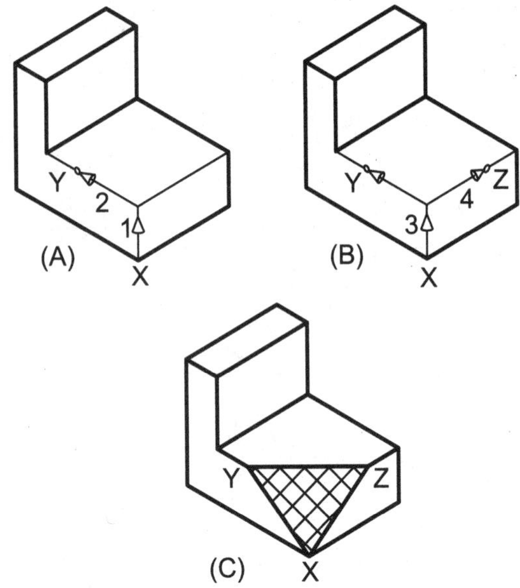

Figure ISO 17: A) Locating point Y using the two steps 1 and 2, B) finding point Z using the two steps 3 and 4, C) points X, Y, and Z connected to form the isometric oblique plane.

Isometric Circles and Arcs

Holes and cylinders that lie on normal planes appear as ellipses in isometric drawings. Figure ISO 18 illustrates a multiview drawing with circles representing holes in the vertical and horizontal planes and arcs representing curved surfaces in the vertical, horizontal, and the profile planes. Figure ISO 19 illustrates an isometric sketch of the multiview Figure ISO 18. Figure ISO 20 illustrates isometric circles (ellipses) in the top (horizontal), front (vertical), and right (profile) normal isometric planes. Each ellipse is contained in an isometric square (or rhombus) whose sides are equal to the circle's diameter. The minor axis of each ellipse is perpendicular to the corresponding normal isometric plane. Another way of describing the orientation of the minor axis is to imagine the minor axis being in the same orientation of a drill bit as if you were drilling a hole into each of the isometric planes. Note that when drawing an isometric ellipse it is necessary to draw three construction center lines, one for the X, Y, and Z axes. The most common error when drawing an isometric ellipsis is to incorrectly orient the minor axis.

> ***Isometric Rule #2***
>
> *When drawing ellipses on normal isometric planes, the minor axis of the ellipse is perpendicular to the plane containing the ellipse.*

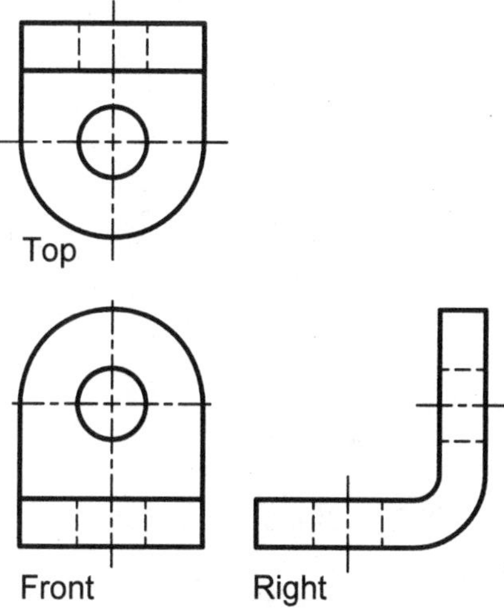

Figure ISO 18: Multiview with circles and arcs in the front, top, and right views.

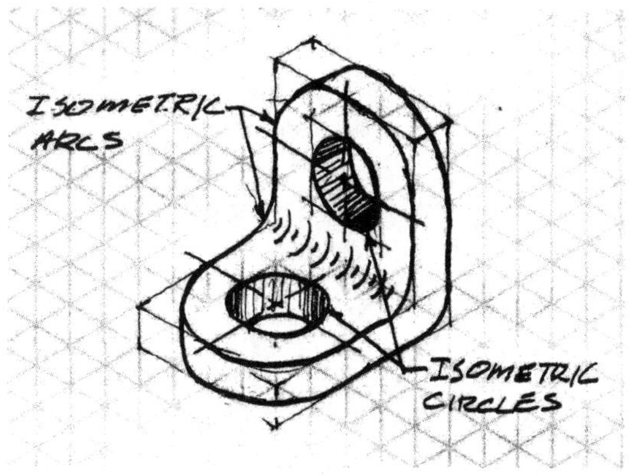

Figure ISO 19: A sketch illustrating circles and arcs in isometric planes.

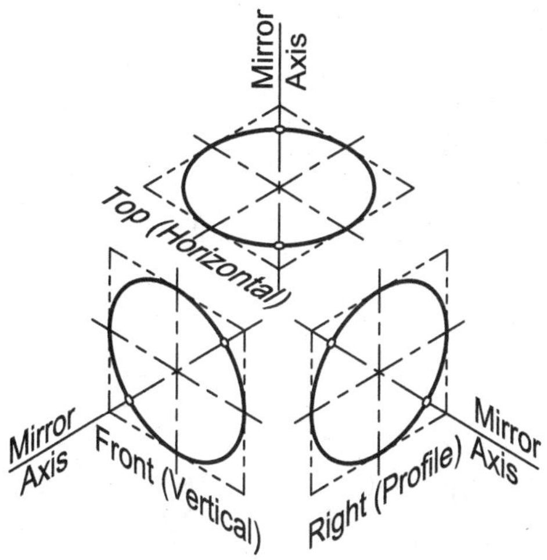

Figure ISO 20: Isometric ellipses in three isometric planes with a minor axis at either 30° to the right, 30° to the left, or vertical.

Approximation Method:

To sketch isometric circles and arcs, it is helpful to know how the isometric circle is constructed using traditional manual drafting instruments. The construction technique is called the *four center approximation method.* It is approximate because it does not result in a true ellipse. The four center approximation method has four steps:

1. Draw an isometric square (rhombus) whose sides are equal to the diameter of the circle as illustrated in Figure ISO 21A.
2. From the large angles W and X, draw lines to the midpoints 1, 2, 3, and 4 of the opposite sides forming perpendicular bisectors, as illustrated in Figure ISO 21B.
3. Draw the two large arcs using center points W and X as illustrated in Figure ISO 21C.
4. Draw the two small arcs using center points Y and Z as illustrated in Figure ISO 21D.

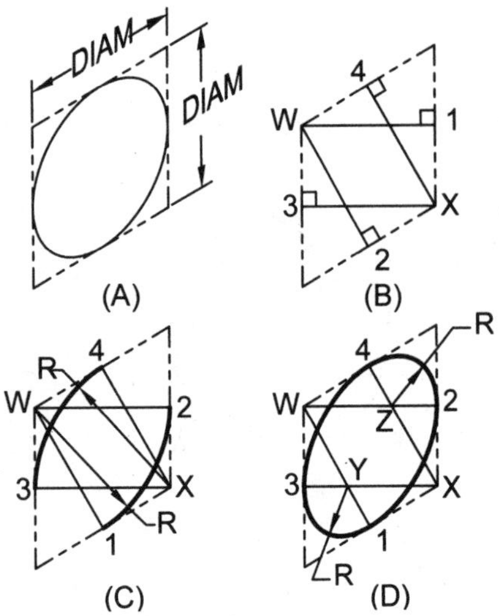

Figure ISO 21: Four steps to the four center approximation method for drafting an isometric ellipse.

Sketching an Isometric Circle:

Sketching isometric circles and arcs follows the same method as the four center approximation method except you do not draw the perpendicular bisectors.

Figure ISO 22 illustrates the steps used to sketch an isometric circle:

1. Draw an isometric square (rhombus) whose sides are equal to the circles diameter as illustrated in Figure ISO 22A.
2. Mark off the points of tangency at the midpoint of each side of the square as illustrated in Figure ISO 22B.
3. Sketch the two long arcs as illustrated in Figure ISO 22C.
4. Sketch the two short arcs as illustrated in Figure ISO 22D. Pay particular attention to the orientation of the minor axis.

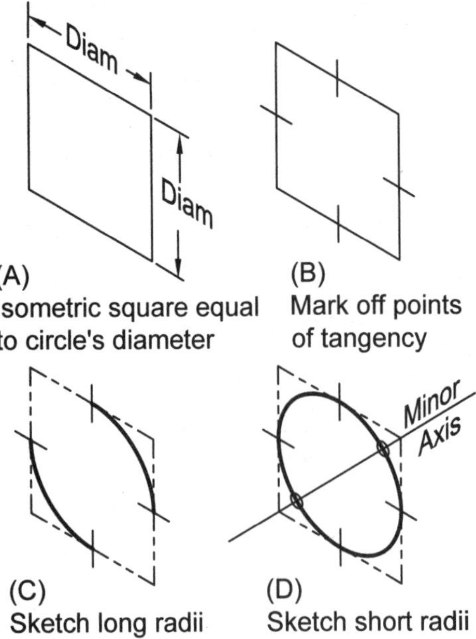

Figure ISO 22: Steps for sketching an isometric circle (ellipse).

The next logical step once you know how to sketch a circle is to sketch an isometric cylinder. Figure ISO 23 illustrates how to sketch an isometric cylinder. In ISO 23A you sketch the front circle and then layout and sketch the rear parallel arcs. In ISO 23B you sketch the two 30° tangent lines and then trim the back arcs as required.

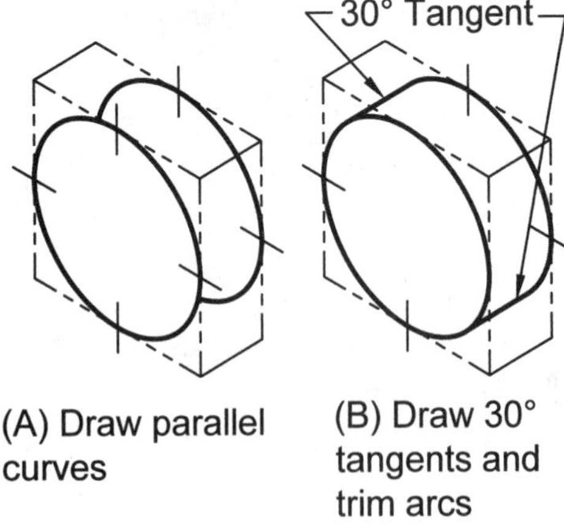

Figure ISO 23: Sketching an isometric cylinder.

When sketching isometric arcs, it is best to very lightly lay out the entire circle not just the portion used to construct the arc. As illustrated in Figure ISO 24 sketching the entire circle helps you visualize the arc and see if your arc looks proportional and the minor axis has the correct orientation. Figure ISO 25 illustrates the construction of a drill jig having isometric circles and arcs drawn in the front, top and right planes. Tangent lines are also illustrated.

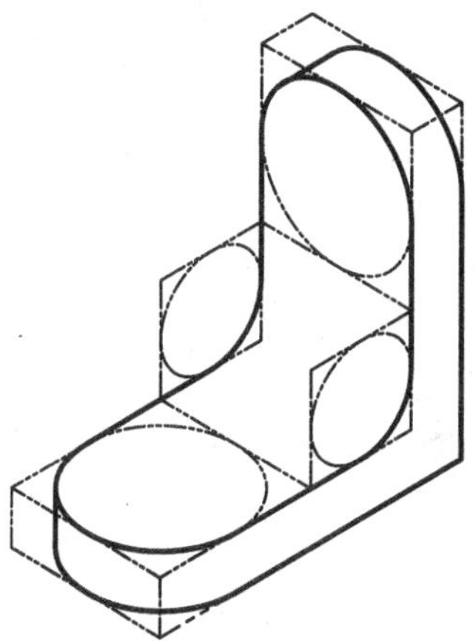

Figure ISO 24: Isometric arcs sketched using full circle construction.

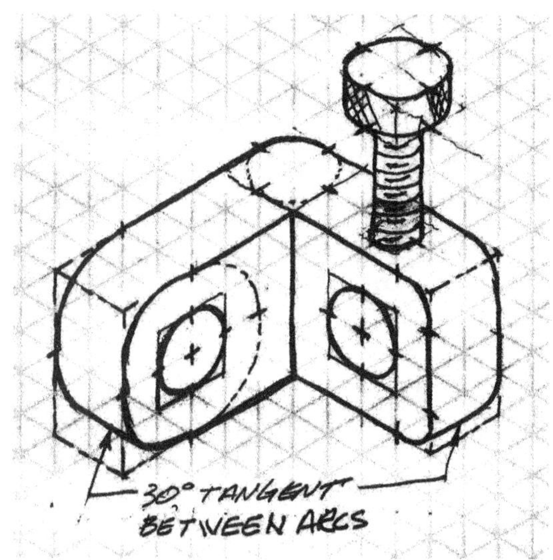

Figure ISO 25: Examples of sketched isometric circles and arcs.

Isometric Elliptical Templates:

If you find it difficult to draw isometric ellipses, you may want to use an isometric elliptical template until your free hand skills improve. Figure ISO 26 shows an isometric elliptical template. Since isometric drawing is 3-D, three construction center lines need to be drawn to use the template as illustrated in Figure ISO 27. Figure ISO 28 illustrates the use of an isometric elliptical template. Remember to keep the minor axis perpendicular to the plane containing the ellipse.

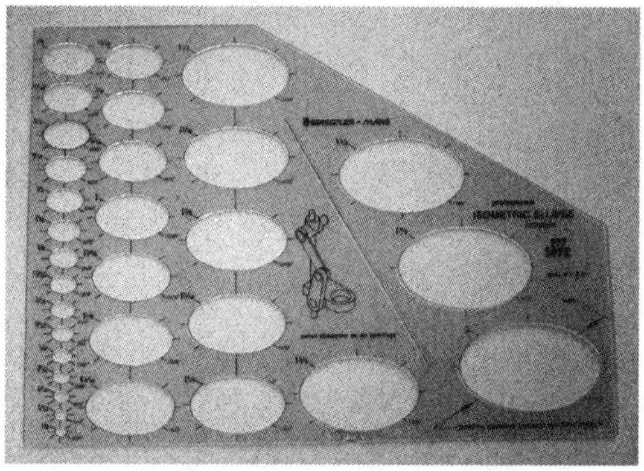

Figure ISO 26: Isometric elliptical template.

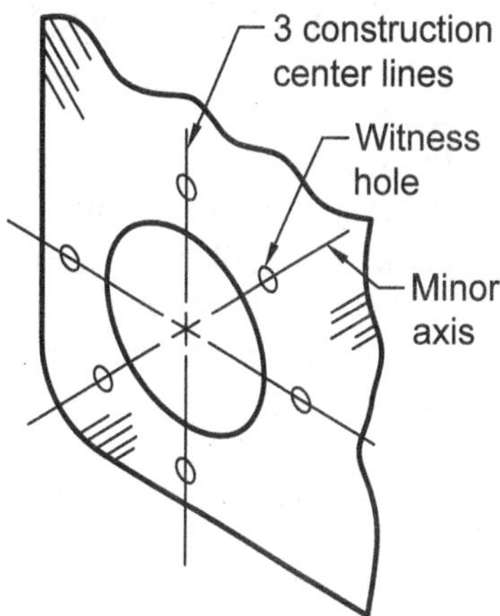

Figure ISO 27: Elliptical template with witness holes.

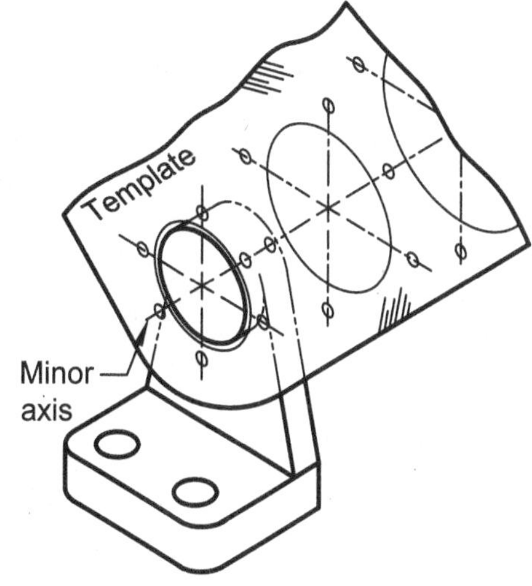

Figure ISO 28: Using an isometric elliptical template. Line up the witness holes (or lines) on all three axes. Be sure the minor axis of the template is in line with the minor axis of the hole or cylinder.

Isometric Irregular Curves:

Irregular curves do not contain true circular arcs. Therefore there is no simple way to lay one out in an isometric drawing. Irregular curves must be constructed by locating the XYZ coordinates of several points on the curve and then connecting the points to form the curve. Figure ISO 29 illustrates the steps used to lay out an irregular curve in an isometric drawing:

1. Divide the vertical axis into the equal increments 1, 2, 3, and 4 as illustrated in Figure ISO 29A. Then project these points onto the irregular curve resulting in points 5, 6, 7, and 8 as illustrated in Figure ISO 29A.
2. Locate points 1, 2, 3, and 4 in the isometric view by laying off the Y axis distances on the vertical axis. Then locate points 5, 6, 7, and 8 by lying off the X axis distances as illustrated in Figure 29B.
3. Connect points 5, 6, 7, and 8 to form the front irregular curves as illustrated in Figure 29B.
4. Locate points 9, 10, 11, and 12 by lying off the Z axis distances from point 5, 6, 7, and 8 and connect points 9, 10, 11, and 12 to form the back irregular curve.

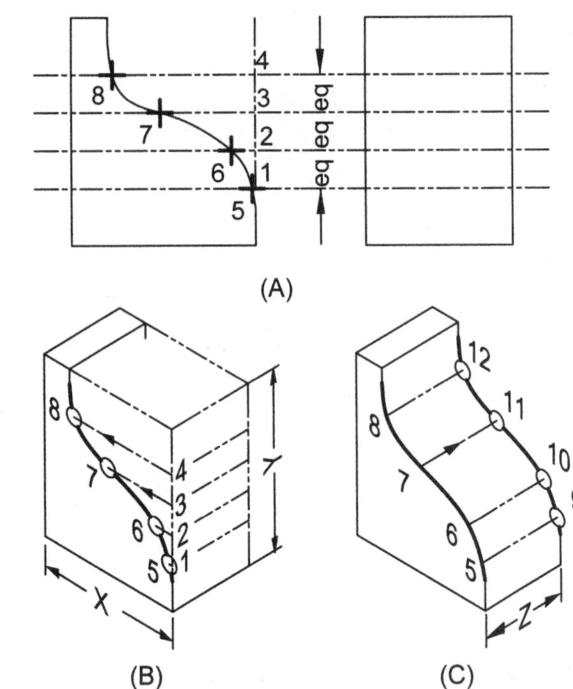

Figure ISO 29: Constructing an isometric irregular curve.

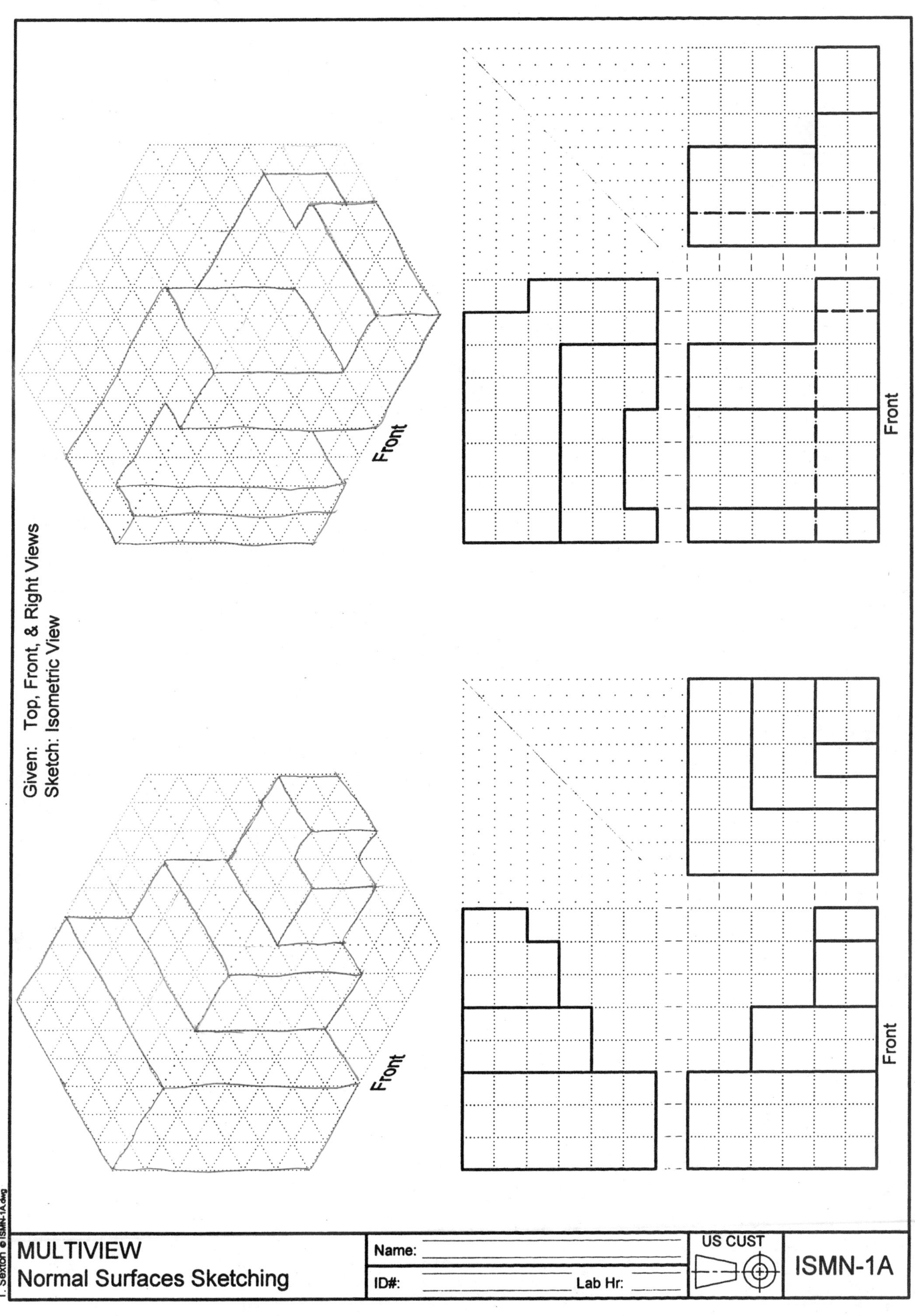
Given: Top, Front, & Right Views
Sketch: Isometric View
Front
Front
Front
Front
T. Sexton © ISMN-1A.dwg
MULTIVIEW
Normal Surfaces Sketching
Name:
ID#:
Lab Hr:
US CUST
ISMN-1A

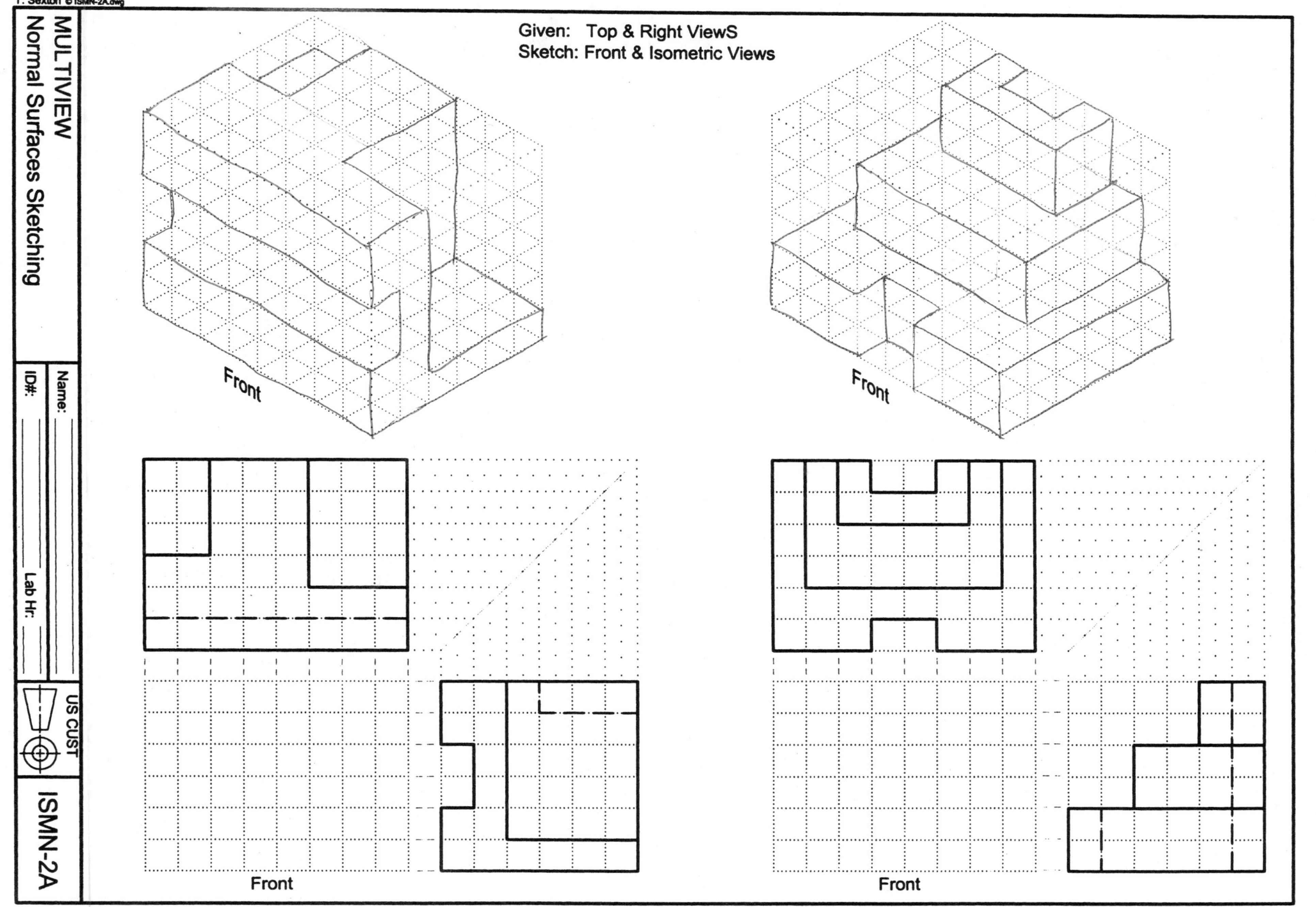
Given: Top & Right ViewS
Sketch: Front & Isometric Views
Front
Front
Front
Front
MULTIVIEW
Normal Surfaces Sketching
Name:
ID#:
Lab Hr:
US CUST
ISMN-2A
T. Sexton © ISMN-2A.dwg

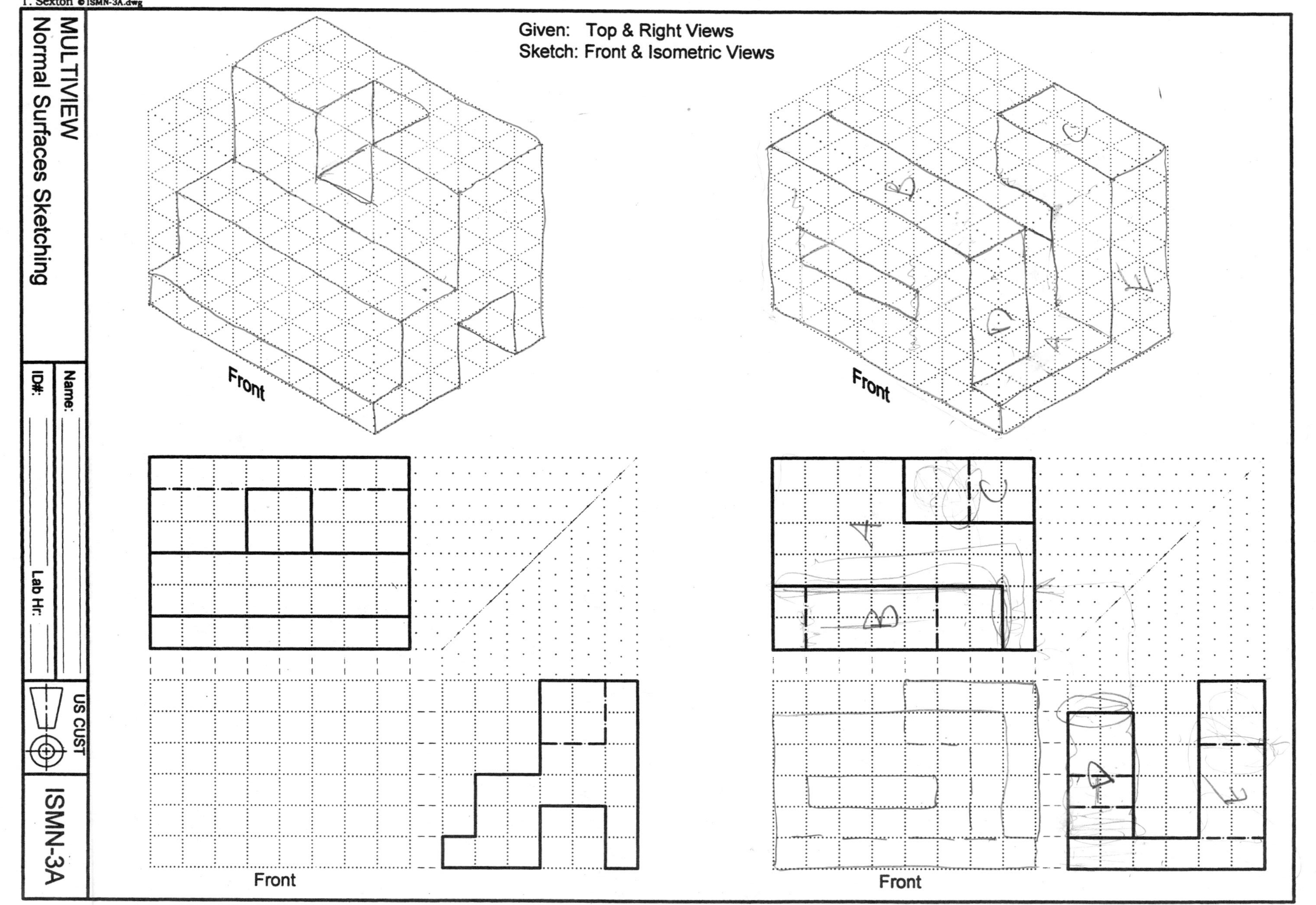
Given: Top & Right Views
Sketch: Front & Isometric Views
Front
Front
Front
Front
MULTIVIEW
Normal Surfaces Sketching
Name:
ID#:
Lab Hr:
US CUST
ISMN-3A
T. Sexton © ISMN-3A.dwg

Given: Top & Front Views
Sketch: Right, Left, Back & Isometric Views
NO HIDDEN LINES RQUIRED!

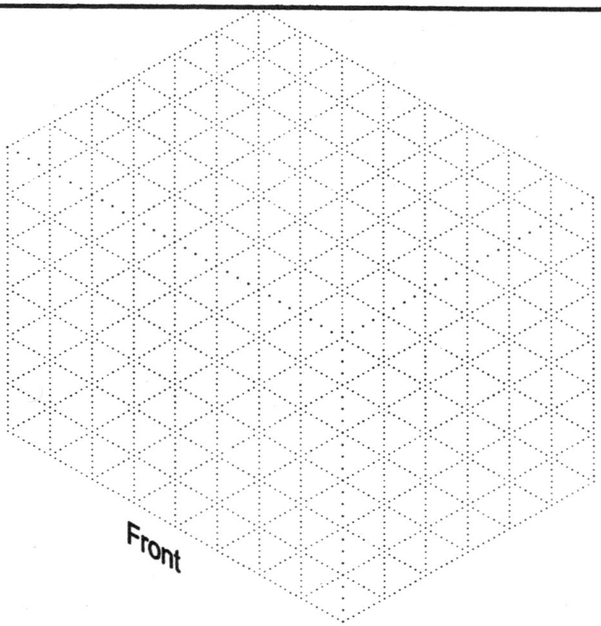

Front

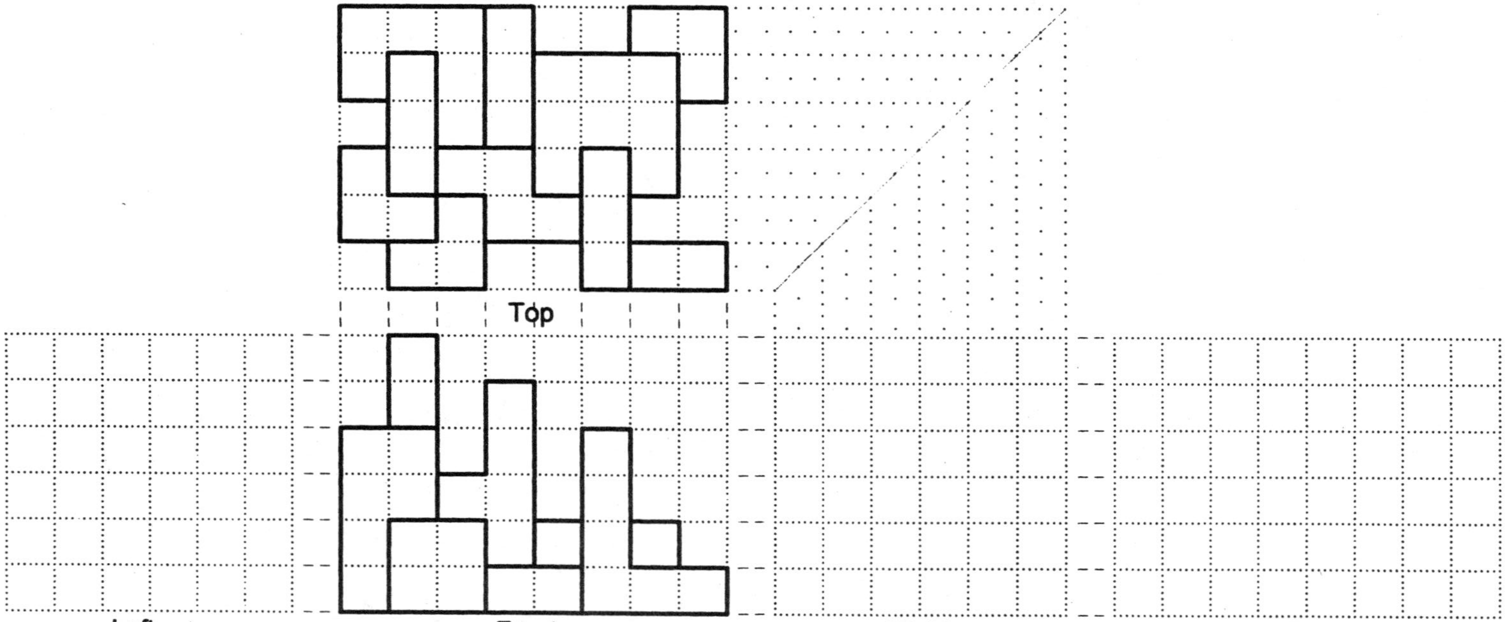

Top

Left

Front

Right

Back

MULTIVIEW Normal Surfaces Sketching	Name: ________ ID#: ________ Lab Hr: ____	US CUST	ISMN-4A

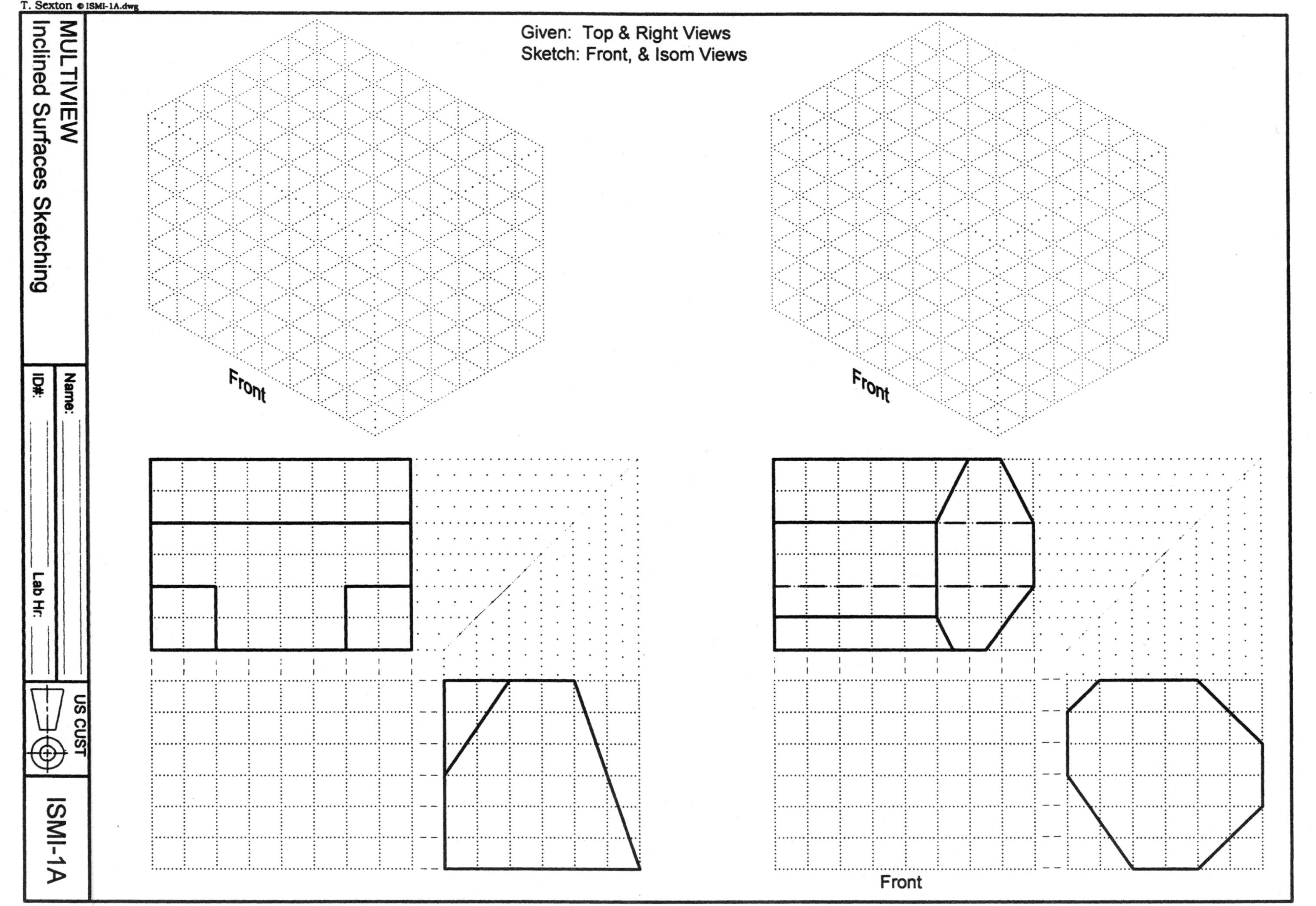
Given: Top & Right Views
Sketch: Front, & Isom Views
Front
Front
Front
T. Sexton © ISMI-1A.dwg
MULTIVIEW
Inclined Surfaces Sketching
Name:
ID#:
Lab Hr:
US CUST
ISMI-1A

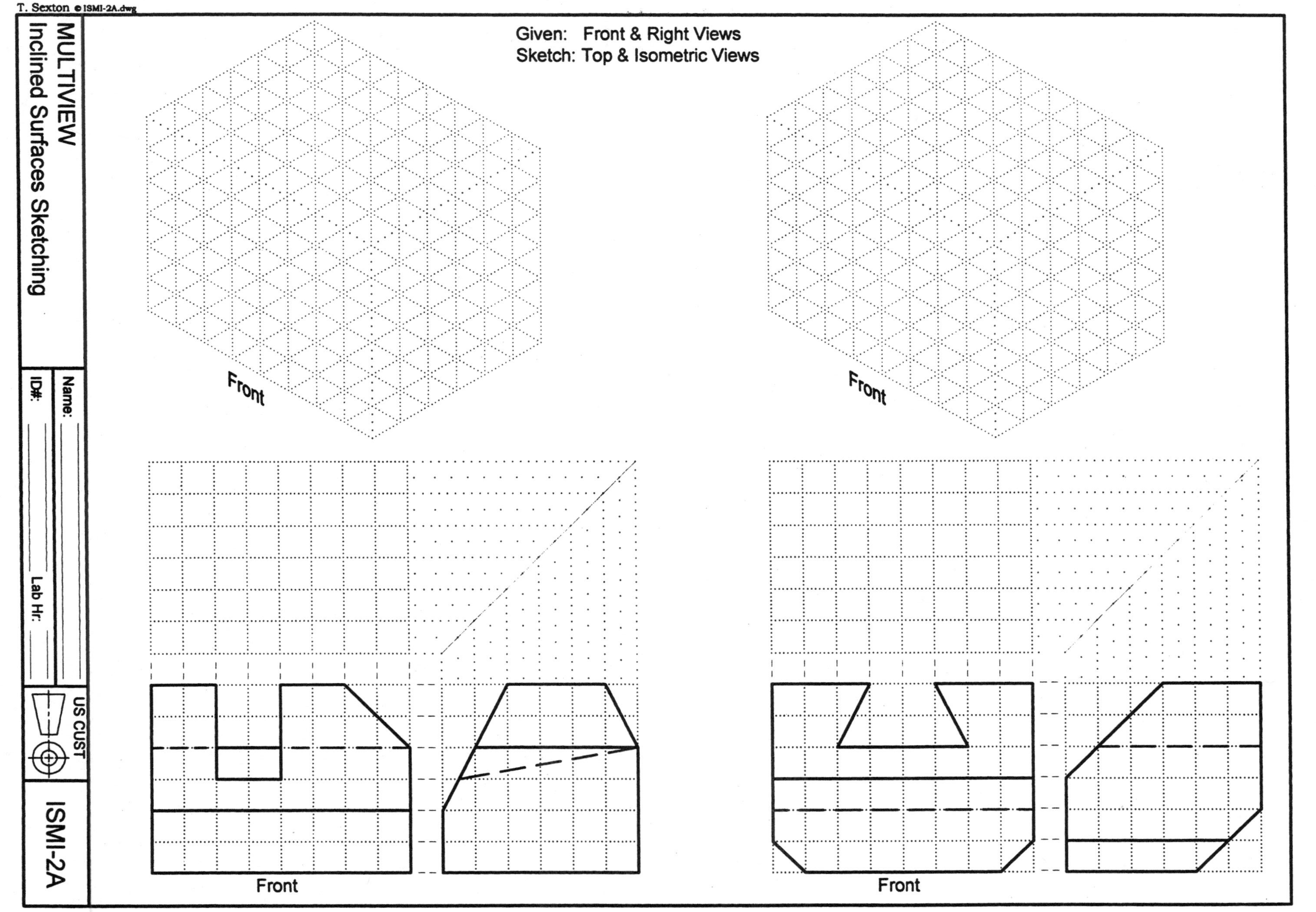
T. Sexton © ISMI-2A.dwg
Given: Front & Right Views
Sketch: Top & Isometric Views
Front
Front
Front
Front
MULTIVIEW
Inclined Surfaces Sketching
Name:
ID#:
Lab Hr:
US CUST
ISMI-2A

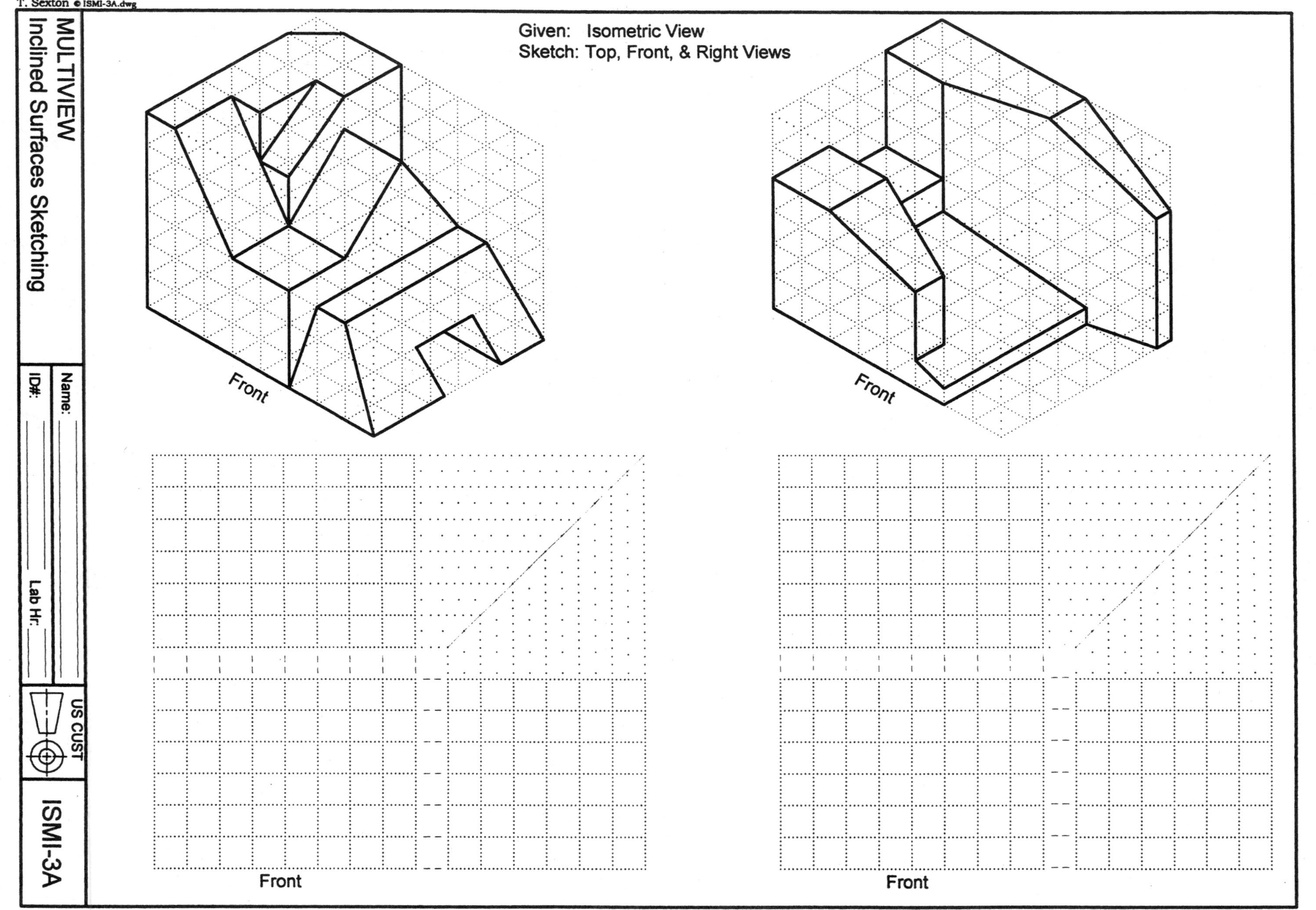
Given: Isometric View
Sketch: Top, Front, & Right Views
Front
Front
Front
Front
MULTIVIEW
Inclined Surfaces Sketching
Name:
ID#:
Lab Hr:
US CUST
ISMI-3A
T. Sexton © ISMI-3A.dwg

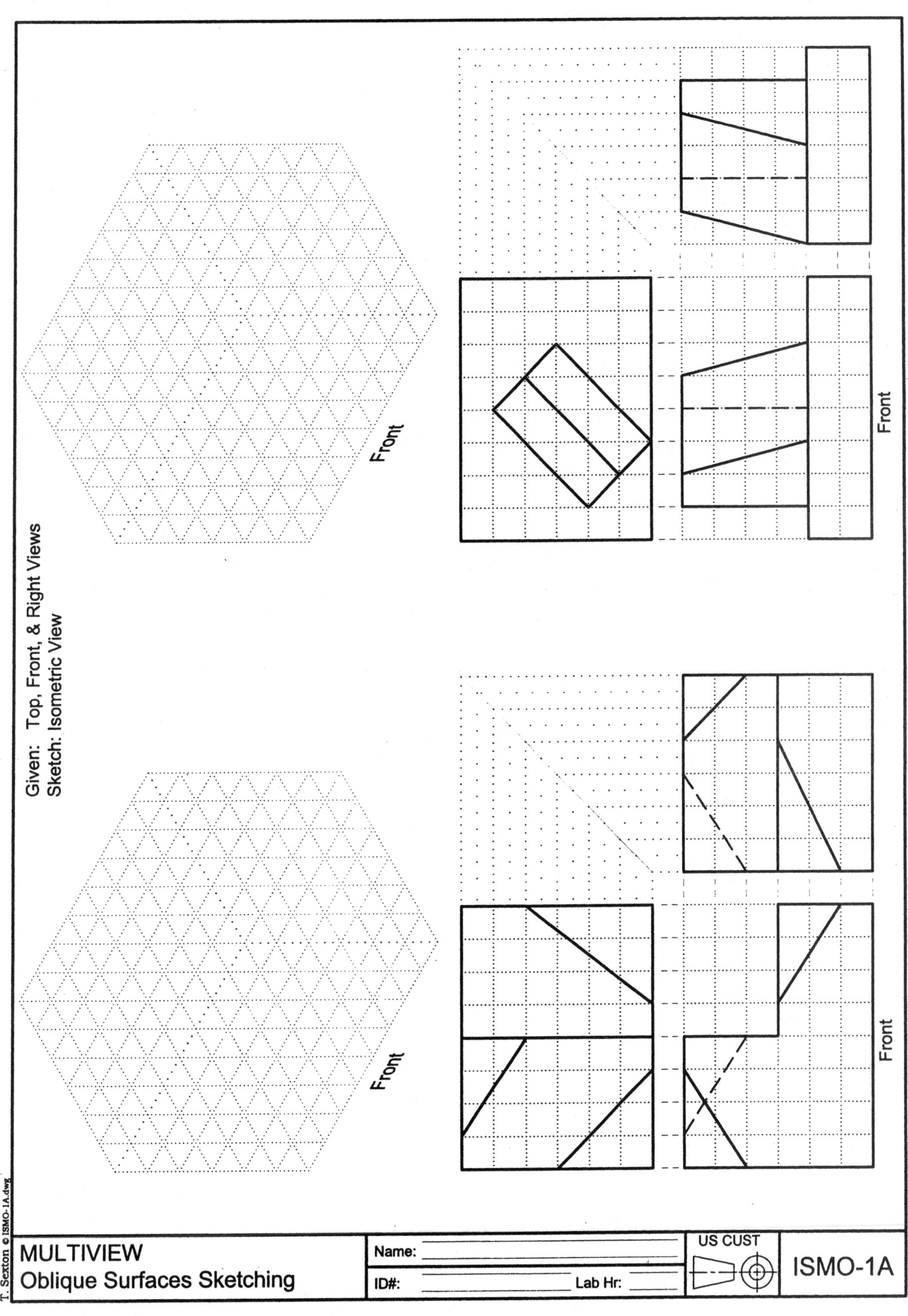
Given: Top, Front, & Right Views
Sketch: Isometric View
Front
Front
Front
Front
MULTIVIEW
Oblique Surfaces Sketching
Name:
ID#:
Lab Hr:
US CUST
ISMO-1A
T. Sexton © ISMO-1A.dwg

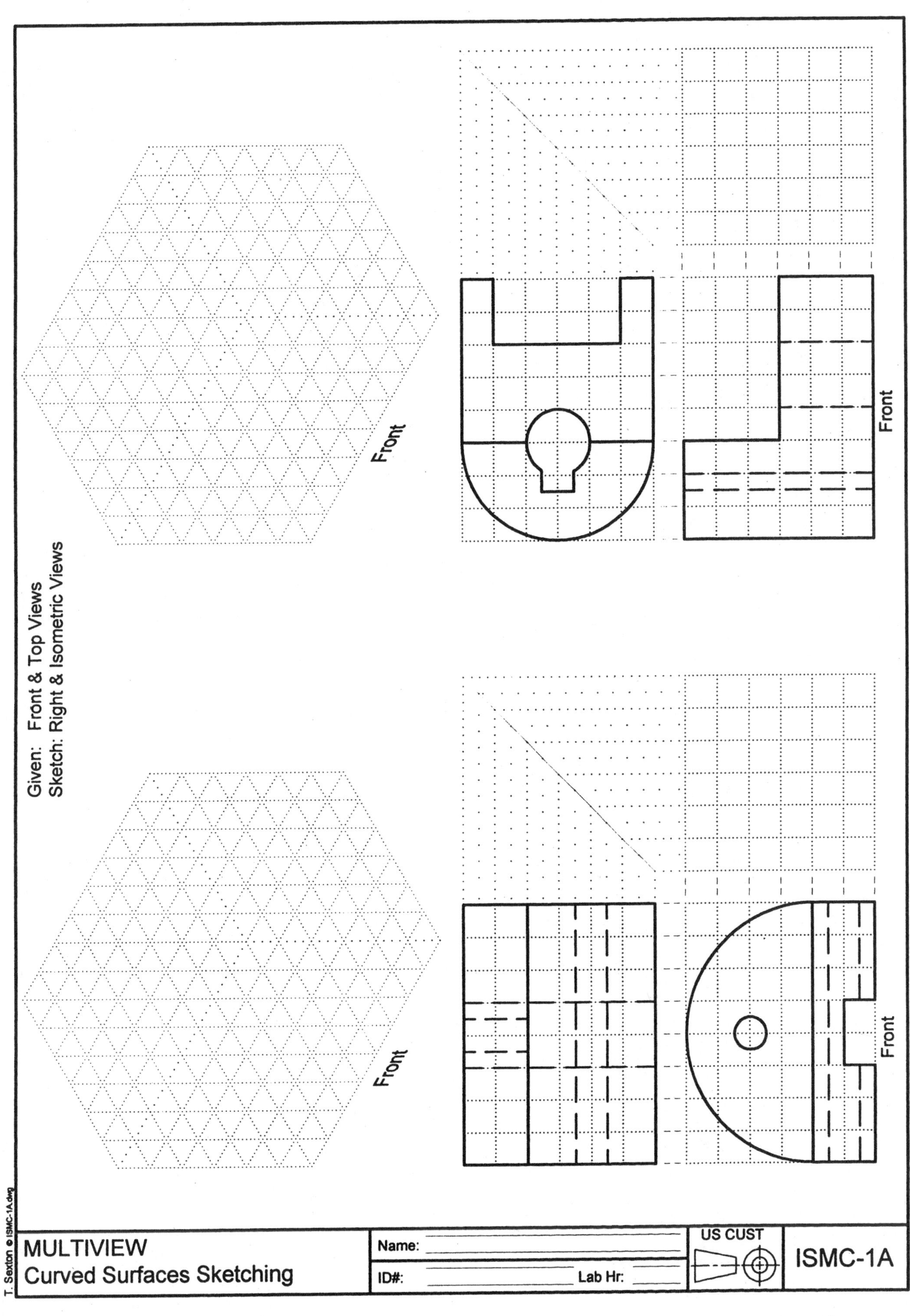
Given: Front & Top Views
Sketch: Right & Isometric Views
Front
Front
Front
Front
MULTIVIEW
Curved Surfaces Sketching
Name:
ID#:
Lab Hr:
US CUST
ISMC-1A
T. Sexton © ISMC-1A.dwg

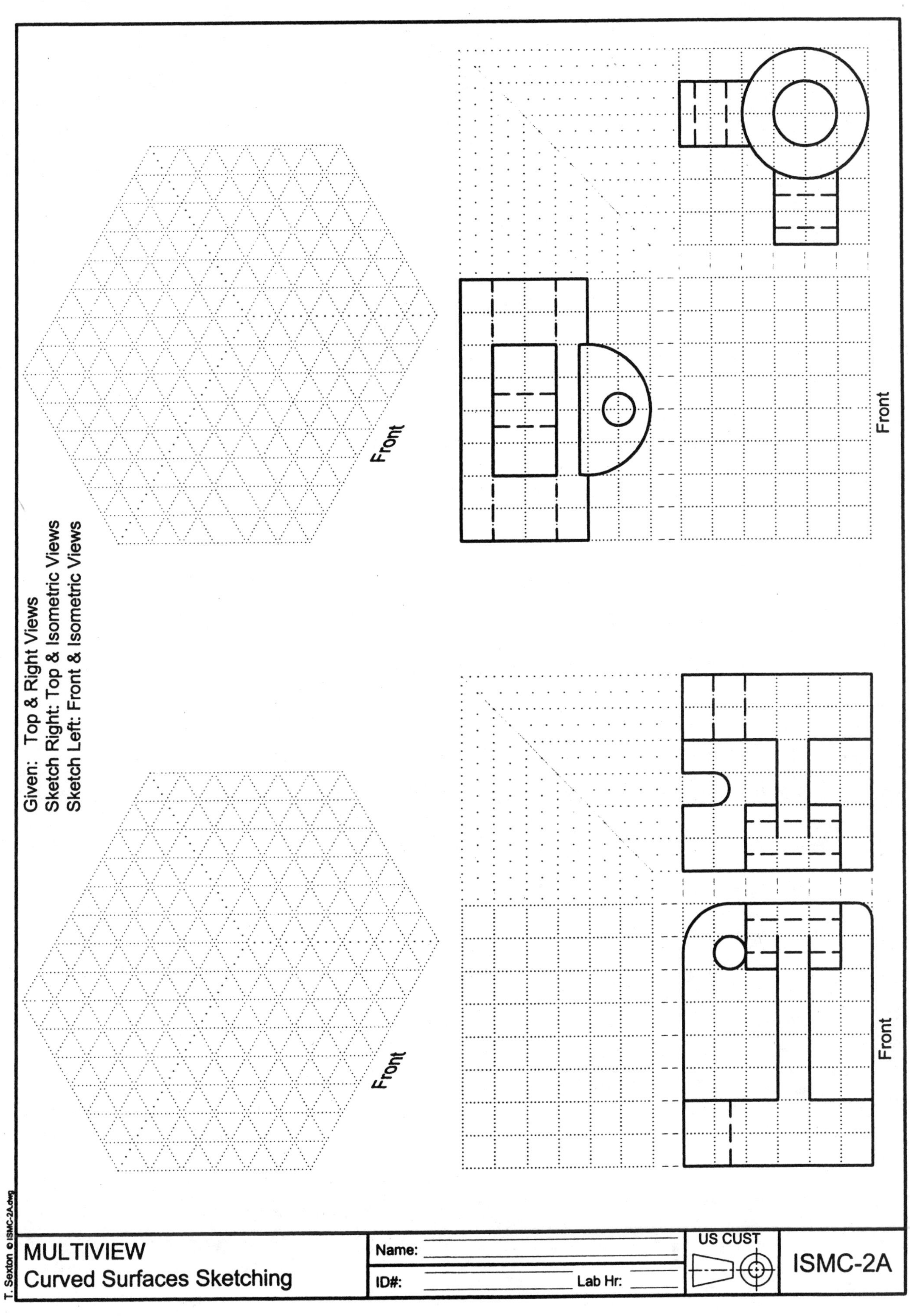
Given: Top & Right Views
Sketch Right: Top & Isometric Views
Sketch Left: Front & Isometric Views
Front
Front
Front
Front
MULTIVIEW
Curved Surfaces Sketching
Name:
ID#:
Lab Hr:
US CUST
ISMC-2A
T. Sexton © ISMC-2A.dwg

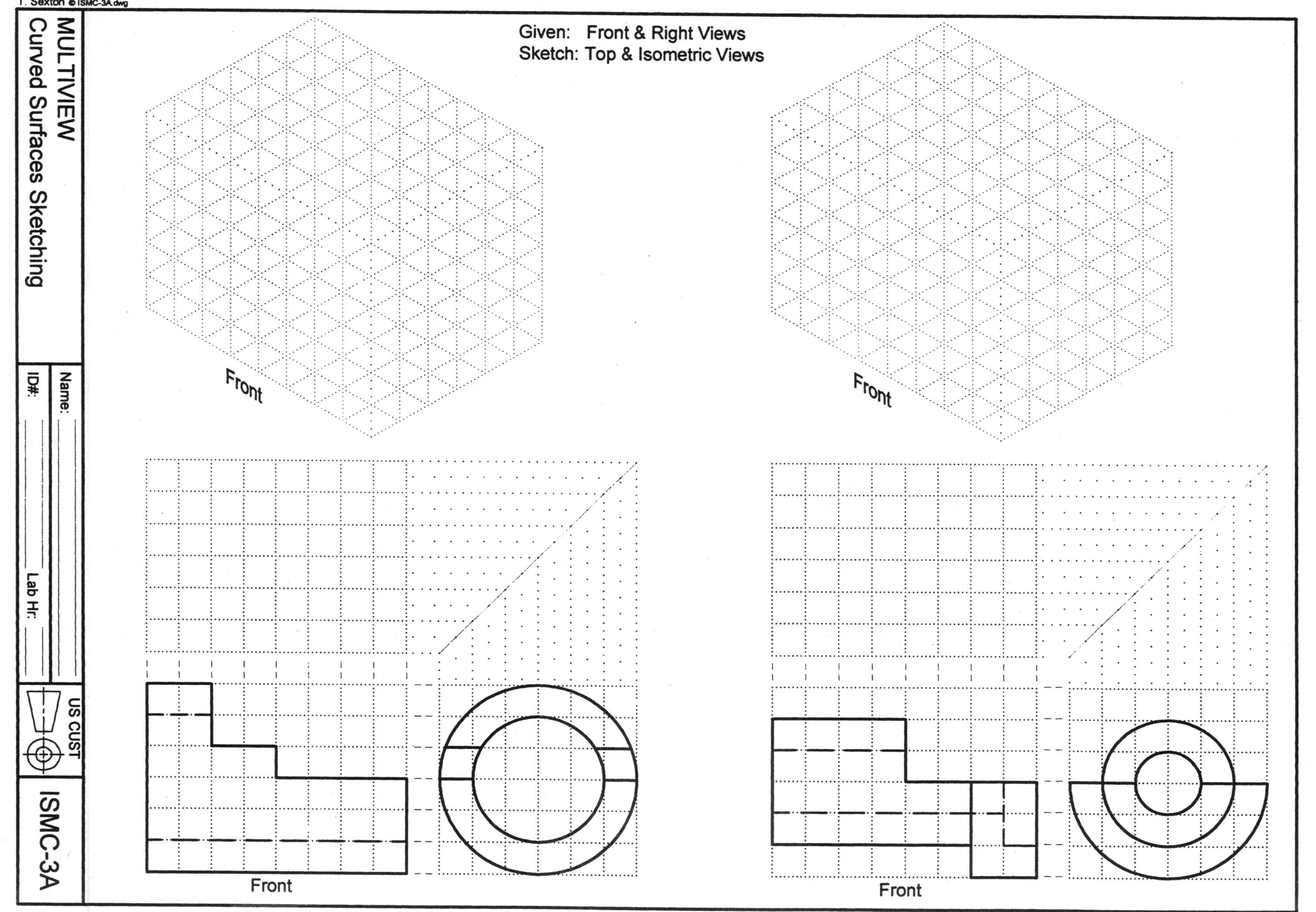
Given: Front & Right Views
Sketch: Top & Isometric Views
Front
Front
Front
Front
MULTIVIEW
Curved Surfaces Sketching
Name:
ID#:
Lab Hr:
US CUST
ISMC-3A
T. Sexton © ISMC-3A.dwg

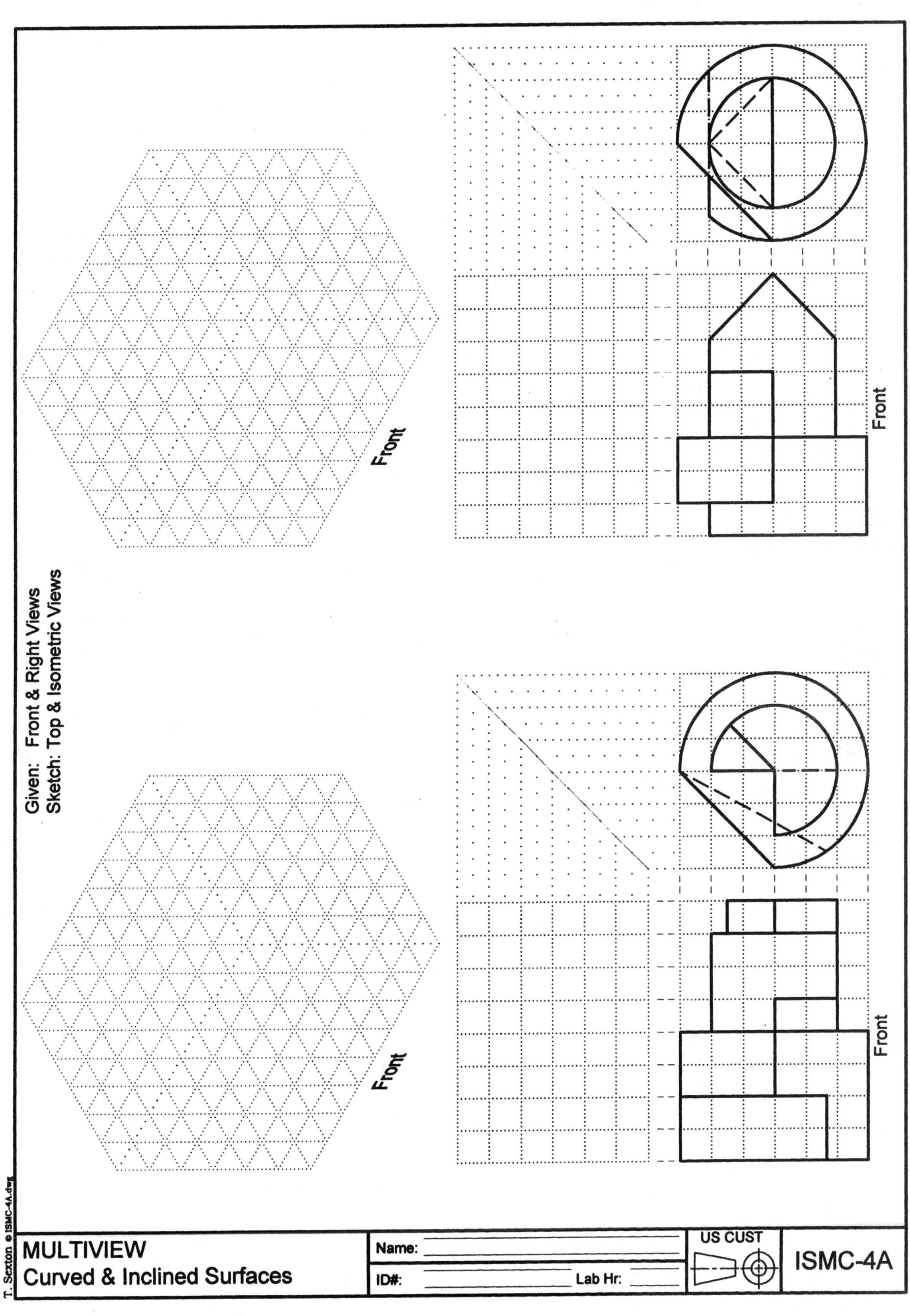
Given: Front & Right Views
Sketch: Top & Isometric Views
Front
Front
Front
Front
T. Sexton © ISMC-4A.dwg
MULTIVIEW
Curved & Inclined Surfaces
Name:
ID#:
Lab Hr:
US CUST
ISMC-4A

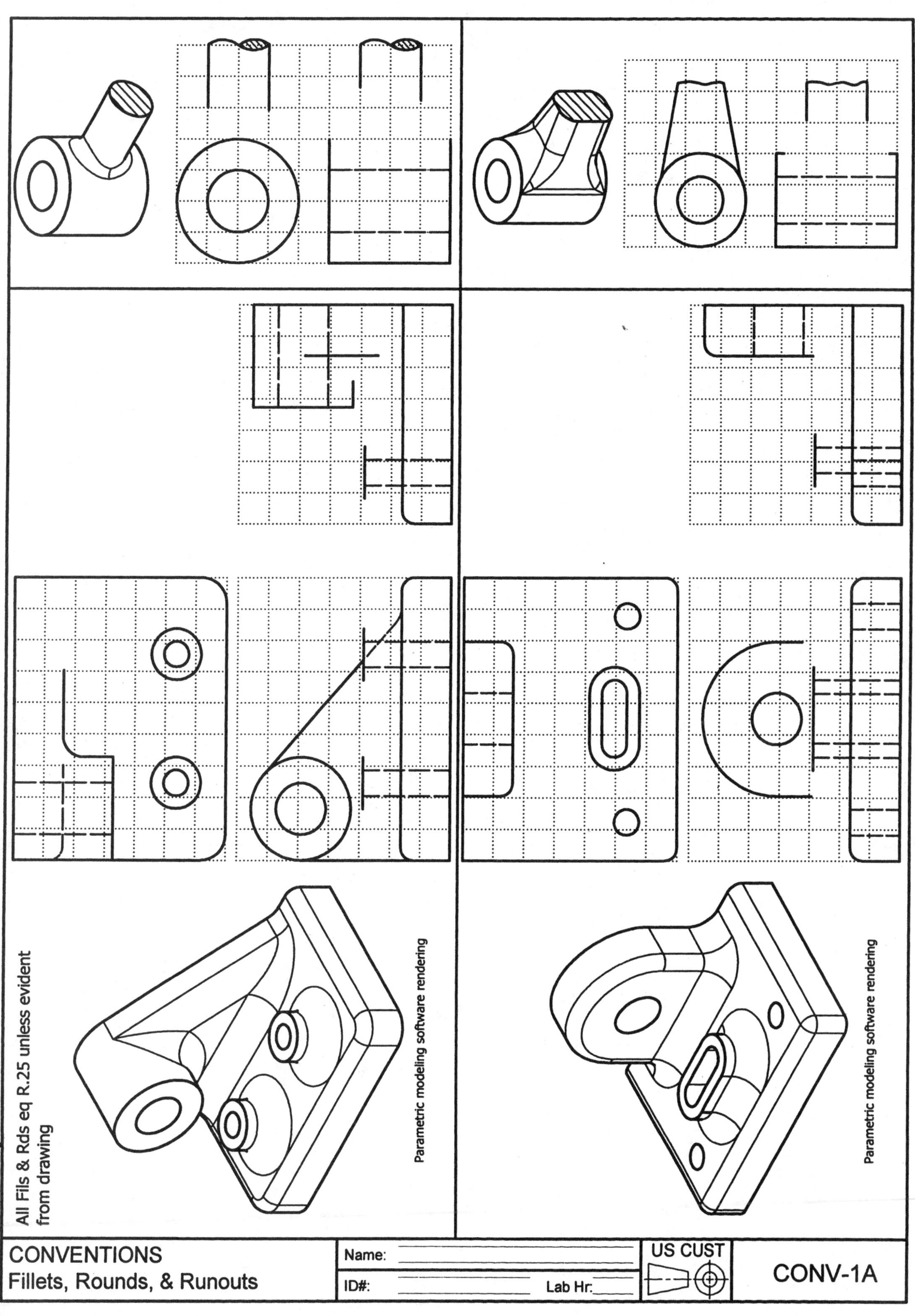
All Fils & Rds eq R.25 unless evident from drawing
Parametric modeling software rendering
Parametric modeling software rendering
CONVENTIONS
Fillets, Rounds, & Runouts
Name:
ID#:
Lab Hr:
US CUST
CONV-1A

Chapter 5

Line Types

Drawing Line Types:

Technical drawings use several different line types to help convey the shape and size of an object or building. Figure LNT 1 shows all the different *American National Standards Institute (ANSI)* approved line types used in technical drawing. Lines are drawn with two different weights (or thickness). Lines are either thick (0.7mm) or thin (0.35mm). If your drawing is done on a computer, you can specify the 0.7mm or 0.35mm line weights. When sketching, it is best to use a 0.7mm pencil for the thick lines and a 0.5mm pencil for the thin lines. A 0.3mm pencil does not show up well when sketching on grid paper. Figure LNT1 notes the line thickness with a thick (0.7) or thin (o.35) designation. Figure LNT 2 shows how the different line types are applied.

Visible object lines (or simply visible lines) are thick solid lines used to represent the edges and outlines of geometric features such as the edge view of a planes, intersection of planes, circular view of holes or cylinders, or the edges of cylinders, cones, or spheres.

Hidden object lines (or simply hidden lines) are thin dashed lines used to represent the edges or outlines of objects that are behind or within an object from the vantage point of the viewer. The length of the dashes and spaces vary with the size of the drawing but a starting point is 3mm (1/8") dashes with 1mm (1/32") spacing. A good rule of thumb is to make the dashes four times longer than the spaces.

Center lines are thin long-dash-long lines used to locate the center of a geometric feature or to represent the central axis of a symmetrical object. A center line should not terminate on another line but should extend slightly beyond other lines as illustrated by the centerline passing through the counterbored hole in the front view and right side sectioned view in Figure LNT 1. The length of the dash, space, and long segment of the center line varies with the size of the drawing. A starting point for the dash and the spaces are 3mm (1/8") for the dash and 1.5mm (1/16") for the space. In other words, the dashes are proportionately twice as long as the spaces.

Phantom lines are thin long-dash-long lines used to show the alternate position of an object. For example, a wall toggle switch could be drawn with visible lines in the off position and with phantom lines in the on position. It can also be used to show how a mating part fits into a part being detailed as illustrated by the dowel pin in Figure LNT 2. A starting point for the dashes and the spaces are 3mm (1/8") for the dashes and 1.5mm (1/16") for the spaces.

Section lines are thin angled lines used to indicate the solid material that would touch a cutting blade when cutting through an object. Figure LNT 2 has section lines in both the front and right sectioned views. The angle of section lines can vary. If you are cutting through a single part, all the resulting section lines will be at the same angle. If you are cutting through several assembled parts, each part will have section lining at a unique direction and angle as illustrated in Figure LNT 3. The space between the section lines varies with the size of the object being sectioned but a rule of thumb is to space the section lines about 3mm (.12") apart.

A ***cutting or viewing plane line*** is a thick hidden line or a thick phantom line terminating with large arrowheads at ninety degrees as illustrated in Figure LNT 1 and Figure LNT 2. If it is desirable to not cover up some details with a thick cutting plane line, draw just a small portion of line and a ninety degree line terminating with an arrowhead, as illustrated in Figure LNT 1. A cutting plane lines indicate the location and viewing direction of a section being cut or special view.

Break lines are used to show that a portion of an object has been left out. It can be either a thick irregular freehand line used with short breaks or a thin line segment connected by a zigzag used for long breaks as illustrated in Figure LNT 1 and 2.

Dimension lines are thin lines terminating in arrow heads with a dimension placed between the straight lines as illustrated in Figure LNT 1 and Figure LNT 2. A dimension line denotes the size or location of a geometric feature. In architectural drawing the dimension lines can terminate with a variety of symbols. Most common are the dot and tic mark as illustrated in Figure LNT 1.

Extension lines are a pair of thin solid lines extending from a geometric feature on the object to slightly beyond the dimension line as illustrated in Figure LNT 1 and Figure LNT 2. Extension lines help clarify exactly what is being dimensioned. Leave a small gap approximately 1/4 the height of the lettering between the object and an extension line.

Leaders are thin solid lines beginning with a short horizontal line then extending on an angled until it touches the object. The angled line terminates in an arrowhead as illustrated in Figure LNT 1 and Figure LNT 2. A leader directs information to a specific area or geometric feature.

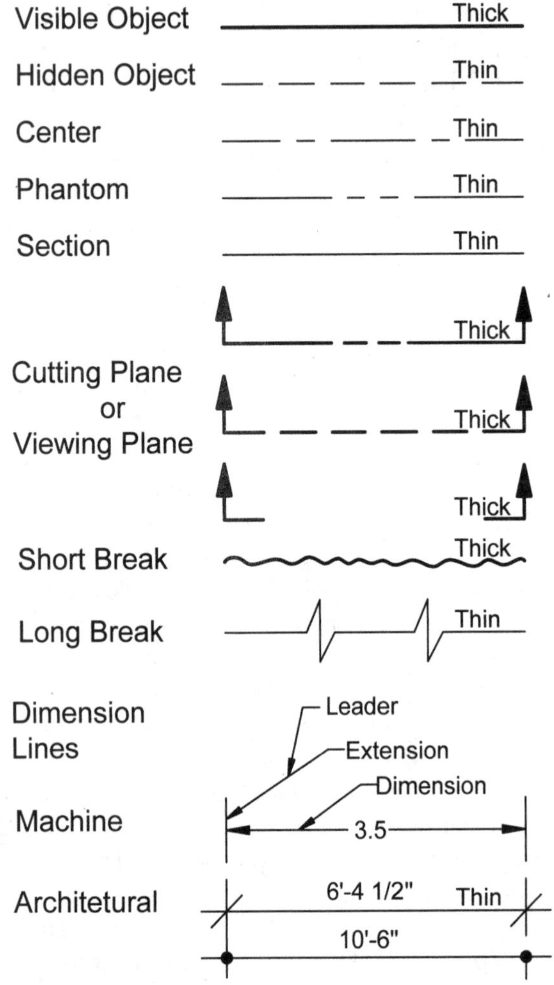

Figure LNT 1: American National Standards Institute (ANSI) approved line types used in technical drawing.

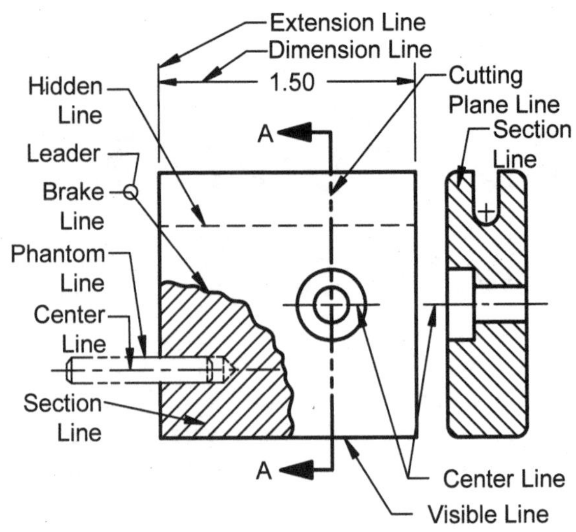

Figure LNT 2: Application of ANSI line types.

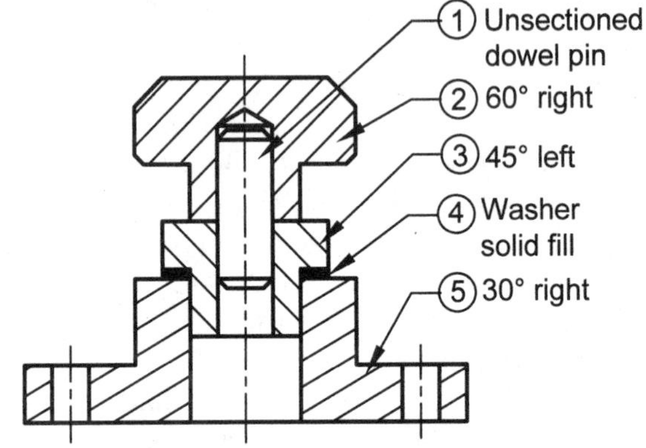

Figure LNT 3: A sectioned assembly showing each part with section lining at a unique direction and angle.

Hierarchy of Lines:

Often line types representing different details of an object will be collinear, meaning they fall on top of one another in a given view. When this happens you draw the line type that is highest on the hierarchy scale. Visible lines are most important, followed by hidden, then cutting plane, and finally center. Figure LNT 4 illustrates several applications of this rule. For example, line D in the right view of Figure LNT 4 would be a visible line representing the edge view of plane A in the front view. It supersedes the hidden line representing the intersection of planes B and C in the front view.

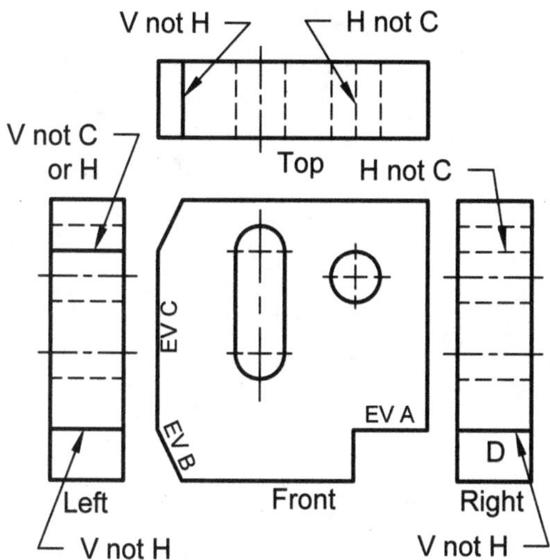

Figure LNT 4: Hierarchy of lines application where V=Visible, H=Hidden, CP=Cutting Plane Line, C=Center, PL=Plane, and EV=Edge Hidden Line Technique View.

Hidden Line Technique:

Hidden lines can be confusing! To make your drawings/sketches easier to read follow the illustrated hidden line techniques in Figure LNT 5. These techniques help delineate where hidden lines form corners, makes it easier to distinguish between visible and hidden lines, or tell which hidden line is closer. Figure LNT 6 illustrates the application of hidden line techniques and the ballooned numbers tie each example back to the specific hidden line technique in Figure LNT 5. Drawings done on a computer do not follow these techniques.

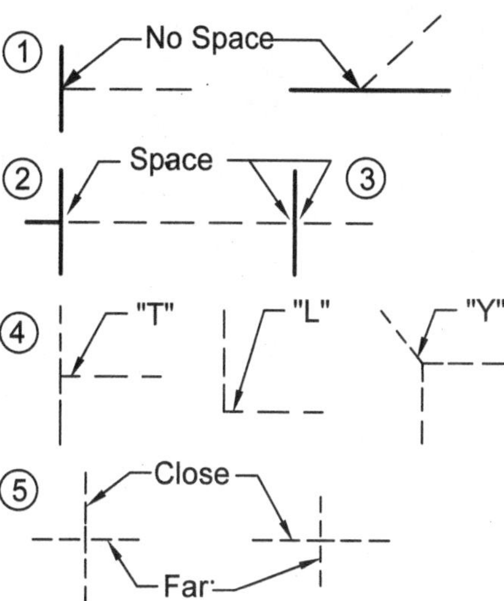

Figure LNT 5: Hidden line techniques

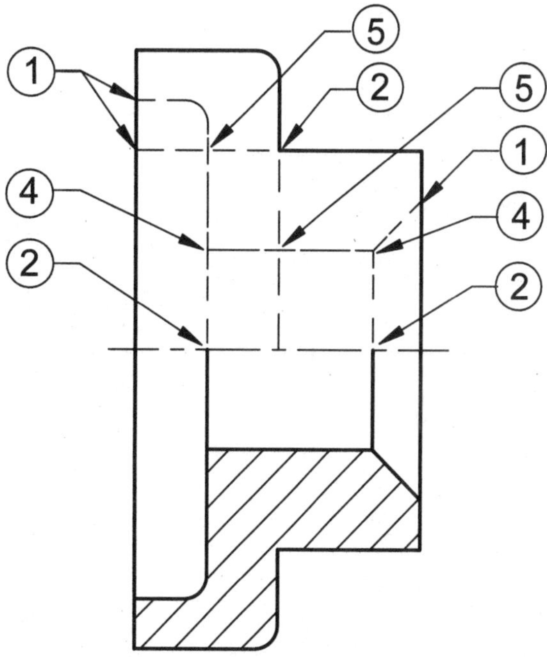

Figure LNT 6: Application of the hidden line techniques from Figure LNT 5.

NOTES:

Chapter 6

Auxiliary Views

Visualizing a Primary Auxiliary View:

For many objects the six principle views do not always describe an object's true shape. If the object contains an inclined surface, you need an additional view called a primary auxiliary view. Figure AUX 1 illustrates an example of an object with the inclined surface A. Surface A appears as an edge view in the front view and as a foreshortened plane in both the top and right views. But the true size and shape (TSS) of surface A is missing. In order to see the TSS of surface A you need to view surface A orthographically (i.e., at a right angle). Figure AUX 2 illustrates the line of sight necessary to obtain the TSS of surface A.

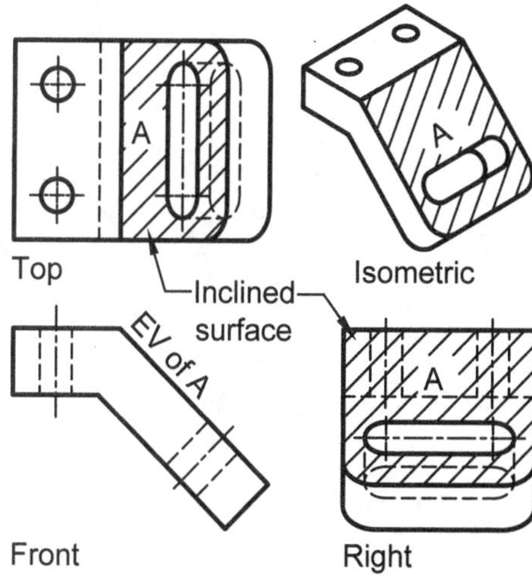

Figure AUX 1: Surface A is in edge view in the front and foreshortened in the top and right views.

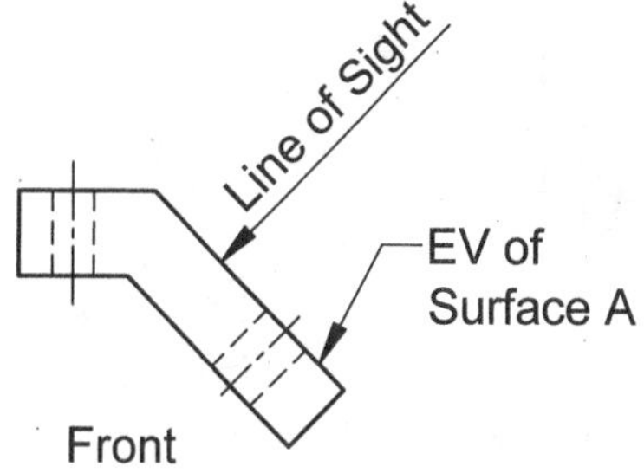

Figure AUX 2: The front view shows surface A in edge view and the line of sight necessary to produce an orthographic TSS view of surface A.

Auxiliary views should be visualized using the "plastic box" as was done when visualizing the principle views of a multiview drawing. Figure AUX 3A shows our object in a plastic box made up of only three principal views. To obtain the TSS of the inclined surface, a plastic plane must be parallel to the inclined plane as illustrated in Figure AUX 3B. Figure AUX 3C shows the auxiliary plane of projection partially rotated about the front view. Figure AUX 4 shows the auxiliary view after a complete 90° rotation. As illustrated in Figure AUX 4, the auxiliary view must stay aligned with the view it is rotated about, i.e., it must maintain projectability.

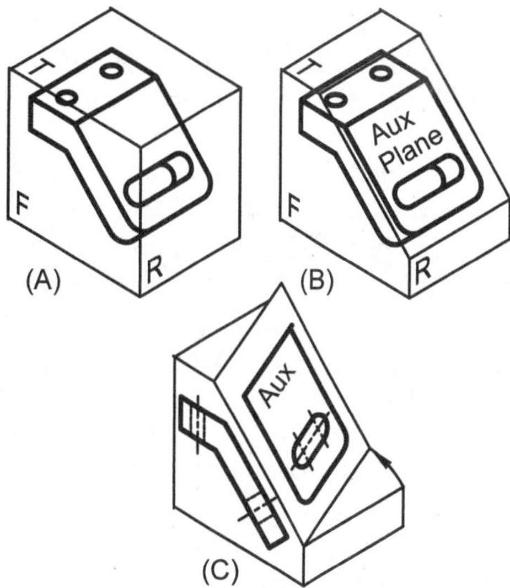

Figure AUX 3: A) Planes of projection for the Top, Front, and Right views, B) a plane of projection is placed parallel to the inclined surface, C) the auxiliary plane is revolving about the front view.

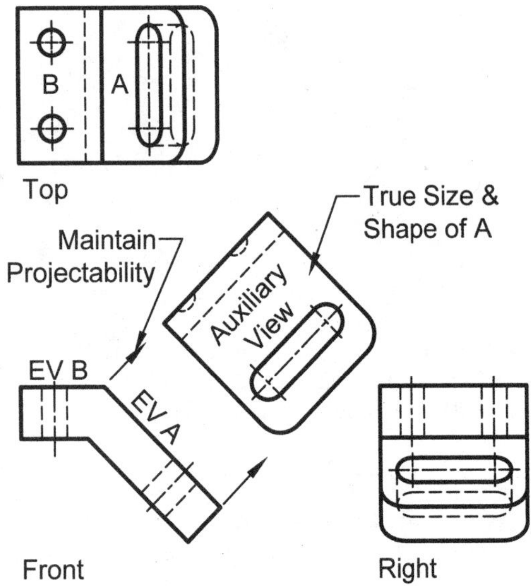

Figure AUX 4: An auxiliary view showing the true size and shape of the inclined surface.

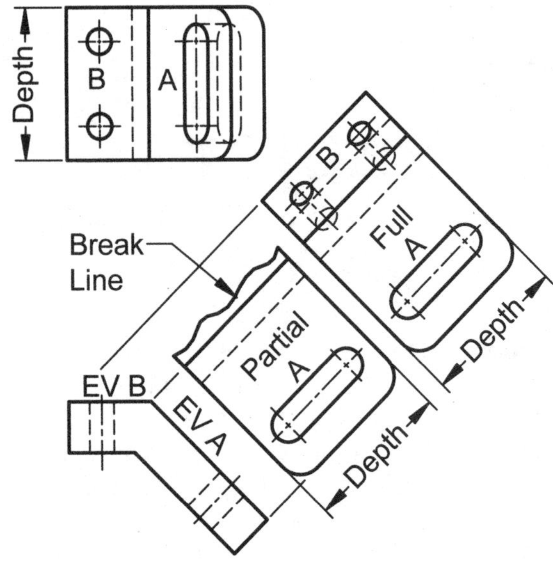

Figure AUX 5: A partial versus a full auxiliary.

Full vs. Partial Auxiliary Views

In Figure AUX 4, the auxiliary view shows only the inclined surface A. This is called a *partial auxiliary*. Figure AUX 5 illustrates both a partial auxiliary, showing only surface A, and a full auxiliary showing both the inclined surface A and the foreshortened top plane B. Since the TSS of surface B is displayed in the top view, the foreshortened view in the auxiliary can be considered unnecessary. In the partial auxiliary of Figure AUX 5, a break line is used to show that the view is incomplete. In Figure AUX 6 both the auxiliary view and the top view are partial views. Since the inclined surface A is foreshortened in the top view it can be considered unnecessary. The use of partial views is a judgment call. If you are sketching or manually drafting, it can save a lot of time. If you have a 3-D computer generated model, then full auxiliary views are effortless to generate. But you must ask yourself if the full view clarifies the inclined plane or makes it more difficult to read.

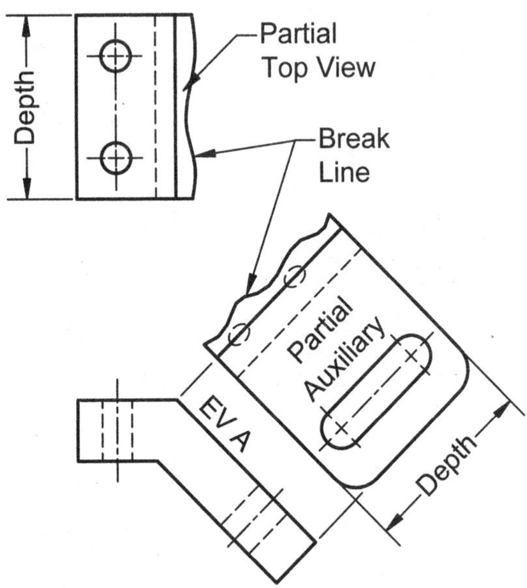

Figure AUX 6: Partial top and partial auxiliary views.

The most common break lines used to indicate that a view is incomplete are illustrated in Figure AUX 7. Figure AUX 7A illustrates a free hand wavy line used as a break line for short distances. The "lightning bolt" (author's terminology) break line illustrated in Figure AUX 7B is used for long distances and in architectural drawings.

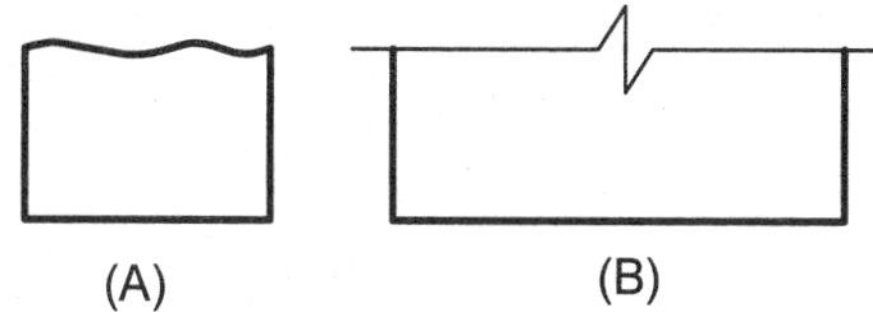

Figure AUX 7: A) short freehand break line, B) long break line.

Type of Primary Auxiliary Views

In Figure AUX 4 the auxiliary was hinged off the front view. This type of auxiliary is referred to as a *front auxiliary* or a *depth auxiliary*. It is called this because the edge view of the inclined plane is in the front view and the auxiliary plane is rotated about this edge view. A front auxiliary can also be called a depth auxiliary because depth cannot be measured in the front view but is needed to complete the auxiliary rotated about the front view.

Figure AUX 8 shows the top, front, and right views of an object that includes the EV of an inclined plane in the top view. Figure AUX 9 shows the top and front views and the primary auxiliary view which has been rotated about the top view. This type of auxiliary is called a *top auxiliary* or *height auxiliary*.

In Figure AUX 8 and 9 the lines labeled T/F and T/A are reference lines that can be envisioned as the hinges or "folding lines" where the theoretical "plastic" planes of projection intersect. Notice how the height measurements are made with reference to these folding lines. A second way to envision the T/F and T/A lines is to regard them as reference planes from which measurements are taken. In Figure AUX 8 the T/F line represents a horizontal plane when looking at the front view and a vertical plane when looking at the top view.

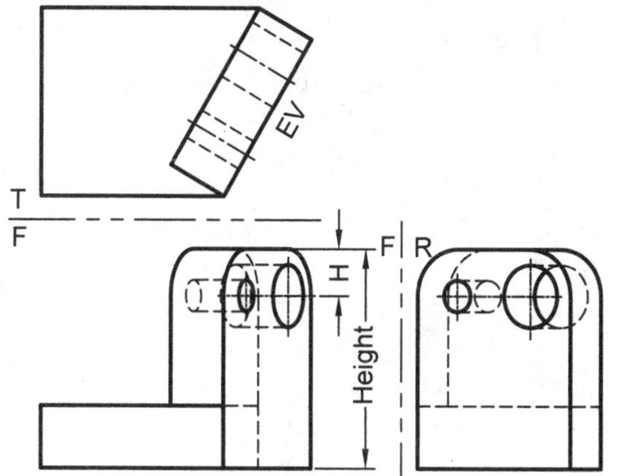

Figure AUX 8: Inclined surface with its edge view in the top view.

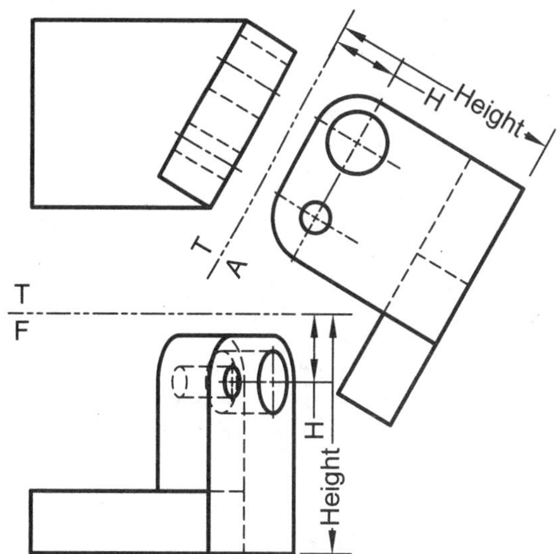

Figure AUX 9: To and height auxiliary.

Figure AUX 10 shows the top, front, and right views of an object with the edge view of an inclined surface in the right view. Figure AUX 11 shows the right view and a *width auxiliary* or *right auxiliary*. The width measurements should be taken in the front view with respect to the reference plane F/R.

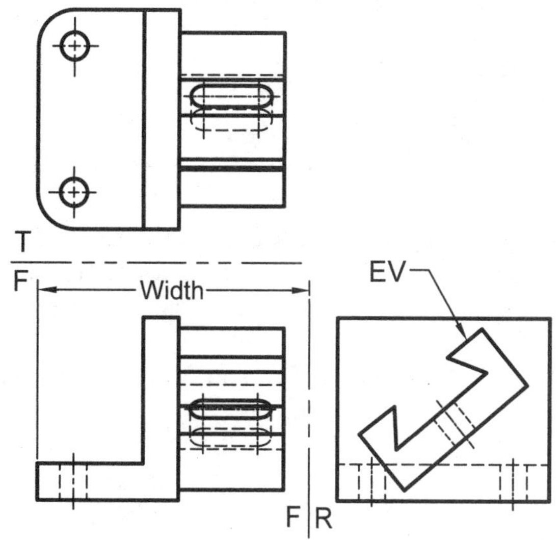

Figure AUX 10: edge view of an inclined surface in the right view.

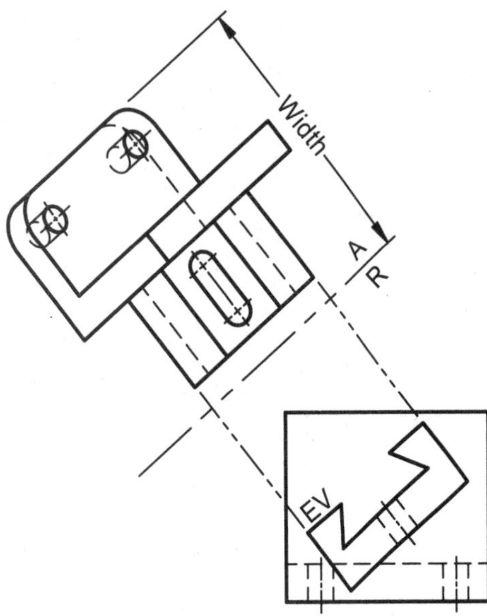

Figure AUX 11: Right or depth auxiliary.

Reverse Construction:

When drafting by hand or using a 2-D CAD system, it is often easier to draw the auxiliary first and then project points back to the principle views. Figure AUX 12 illustrates the procedure:

1. draw the front view, partial auxiliary view, and the right view except the irregular curve;
2. in the auxiliary view, layout points a, b, c, d, and e (every 15°);
3. project points a – e back onto the edge view of the inclined surface in the front view;
4. project points a – e from the front view to the right view;
5. lay out points a – e using their corresponding measurements, point a and e use distance x, point b and d use distance y, and point c uses distance z.
6. connect points w, a-e, and V with a French curve or elliptical template

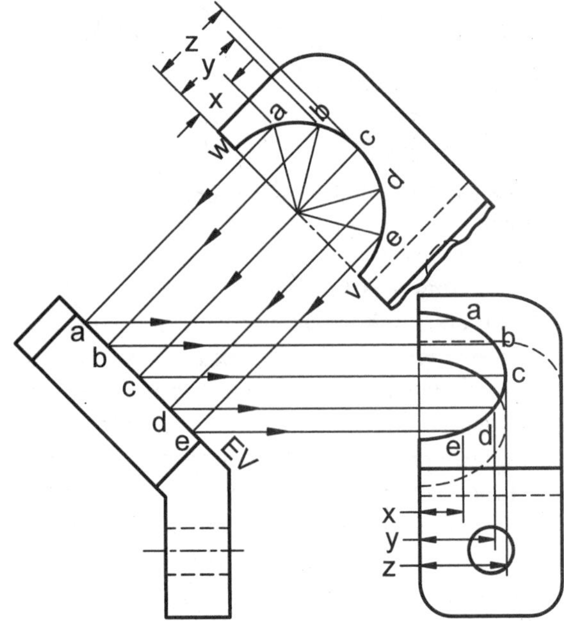

Figure AUX 12: The curve in the right view is generated by projecting the points a, b, c, d, and e from the auxiliary view to the EV in the front view and then to the right view. Then the distances x, y, and z are transferred from the auxiliary view to the right view.

Successive Auxiliaries:

The true size and shape (TSS) of an inclined plane can be found by a single 90° rotation about an edge view located in one of the principle views. But finding the TSS of an oblique plane takes one additional 90° rotation or *successive auxiliary* called a *secondary auxiliary.*

In Figure AUX 13 there is the front, top, primary auxiliary A1 and secondary auxiliary A2. The elliptical holes and arcs of the top and front views will need to be constructed after the TSS oblique plane in view A2 is laid out. The following steps are a guide to construct the four views:

1. layout as much of the front and top views as possible (the elliptical portions will have to wait for now),
2. draw the folding line T/A1 perpendicular to the TL line 1,2 in the top view. This gives you the edge view of both surfaces and the true angle between the surfaces. Then layout primary auxiliary A1 using height measurements X from the folding line T/F into the front view,
3. draw the folding line A1/A2 parallel to the EV of the oblique surface, then layout the true circle and arc in the secondary auxiliary A2

using measurements Y from the folding line A1/T into the top view,

4. from the secondary auxiliary view A2 project the points 3, 4, 5, 6, and 7 and the center hole back onto the EV in the primary auxiliary A1 and complete the primary auxiliary A1,
5. project point 3, 4, 5, 6, and 7 from the primary auxiliary A1 into the top view and locate these points using Y dimensions taken from folding line A1/A2 into the secondary auxiliary A2,
6. project points 3, 4, 5, 6, and 7 from the top view into the front view and locate these points using the height X dimensions taken from folding line T/A1 into the primary auxiliary A1,
7. complete the hole in a similar manner,
8. use French curve or elliptical template to construct the elliptical hole and arcs.

To find the dimensions needed to plot points on auxiliary views, place your finger on the auxiliary view then jump back over two successive folding lines putting your finger on each view as you jump. Obtain your measurements by measuring from the second folding line jumped to the view your finger is touching.

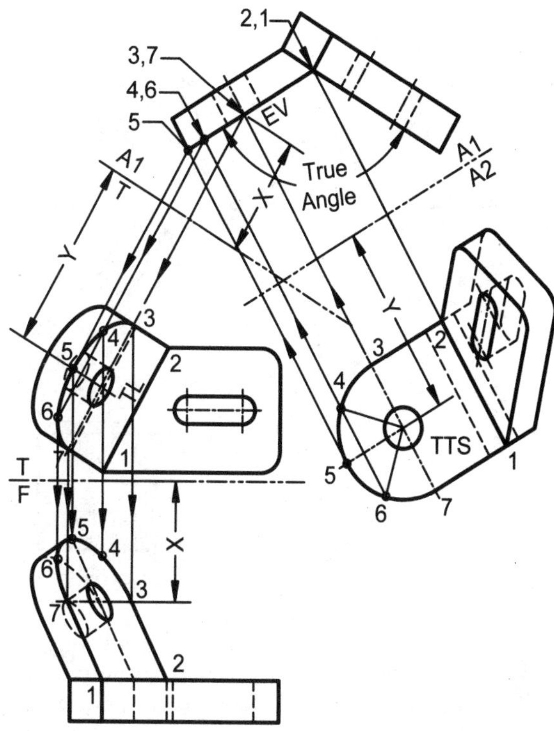

Figure AUX 13: Primary auxiliary A1 displays the EV of the oblique surface and secondary auxiliary A2 displays the TSS of the oblique plane.

Dihedral Angles:

A dihedral angle is the angle formed when two planes intersect. Figure AUX 14A shows an example of a V block (used to cradle cylindrical stock in a machine shop when the cylinder, for example, is to receive a drilled hole) with the line 1, 2 forming the line of intersection between the planes W and X. In Figure 14A the line of intersection 1,2 is true length in the top view and shows as a point projection 1,2 in the front view. When you have the point projection of the true length line of intersection (line 1,2) between planes X and W, the view will display the edge view of the two intersecting planes allowing you to see the true dihedral angle.

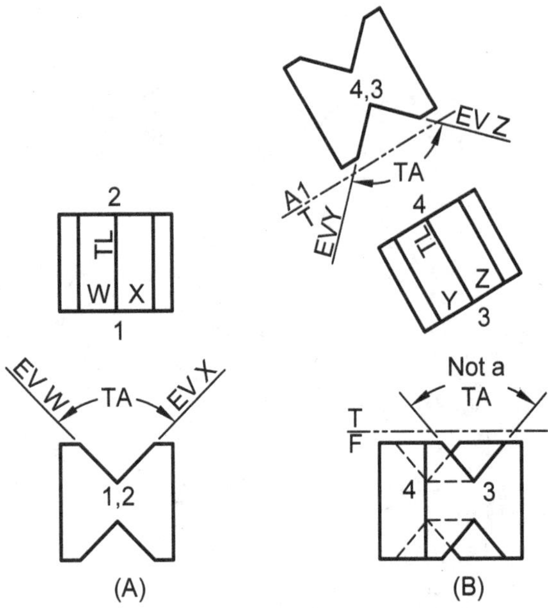

Figure AUX 14: Finding the true dihedral angle A) when the point projection of the line of intersection is given, and B) when finding the point projection of the line of intersection must be found.

When the given views do not show both intersecting planes in edge view, the true dihedral angle will not be displayed. The front view in Figure 14B does not show the edge views of planes Y and Z and thus does not display the true dihedral angle. To find the true angle the point projection of the true length line of intersection 3,4 must be found. First locate line 3,4 in true length as in the top view of Figure 14B. Then set up the folding line T/A perpendicular to the true length line 3,4. Project points into the auxiliary and complete the view by obtaining height measurements from the front view (remember to jump back two successive views.) The auxiliary A1 shows the point view of 3,4 and the edge views of surfaces Y and Z and thus the dihedral angle as a true angle.

When the line of intersection forming the dihedral angle is an oblique line, two successive auxiliaries are required to find the true dihedral angle. The first auxiliary finds the TL of the line of intersection and a second auxiliary finds the point projection of the line of intersection and thus the true angle. In Figure AUX 15 the line of intersection DF is an oblique line. The following steps are required to find the true angle between the planes ABDF and DEFG:

1. locate the primary auxiliary line R/A, parallel to the line DF in the right view. This results is a primary auxiliary showing DF in true length. Any line parallel to a folding line will show true length in the next successive view.
2. set up the folding line A1/A2 perpendicular to the TL line DF in the primary auxiliary A1. This result in a secondary auxiliary with the point projection of line DF and the accompanying edge views and the true dihedral angle. Any time you find the point projection of a true length line of intersection between planes you have the edge views of the planes forming the angle and the true dihedral angle between the planes.

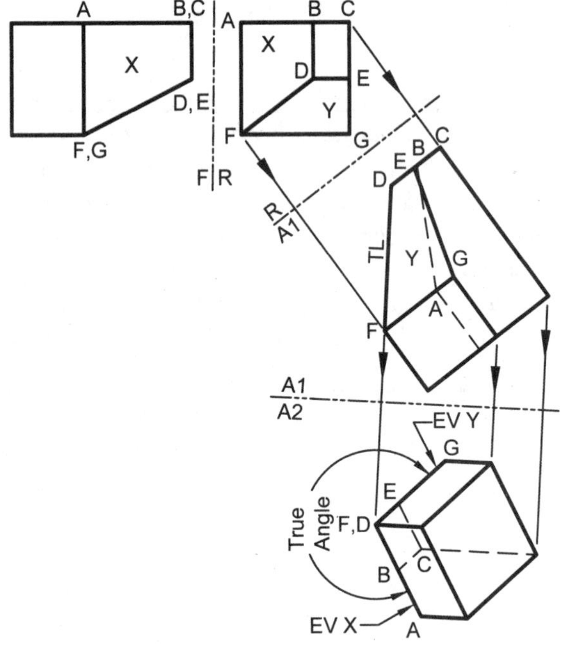

Figure AUX 15: steps to find the true dihedral angle when the line of intersection is an oblique line.

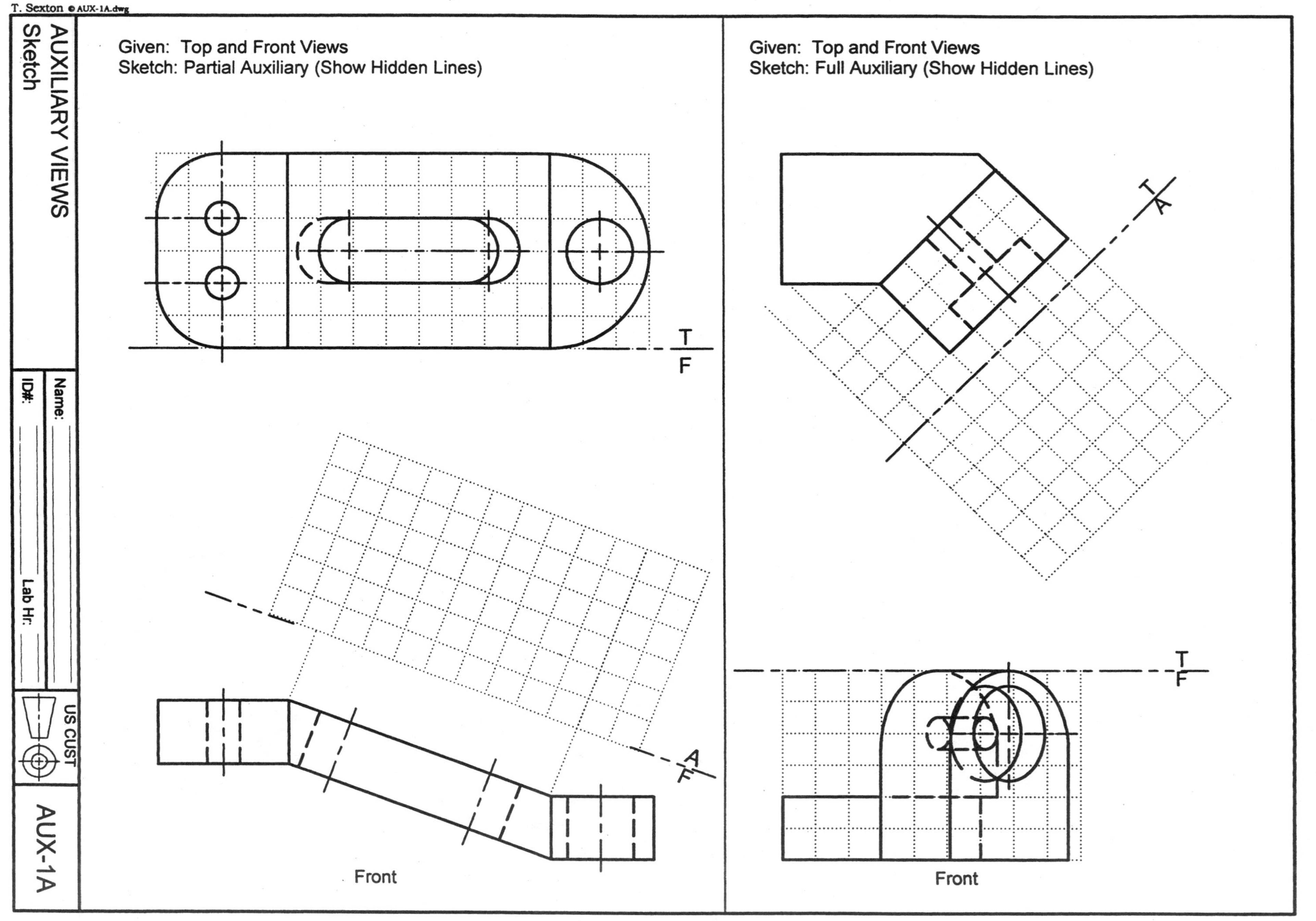
Given: Top and Front Views
Sketch: Partial Auxiliary (Show Hidden Lines)
T
F
A
F
Front
Given: Top and Front Views
Sketch: Full Auxiliary (Show Hidden Lines)
T
A
T
F
Front
AUXILIARY VIEWS
Sketch
Name:
ID#:
Lab Hr:
US CUST
AUX-1A
T. Sexton © AUX-1A.dwg

AUXILIARY VIEWS

Name: ______ ID#: ______ Lab Hr: ______

US CUST

AUX-2A

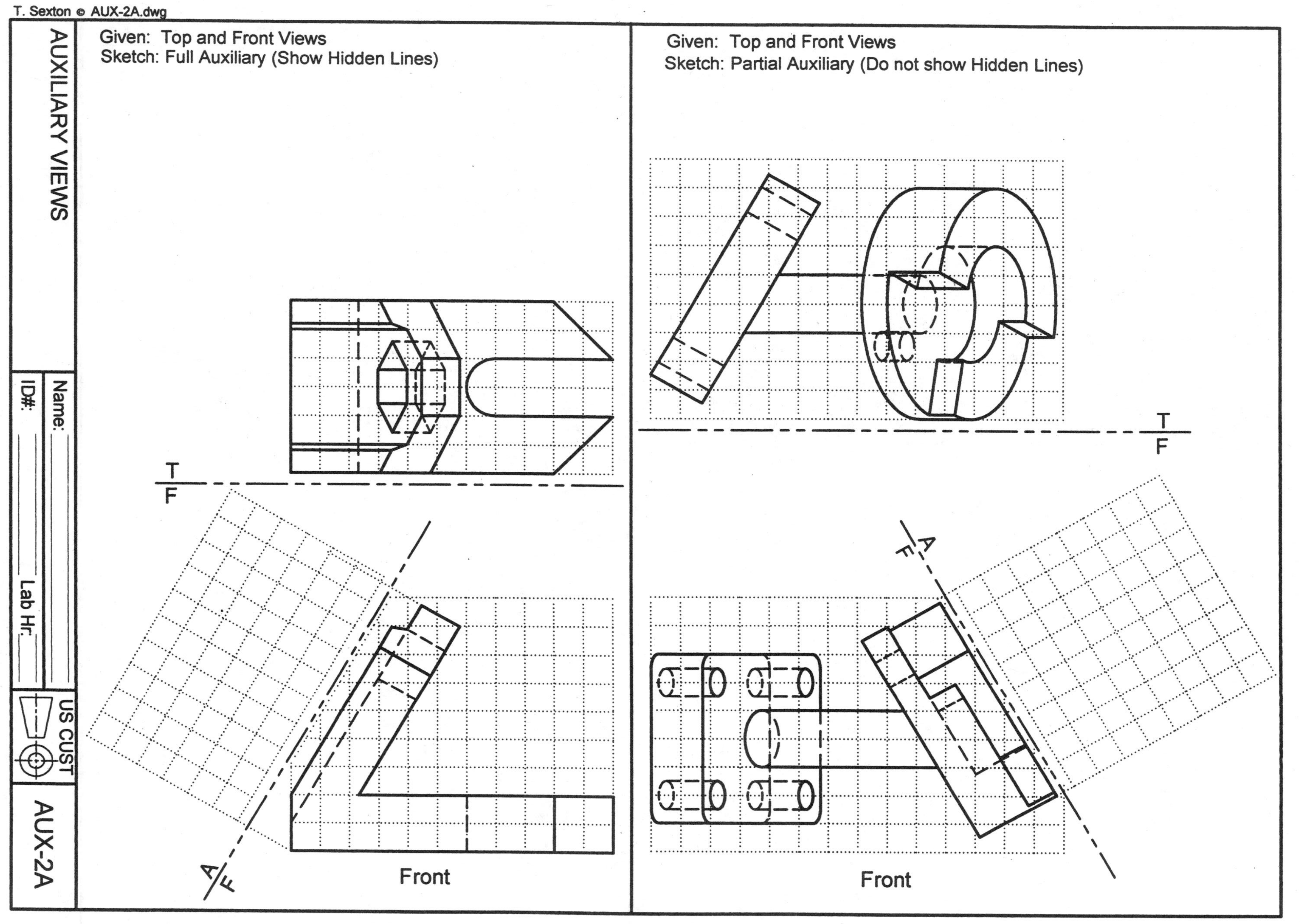

Chapter 7

Sectional Views

Sectional Views:

Multiview drawings do an excellent job describing the shape of objects. But when an object contains complex internal features, the hidden lines describing these features can get confusing as illustrated in Figure SEC 1. To clarify the internal features of an object, sectional views are drawn. A *sectional view* is created by cutting away part of the object in order to expose internal details.

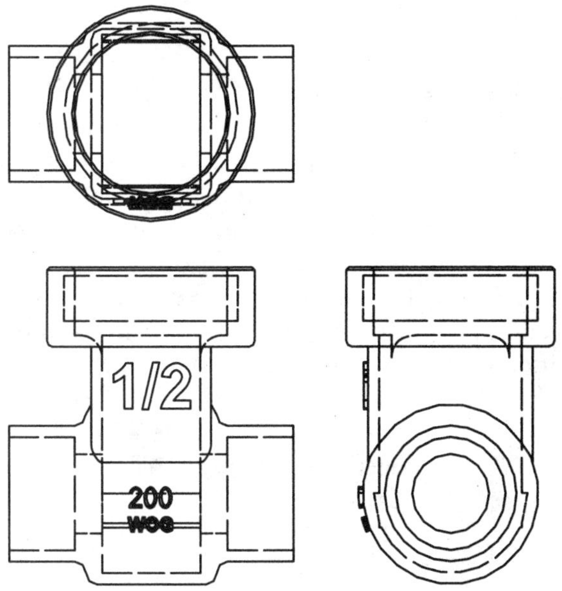

Figure SEC 1: Sample of a drawing with hidden lines indicating the geometry inside the object.

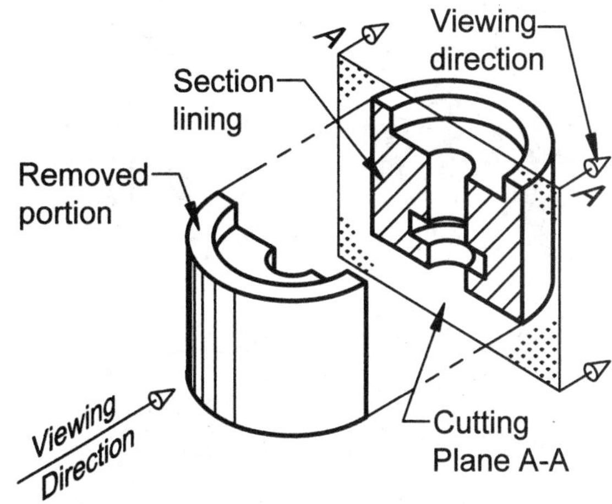

Figure SEC 2: Pictorial drawing showing the cutting plane, viewing direction, and portion of the object removed.

Figure SEC 2 is a pictorial drawing that illustrates how a sectional view is formed. Imagine the cutting plane A-A as a sheet of plastic that cuts the object in half. The portion of the object on the viewer's side of the cutting plane is removed. The portion of the object touching the cutting plane receives angled section lining.

Figure SEC 3A shows the standard front and top views of a cylindrical object containing internal features. To clarify these internal details a portion of the object needs to be cut away. Figure SEC 3B shows the top view of cutting plane line A-A which indicates the cylindrical object is to be cut in half. The arrowheads in SEC 3B indicate the observer's viewing direction and the portion of the object to be removed. In Figure SEC 3B, the standard front view is replaced by Section A-A which exposes the internal details. Any portion of the object "in" the plastic cutting plane sheet receives angled section lining (CAD programs may refer to them as x-hatching or crosshatch).

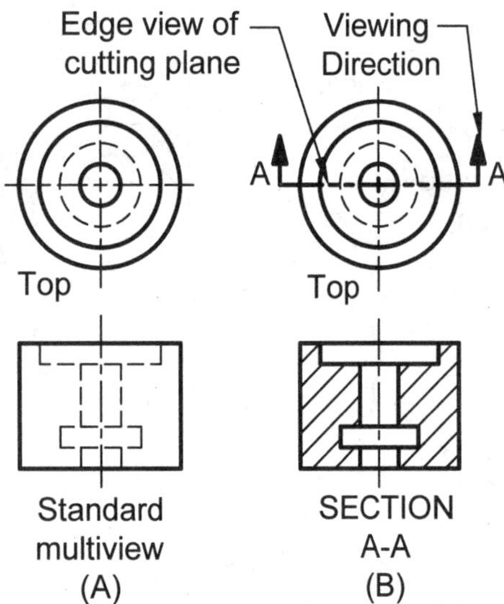

Figure SEC 3: A) Standard top and front views; B) top view with cutting plane line and resulting full SECTION A-A.

Cutting Plane Line:

The cutting plane line, illustrated in Figure SEC 3B, shows exactly where and how the object is to be sectioned or cut apart.

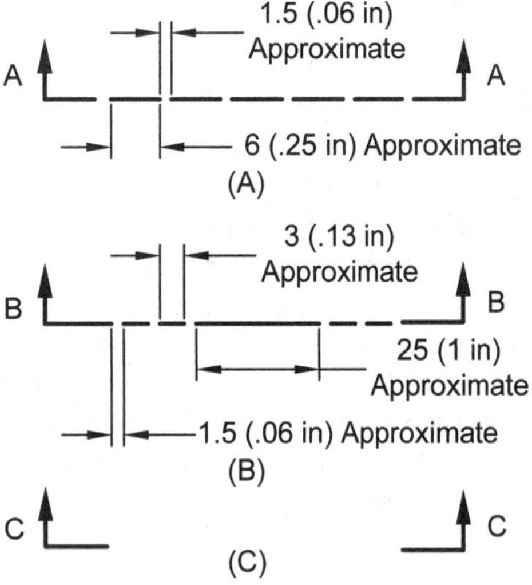

Figure SEC 4: All dimensions given above are approximate as their actual length depends on the size of the drawing. The three styles of cutting plane lines are: A) hidden line style, B) phantom line style, and C) ends only style used when the cutting plane line would hide important details.

Figure SEC 4 illustrated three different styles of the cutting plane lines. All are drawn using a line thickness equal to a visible line which is 0.7mm. Note that all the dimensions given in Figure SEC 4 are approximate as their actual lengths depend on the size of the drawing. For example, if the cutting plane line is just 40mm (1.5 in.) long, then the dashes may need to be less than the 6mm dashes called for in Figure SEC 4A. But if you're cutting plane line is 450mm (6 in.) long, the dashes could be longer than the 6mm (.25 in.) called for in Figure SEC 4A. Just be sure to keep the dashes and spaces uniform. The "hidden line" style in Figure SEC 4A uses a hidden line to indicate the location of the cutting plane and arrowheads to indicate the viewing direction. The "phantom line" style in Figure SEC 4B uses a phantom line to indicate the location of the cutting planes and arrowheads to indicate the viewing direction. The "ends only" style in Figure SEC 4C uses only a short line segment on both ends to indicate the location of the cutting plane along with arrowheads to indicate the viewing direction. The ends only method is used when a cutting plane line would hide important details.

Section Lining:

Section lining is used to show what material is in the cutting plane. A second method for visualizing what portion of the sectioned object receives section lining is to imagine using a saw to cut along the cutting plane line. Whatever portion of the object touching the saw blade receives section lining. In the past, machine drawings used different section lining patterns to indicate different materials as illustrated in Figure SEC 5. But today machine drawings use only the pattern for cast iron or ANSI 31 (American National Standards Institute Number 31) illustrated in Figure SEC 5.

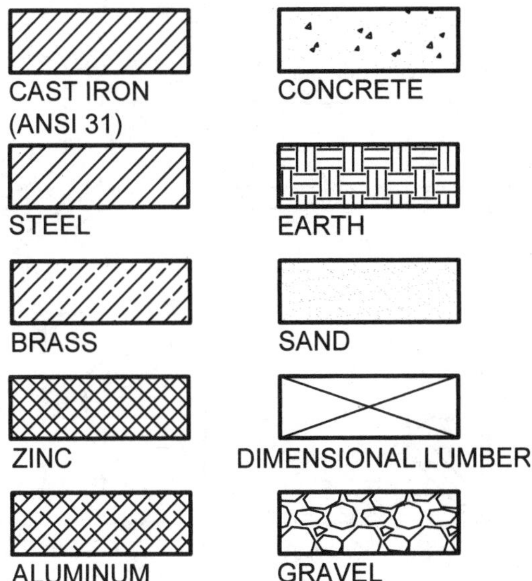

Figure SEC 5: A small sample of cross hatching patterns on machine and architectural drawings. Machine drawings use only the cast iron (ANSI 31) symbol for all materials. Architectural drawings use those listed and several more.

General purpose section lining (ANSI 31) is normally drawn at 30°, 45°, or 60° angles to the left or to the right. The lines are spaced from 1.5mm (.06 in) to 6mm (.25 in.) apart depending on the size of the drawing as illustrated in Figure SEC 6.

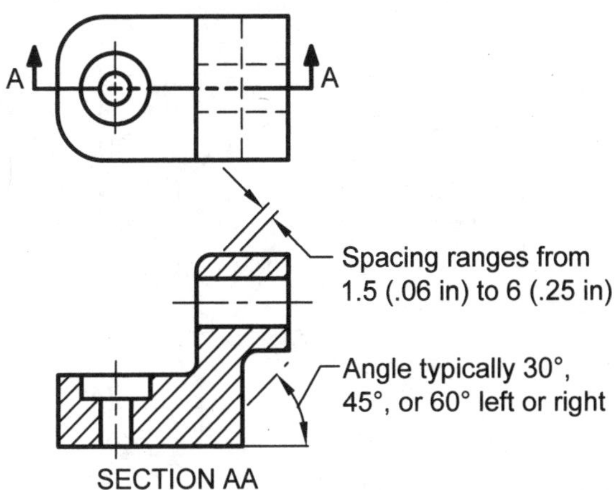

Figure SEC 6: Line spacing for general purpose section lining ranges from 1.5mm (.06 in) to 6mm (.25 in) and are typically drawn at 30°, 45°, and 60° left or right.

To avoid having a section line look like an object line on the drawing, avoid having section lining run parallel to the borders of the sectioned area as illustrated in Figure SEC 7.

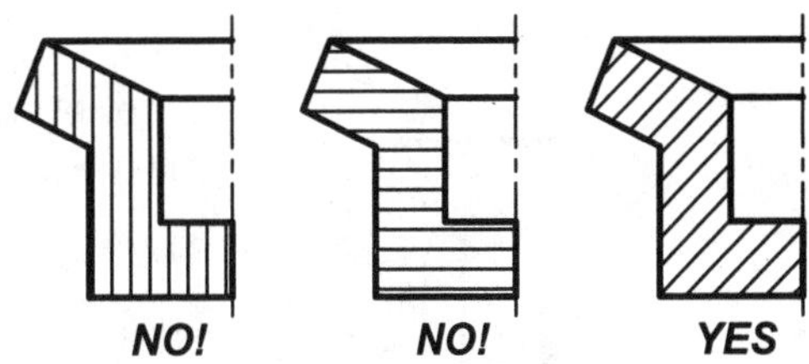

Figure SEC 7: Section lining should not be parallel to the boundaries of the sectioned area.

When a very large area requires section lining, the section lines can be drawn just around the borders of the sectioned area as illustrated in Figure SEC 8.

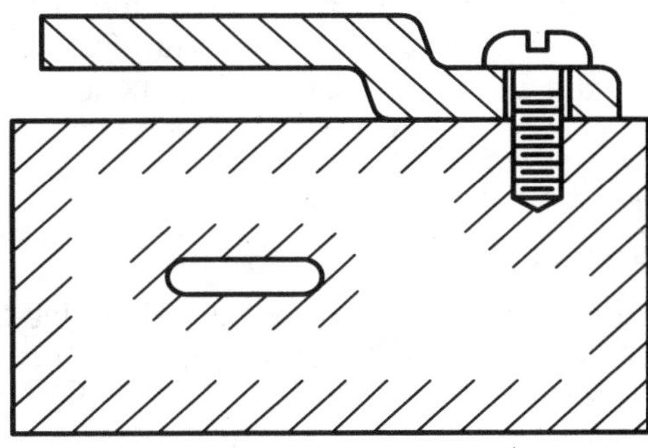

Figure SEC 8: A large area needing section lining can have its section lining just around its perimeter.

An assembly drawing illustrates several individual parts put together. When sectioning through an assembly, section lining methods must be used so that each part can be uniquely recognized. Figure SEC 9 illustrates the required methods. First, each sectioned part should have section lining at a unique angle and direction as illustrated by parts 2, 3, and 5 in Figure SEC 9. Second, when sectioned areas of the same part are separated, the section lining style must have the same angle and direction in all of the separated areas as illustrated by parts 3 and 5 in Figures SEC 9. Third, when a sectioned area is too small to practically use section lining, the area is completely filled in as with part 4 (washer) in Figure SEC 9. As previously mentioned, in machine drawing the section lining pattern cast iron or ANSI 31 is used exclusively.

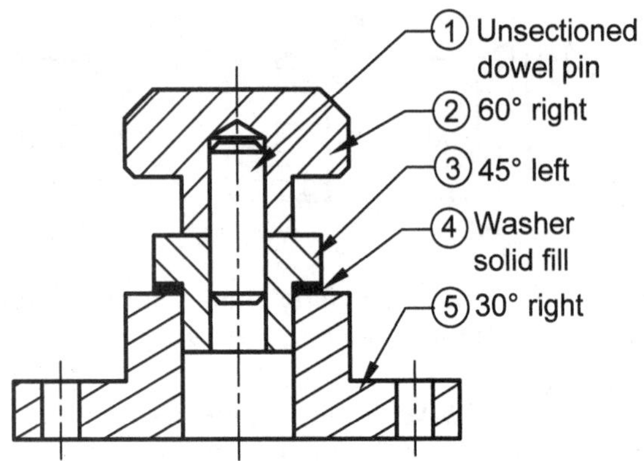

Figure SEC 9: A sectioned assembly showing each part with section lining at a unique direction and angle. Part 4 is a washer that is too small for section lining so it is filled solid. Part 5 has sectioned areas on both sides of the center line but each sectioned area receives the same style of section lining.

Full Sections:

A *full section* is created when the cutting plane line passes entirely through an object. Figure SEC 10 illustrates a full section with the cutting plane line passing through the top view. The pictorial view in Figure SEC 10 is supplied for visualization purposes only. The arrowheads attached to the cutting plane line indicate the viewing direction. The half of the object closest to the viewer (opposite the viewing direction arrows) is removed. Labeling the sectioned view "Section A-A" is probably not necessary for this simple drawing but it illustrates the syntax for labeling sections. The letters used with the cutting plane lines are separated with a dash and accompanied by "SECTION". Note that the section lining is not applied to the slot on the right or the countersunk hole on the left because there was no material for the cutting plane to pass through at those locations.

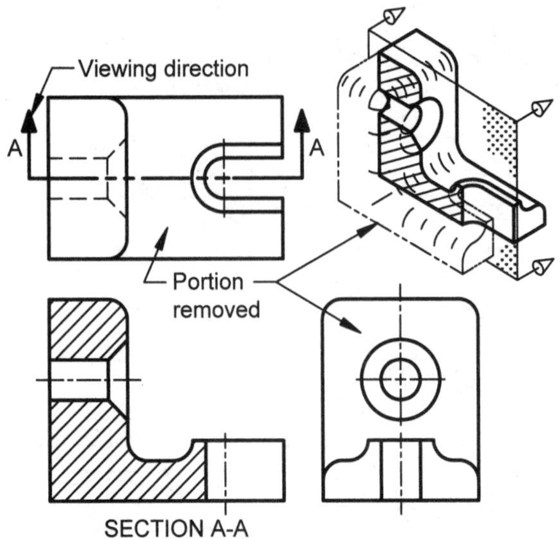

Figure SEC 10: Full section with the cutting plane line A-A in the top view.

Figure SEC 11 illustrates a full section with the cutting plane passing through the front view. The resulting Section B-B is substituted for the right view. Note that: 1) the top view is unaffected by the section, 2) the portion of the object closest to the viewer is removed, and 3) areas not receiving section lines include the threaded hole, the large and small through holes, and the keyway at the top of the large hole.

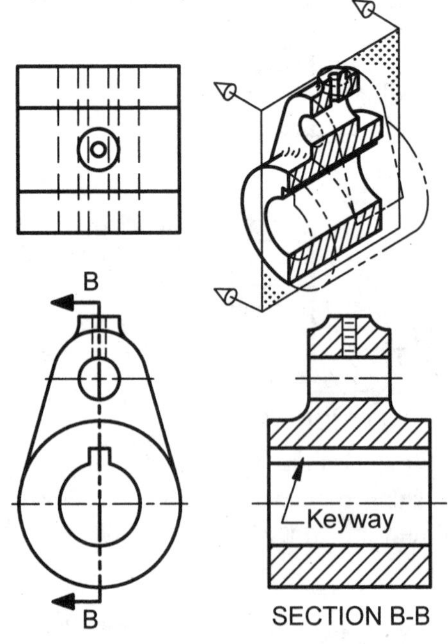

Figure SEC 11: Full section with the cutting plane line B-B in the front view.

Figure SEC 12 illustrates a full section with the cutting plane in the right view. The resulting section C-C is substituted for the front view. Note that: 1) the top view is unaffected by the section, 2) the portion of the object closest to the viewer is removed, 3) the areas not receiving section lines include the mirrored imaged counterbored holes near the top and the 60° dovetailed slot along the bottom.

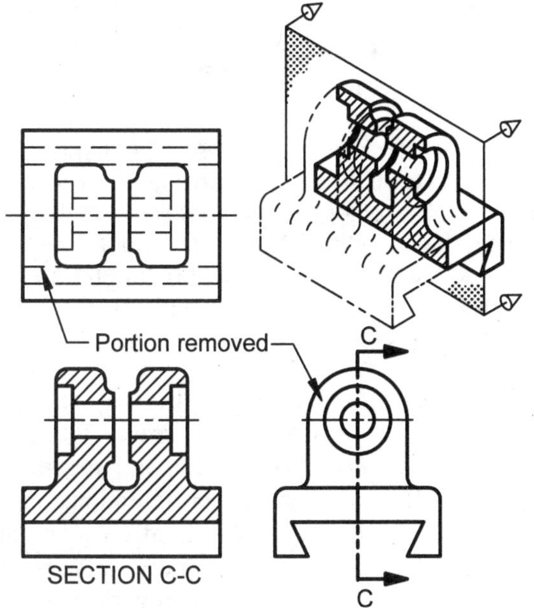

Figure SEC 12: Full section with the cutting plane line C-C in the right view.

Offset Sections:

An *offset section* is a type of full section. Where the cutting plane includes 90° bends in order to pass through a number of important details. The pictorial view in Figure SEC 13 shows the cutting plane with two 90° bends in order to pass through three details. The top view in Figure SEC 13 shows a cutting plane line with two 90° bends. It is important to note that a line at each 90° bend is NOT shown in the sectioned view as illustrated in SECTION A-A in Figure SEC 13.

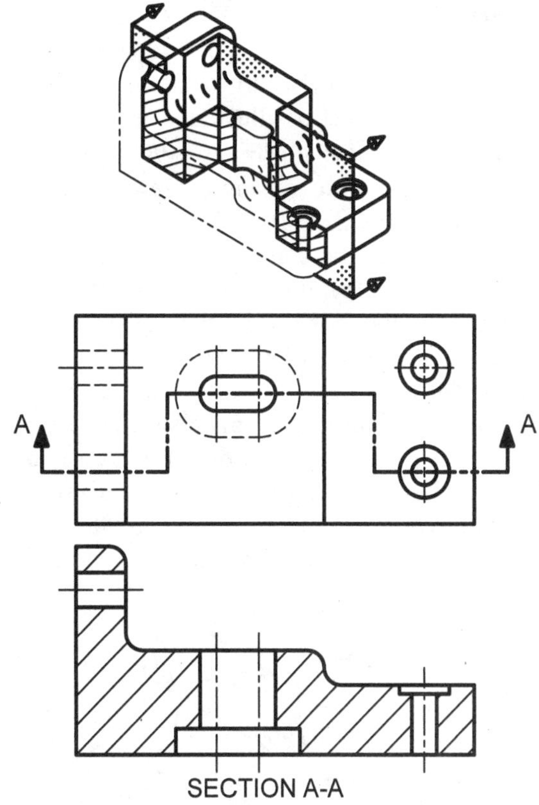

Figure SEC 13: Offset section A-A. Note no lines are drawn in the sectioned view where the cutting planes changes direction.

Figure SEC 14 illustrates the use of multiple offset sections. Section A-A is substituted for the front view and section B-B for the left view.

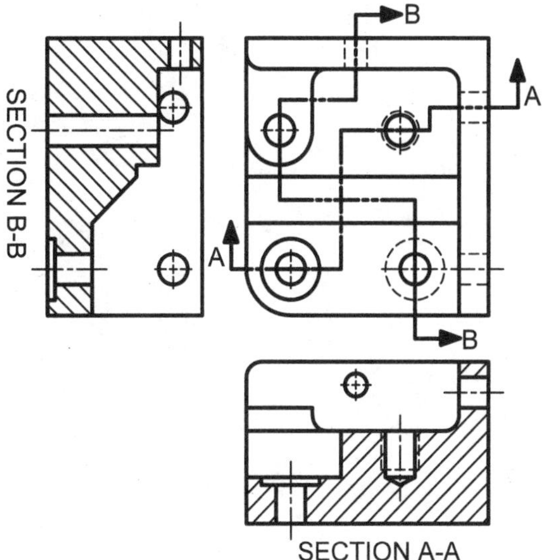

Figure SEC 14: Multiple offset sections in a single view.

Half Sections:

A *half section* is created by cutting into a symmetric cylindrical part and removing one quarter of the object as illustrated in Figure SEC 15. Figure SEC 16A illustrates a 90° bent cutting plane line with only a single viewing direction arrowhead. Figure SEC 16B illustrates the resulting half section. It is the conventional practice for the unsectioned half to not receive hidden lines as illustrated in the bottom portion of Figure SEC 16B. Hidden lines can be used in the unsectioned portion if it helps clarify the object's shape but is not common. Figure SEC 16B also illustrates how the sectioned and unsectioned halves are separated by a centerline not a solid object line. An advantage of a half section is the ability to see the interior detail and exterior shape of the object in a single view.

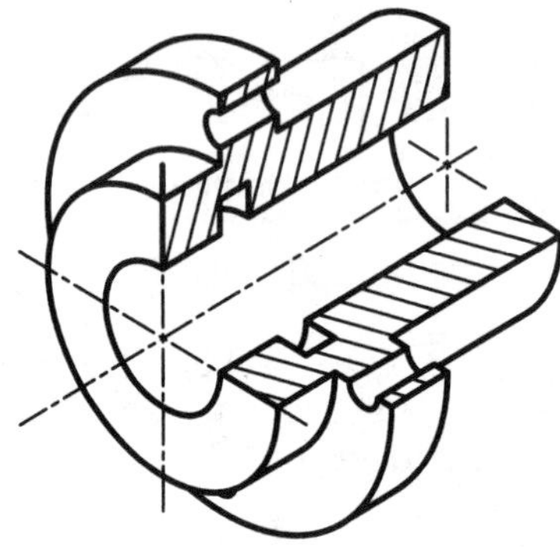

Figure SEC 15: A half section has one fourth of the object removed.

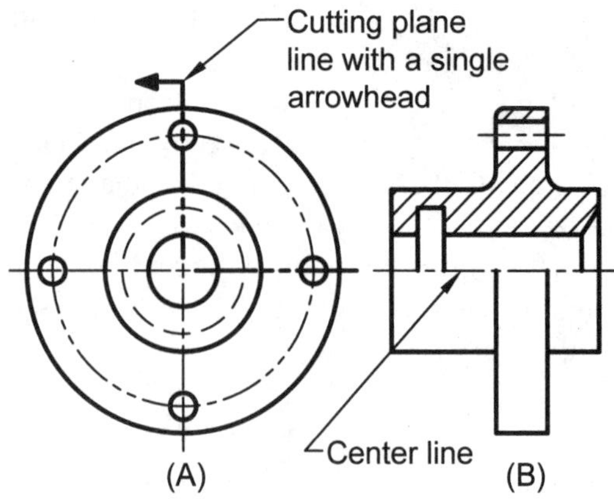

Figure SEC 16: Half section A) The bent cutting plane line with a single arrowhead shows the quarter of the object to be removed, and B) the resulting half section. Note, no hidden lines are used in the lower unsectioned portion, and a center line, not a visible line, is used to divide the sectioned and unsectioned portions.

The most effective use of the half section is using them as half sectioned assembly views. An assembly is simply several parts put together in their normally functioning configuration. Figure SEC 17 is a half section sketch of an assembled globe valve having six parts. A globe valve is used to control the water supply below a bathroom vanity or next to a toilet. Notice that: 1) each part's section lining is at a unique angle and /or direction, 2) the sectioned and unsectioned portions are separated by a center line, 3) part number 5 (washer) is filled in solid, because of its small size, and 4) no hidden lines are used in the unsectioned portion.

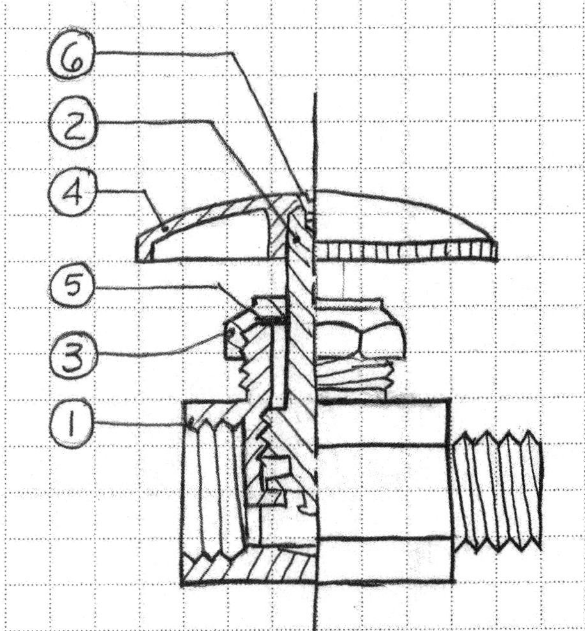

Figure SEC 17: An assembly drawing of a globe valve in half section. Note that: 1) each part sectioned has unique angle and direction section lining, 2) the sectioned and un-sectioned portions are separated by a center line, 3) part number 5 (washer) is filled solid because of its small size, and 4) no hidden lines are used in the un-sectioned portion.

Figure SEC18 is a half section of an assembled gate valve having nine parts.

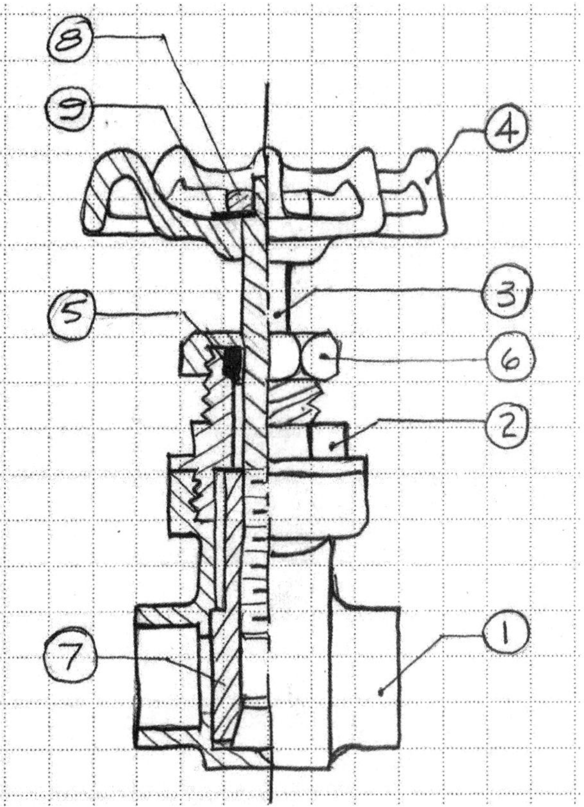

Figure SEC 18: An assembly sketch of a gate valve in half section.

Revolved Sections:

A *revolved section* is developed when a cutting plane passes perpendicular to the axis of an elongated symmetrical feature and rotated 90° into the viewing plane

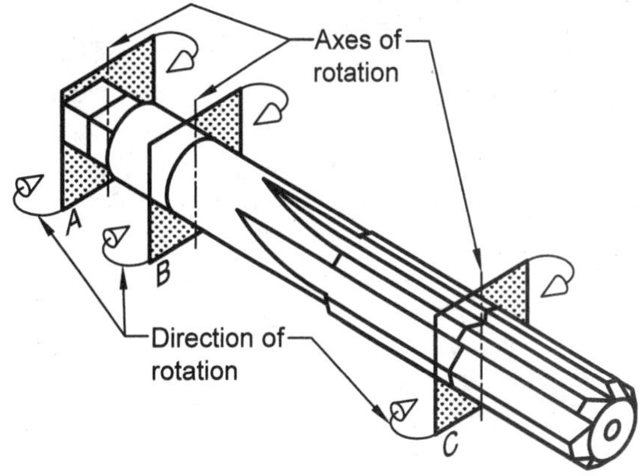

Figure SEC 19: A ream with three imaginary cutting planes that define the three revolved sections in Figure SEC 20.

In Figure SEC 19 three imaginary planes A, B, and C pass through a ream which is a tool that is inserted into a previously drilled hole then rotated to remove a small amount of material and make the hole truer in size and shape. If the cross sectioned areas of each plane in Figure SEC 19 are rotated 90° about the labeled axes of rotation in the direction indicated, three revolved sections are generated. Figure SEC 20 illustrates the three revolved sections A, B, and C.

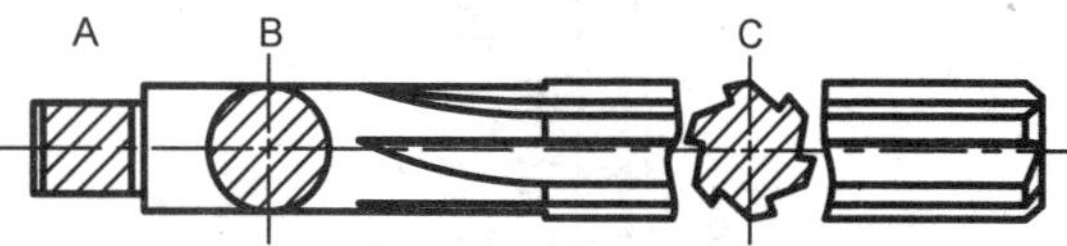

Figure SEC 20: Three revolved sections of a ream. A and B are revolved using the "in-line" method that does not alter the original view. C uses the "break-away" method which breaks open the original view.

Revolved sections are drawn directly on the view using one of two methods. Revolved sections A and B in Figure SEC 20 illustrate the first style which this author calls the "in-line" style. The in-line style leaves the original view unaltered. The advantage of this style is that it allows you to trace details on the normal view directly into the sectioned view. Revolved Section C in Figure SEC 20 is what this author calls the "break-away" style. The break-away style opens up the original view so the rotated section stands alone. The advantage of this style is that it allows the revolved section to stand out and its outline can be easily read.

Figure SEC 21 illustrates a common error when creating revolved sections. Figure SEC 21A illustrates the head of a small woodworking clamp casting and the axis of rotation for the revolved section. Figure SEC 21B shows the correct section while SEC 21C illustrates a common mistake.

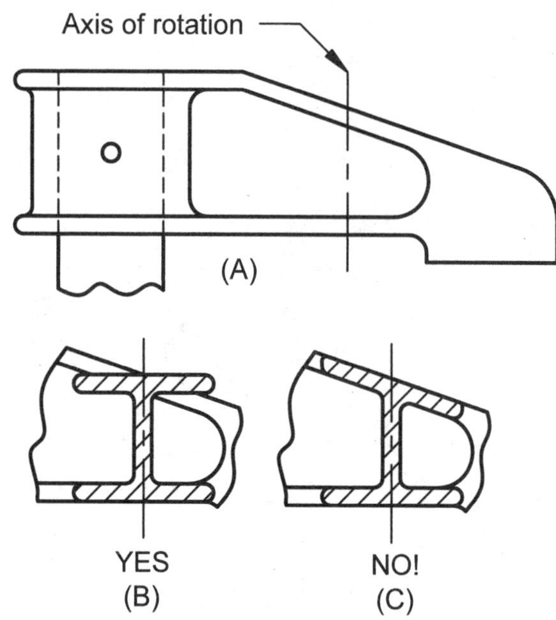

Figure SEC 21: A) locates the axis of revolution on the head of a woodworking clamp casting, B) illustrates the correct method for this revolved section, C) illustrates a common error.

Removed Sections:

The process of creating a removed section is exactly the same as for a revolved section except the resulting revolved cross section is displayed off the original view. Figure SEC 22 illustrates three removed sections of a ream. A cutting plane line is drawn at the location of each section with labels A-A, B-B, and C-C. Each removed section is accompanied by a label Section A-A, Section B-B, etc. If several sections are required, they are lined up and labeled alphabetically from left to right or top to bottom. Do not use letters I, O or Q as they can be confused with numbers. A removed section does not need to be lined up with its accompanying cutting plane line.

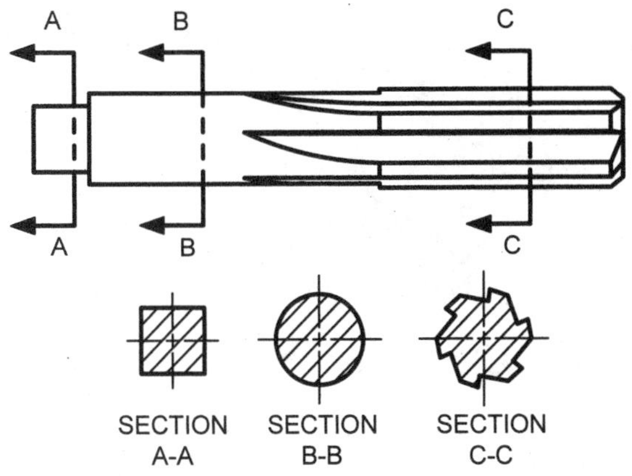

Figure SEC 22: A ream with three removed sections alphabetically listed from left to right.

Removed sections should be placed on the same sheet as the cutting plane line whenever possible. But if the removed section needs to be located on another sheet, the cutting plane line must indicate the sheet number and general location or zone where the section is found. One method illustrated in the upper left half of Figure SEC 23A includes a circle with a letter over a number and is integrated into the cutting plane line. The upper letter "N" is the section callout, i.e., Section N-N while the lower number "4" is the sheet number on which the section is found. A second method, illustrated in the lower left half of Figure SEC 23A, uses a note and leader that specifies the sheet number and zone where the section is located. Figure SEC 23B illustrates the actual Section N-N found on sheet 4. Section N-N needs to be referenced back to the sheet where the cutting plane is located. In our example the section's caption on sheet 4 reads Section N-N from sheet 2 Zone C1.

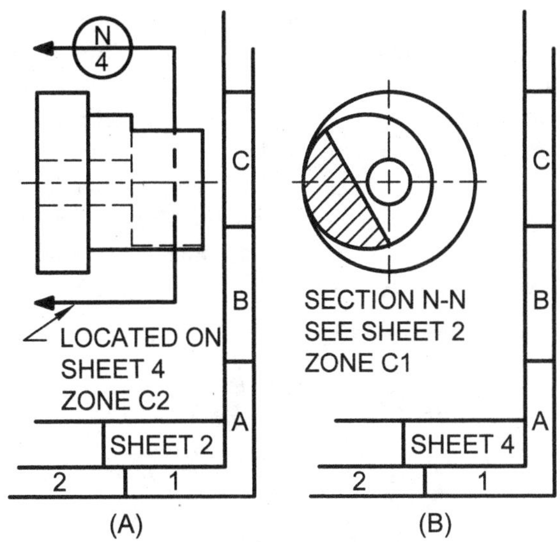

Figure SEC 23: If a removed section will not fit on the same page, its location must be provided. A) The first method in (A) shows a circle on the cutting plane line with the letter "N" representing the detail and the "4" the page on which the section is found. The second method in (A) uses a leader and note specifying the sheet number and its zone. B) The actual section should be labeled with a reference back to the location of the cutting plane line.

With simple objects, removed sections can be displayed by drawing a centerline through the object at the location of the section(s) and extending the centerline through the center of the removed section(s) as illustrated in Figure SEC 24.

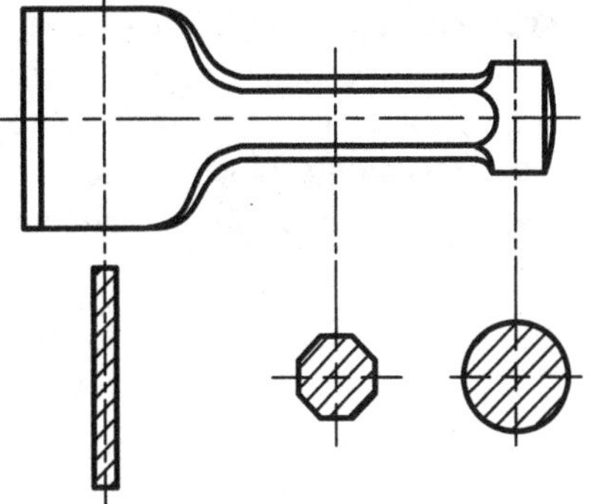

Figure SEC 24: A mason's chisel with three removed sections using the center line extension method.

Broken- out Sections:

If you have ever taken a bite out of an apple to look for a possible worm, you have created a broken-out section. A *broken-out section* removes a small portion of an object or more descriptively takes a small bite out of an object to expose what lies within. Figure SEC 25 illustrates two broken out sections. The upper right broken out section reveals an o-ring in an undercut groove that work together to form a seal. The lower left broken-out section reveals a set screw. The border of the broken-out section is designated by an irregular freehand break line (your teeth marks left after you bite away the removed portion). No cutting plane line is used.

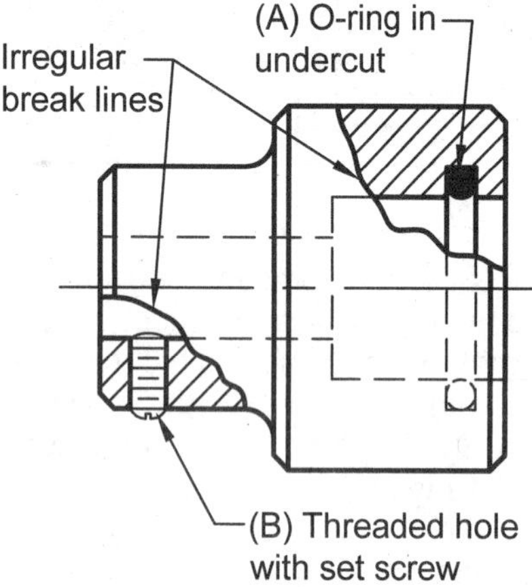

Figure SEC 25: Two broken-out sections A) reveals an o-ring in an undercut groove and B) reveals a set screw.

Auxiliary Sections and Viewing Planes:

Auxiliary sections A-A and B-B in Figure SEC 26 illustrate cutting plane lines and the resulting auxiliary sections. Both sections A-A and B-B are a form of a removed section projected perpendicular to their cutting plane lines. If an auxiliary section must be moved to another location on the drawing, keep it in the same orientation – do not rotate.

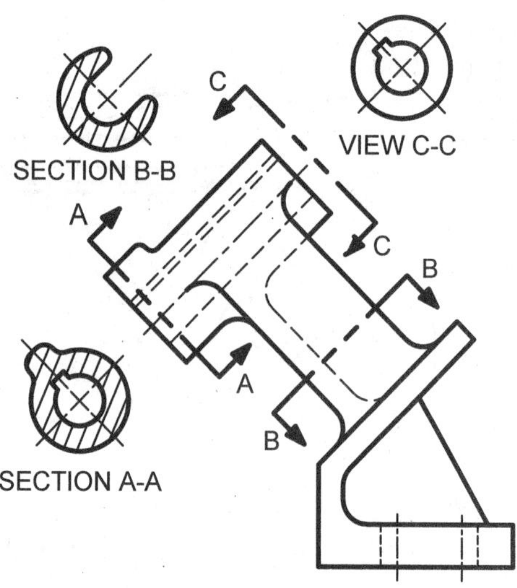

Figure SEC 26: Cutting plane lines A-A and B-B generate the auxiliary views Section A-A and Section B-B. Viewing plane line C-C does not section the object but provides a small auxiliary view.

View C-C in Figure SEC 27 is not a section but is simply a partial auxiliary view. Figure SEC 27 illustrates that the viewing plane line does not pass through the object and no section lining is used in view C-C. A viewing plane line is used when a partial view of the object is desired.

Aligned Sections:

Aligned sections use conventions or standard practices that violate basic projection rules but are used in the interest of describing the part's geometry more clearly.

The object in Figure SEC 27 has a central circular hub with an arm protruding at an angle. The cutting plane line passes through the angled portion. Two steps are needed in order to draw the aligned Section D-D: 1) Revolve the angled portion back to the primary center line. 2) Project the geometric elements marked off on the primary center line from step one down into the Section D-D view. The angled arm in Section D-D is not a true projection of the top view. Rotating the angled arm back to the primary center line then down into the sectional view is a convention that gives a clearer idea of what the object looks like. Note that the length of the angled arm in Section D-D shows its true length.

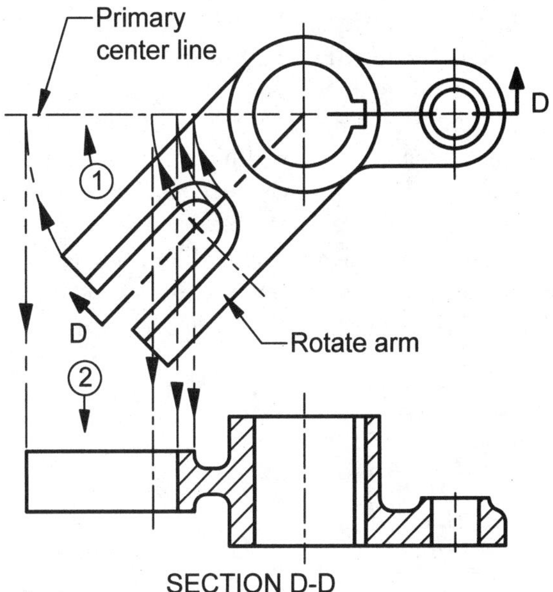

Figure SEC 27: To form aligned Section D-D the slotted arm is rotated back to the primary center line then projected vertically down to locate the slotted arm in the sectional view.

Using traditional conventions, objects with holes and ribs (reinforcing element) distributed about a center axis are revolved back to the cutting plane line when creating an aligned section. In Figure SEC 28A two holes are rotated back to the cutting plane line then into Section E-E. The location of the holes in SECTION E-E are not in a true projected position but their center axes are a true distance from the part's central axis. In Figure SEC 28B a hole is rotated to the left and a rib rotated to the right into the cutting plane line then into Section H-H. Note that both Section E-E and Section H-H are identical and that by convention the ribs do not receive section lining. Figures SEC 28C, 28D, 28E, and SEC 28F illustrate aligned SECTIONS F-F, II, G-G, and J-J. These sections were generated on Autodesk's 3-D parametric software Inventor®. None of these aligned sections follow traditional convention methods but are true projections. The results are not as descriptive as the part's geometry as compared to the sections in Figure SEC 28 and SEC 29. Generating aligned sections using 3-D CAD models can have unpredictable results so you must experiment.

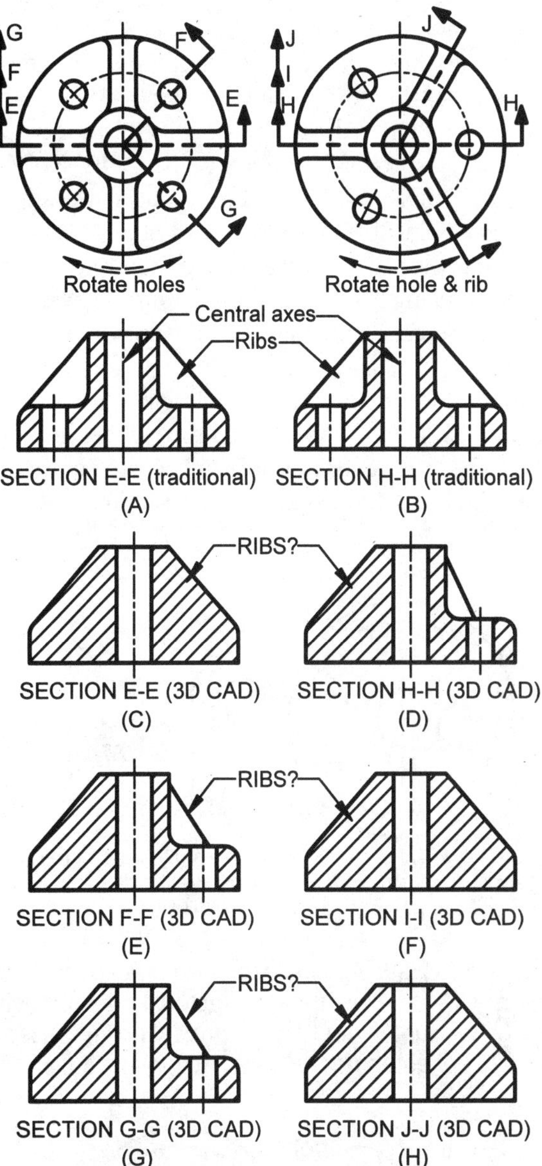

Figure SEC 28: An aligned section of a four and three ribbed object. A) two holes are rotated back to the cutting plane line then projected into Section E-E, B) a hole is rotated left and a rib is rotated right back to the cutting plane then into Section F-F. Note Section E-E and F-F are the same and by convention the ribs are not sectioned.

Spokes are elements that radiate out from a center hub to an outer ring such as spokes on some automobile wheels, and turn wheels on metal and woodworking cutting machinery. Figure SEC 29 illustrates a turn wheel with three spokes. SECTION M-M in Figure SEC 29B shows the spokes in true projection which is considered poor practice. In SECTION N-N in Figure SEC 29C the upper spoke is rotated back to the primary centerline then into the Section G-G. Spokes by convention do not receive section lining. When generating aligned sections on Inventor's 3-D parametric software Inventor®, SECTIONS M-M and SECTION N-N resulted in the same aligned sections as in SECTION N-N in Figure 29C which used traditional conventions. Generating aligned sections using 3-D CAD modeling software can have unpredictable results, so you must experiment.

In Figure SEC 30A the turn wheel has four spokes and does not receive section lining to help distinguish it from the webbed turn wheel in Figure SEC 30B. However, the turn wheel in Figure 30B is webbed and does receive section lining.

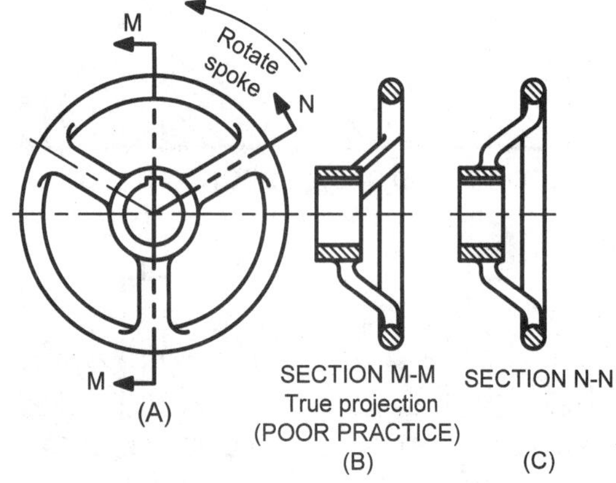

Figure SEC 29: A) a spoke is rotated back to the primary center line then projected into C) Section G-G. The turn wheel's true projection in B) is not used. Spokes do not receive section lining.

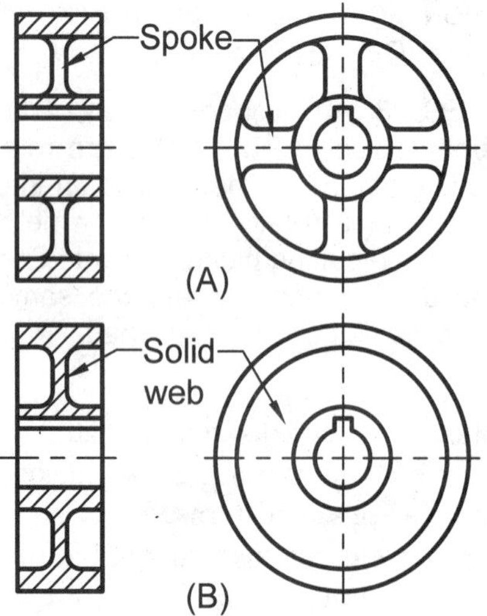

Figure SEC 30: A) spokes do not receive section lining otherwise they would look like a solid web. B) the solid web receives section lining.

Applying section lining to ribs or webs can be confusing. Figure SEC 31 illustrates conventional practice. Cutting plane A-A in Figure SEC 31A passes longitudinally through one of the webs but does not receive section lining as illustrated in Figure SEC 31B. If the web received section lining, as in Figure SEC 31C, the object would look like a solid mass. This misrepresents the geometry of the object. Section B-B in Figure SEC 31 cuts across all the webs, so all the webs receive section lining.

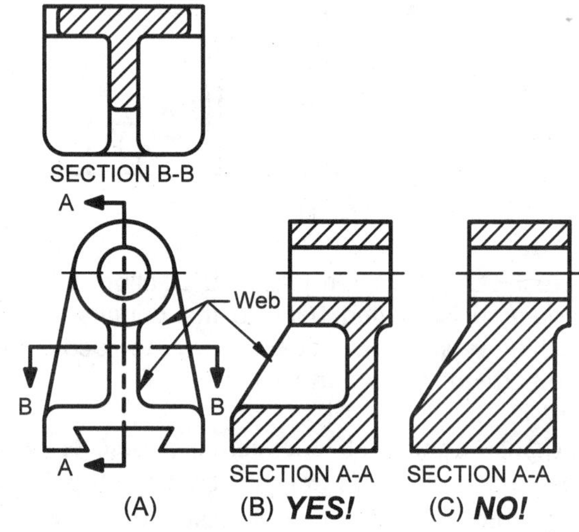

Figure SEC 31: A) SECTION A-A of Figure SEC 31B cuts longitudinally through a web and does not receive section lining, B) SECTION B-B cuts across the webs and receives section lining, C) using section lining as in Figure 31C misrepresents the geometry of the object.

Unsectioned Features:

As illustrated in the previous section, geometric features such as ribs, webs, and spokes do not always receive section lining even when a cutting plane passes through them. Other objects that by conversion do not receive section lining are: shafts, bearings, gear teeth, and fasteners. Shafts and fasteners do not receive section lining when sectioned along their longitudinal axis but do receive section lining when sectioned perpendicular to their longitudinal axis. Figure SEC 32 illustrates a shaft in a journal housing with ball bearings. The shaft and ball bearings do not receive section lining while the housing, bearing casing, retaining clips, and end of shaft do receive section lining. The retaining rings are filled solid because they are so small. Figure SEC 33 illustrates a pulley attached to a shaft with a set screw and rectangular key. The set screw, rectangular key, and most of the shaft do not receive section lining while the housing, the broken out portion of the shaft, and the end of the shaft receive section lining. Figure SEC 34 illustrates a sample of four different types of fasteners that do not receive section lining. A general rule is that shafts and fasteners do not receive sectional lining.

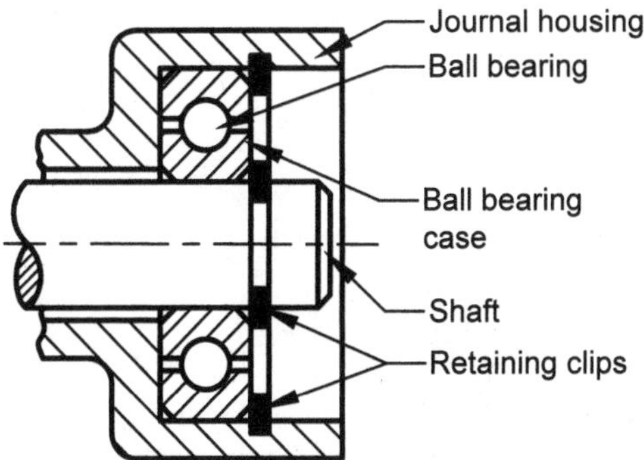

Figure SEC 32: The journal housing, ball bearing case, retaining rings, and end of the shaft receive section lining. The retaining rings are so thin that they are filled solid. The ball bearings and the shaft do not receive section lining.

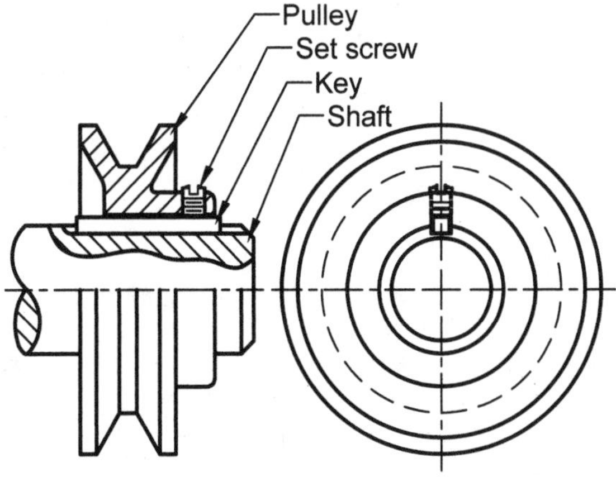

Figure SEC 33: Fasteners such as this set screw and rectangular key are categories of parts that do not receive section lining.

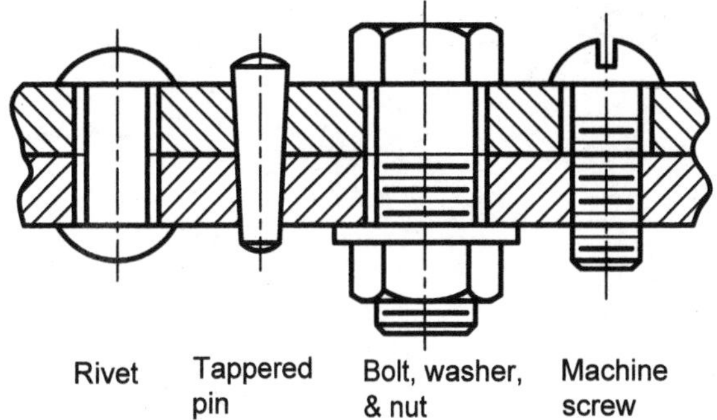

Figure SEC 34: When sectioning through fasteners longitudinally, they do not receive section lining.

Conventional Breaks:

A single break line indicates that a part of the object has been cut off. A pair of break lines indicates that something between them has been removed. Figure SEC 35A illustrates a pair of break lines used with wood and most other architectural applications. Figure SEC 35B illustrates a pair of break lines used with metal objects. Figures SEC 35C and SEC 35D illustrate a pair of break lines used with tubular and solid tubular stock. Figure SEC 35E illustrates break lines when they need to be very long.

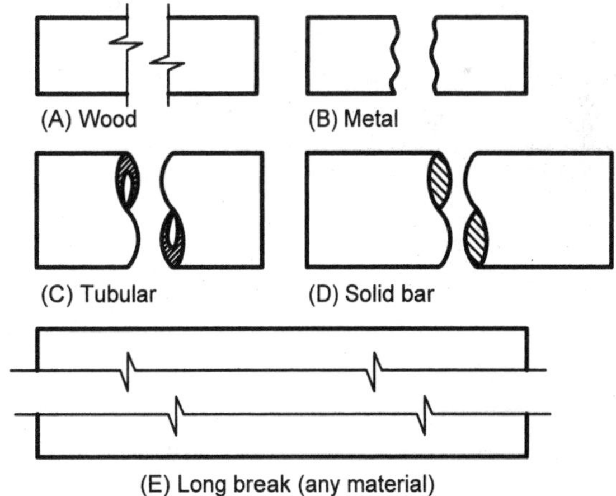

Figure SEC 35: Conventional breaks.

NOTES:

Given: Top and Right views
Sketch: Front view as a full section

1.

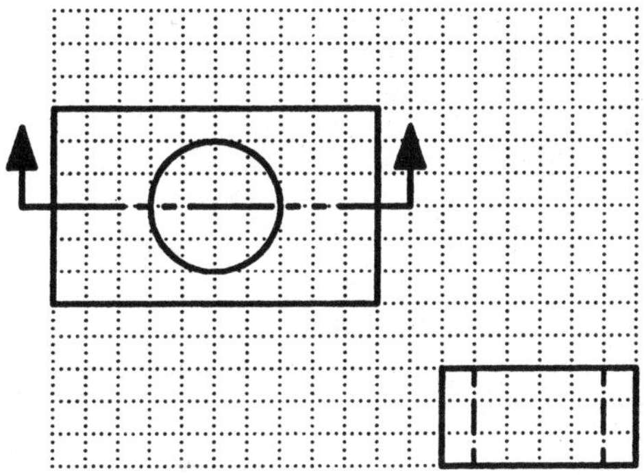

2.

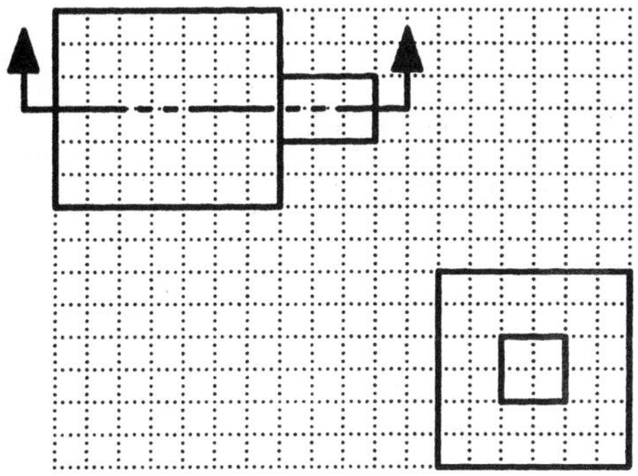

3.

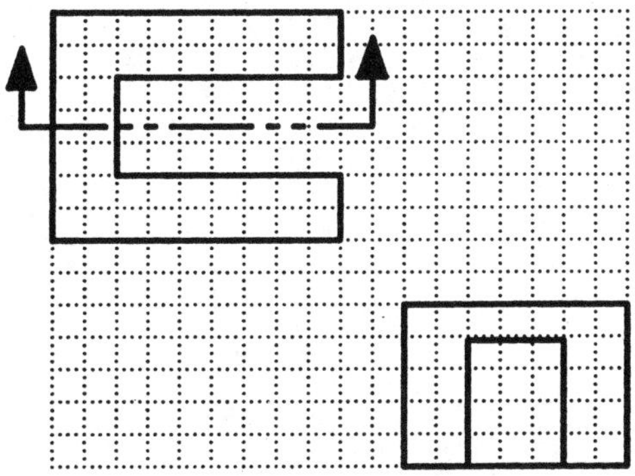

4.

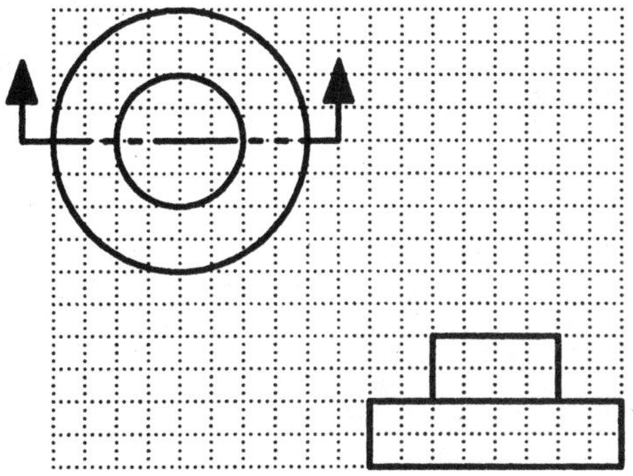

5.

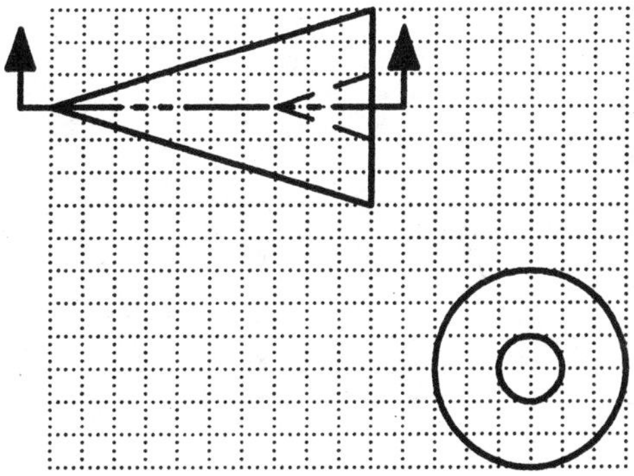

6.

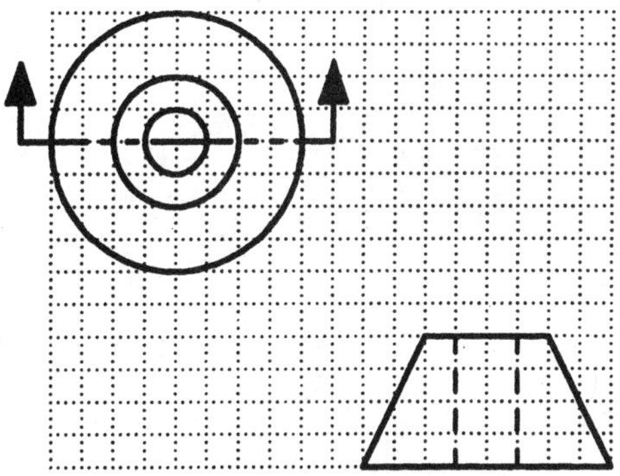

7.

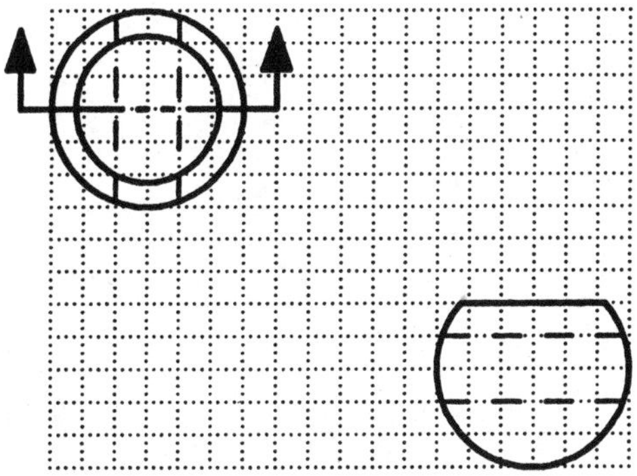

8.

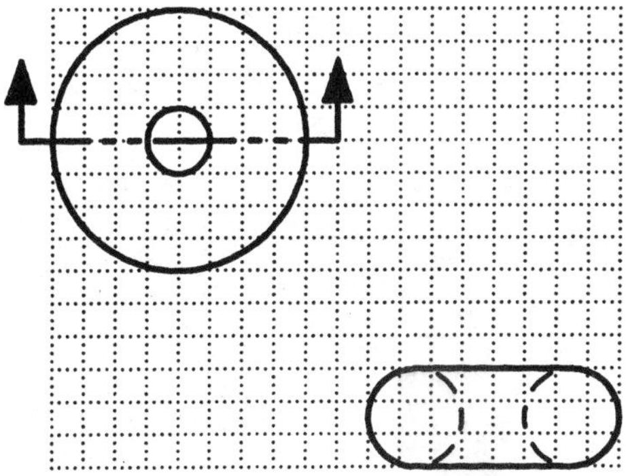

SECTIONING Full Sections Sketching	Name: ______ ID#: ______ Lab Hr: ______	US CUST 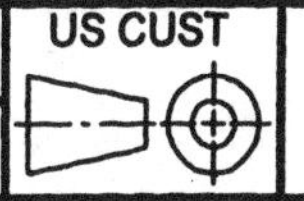	SECF-1A

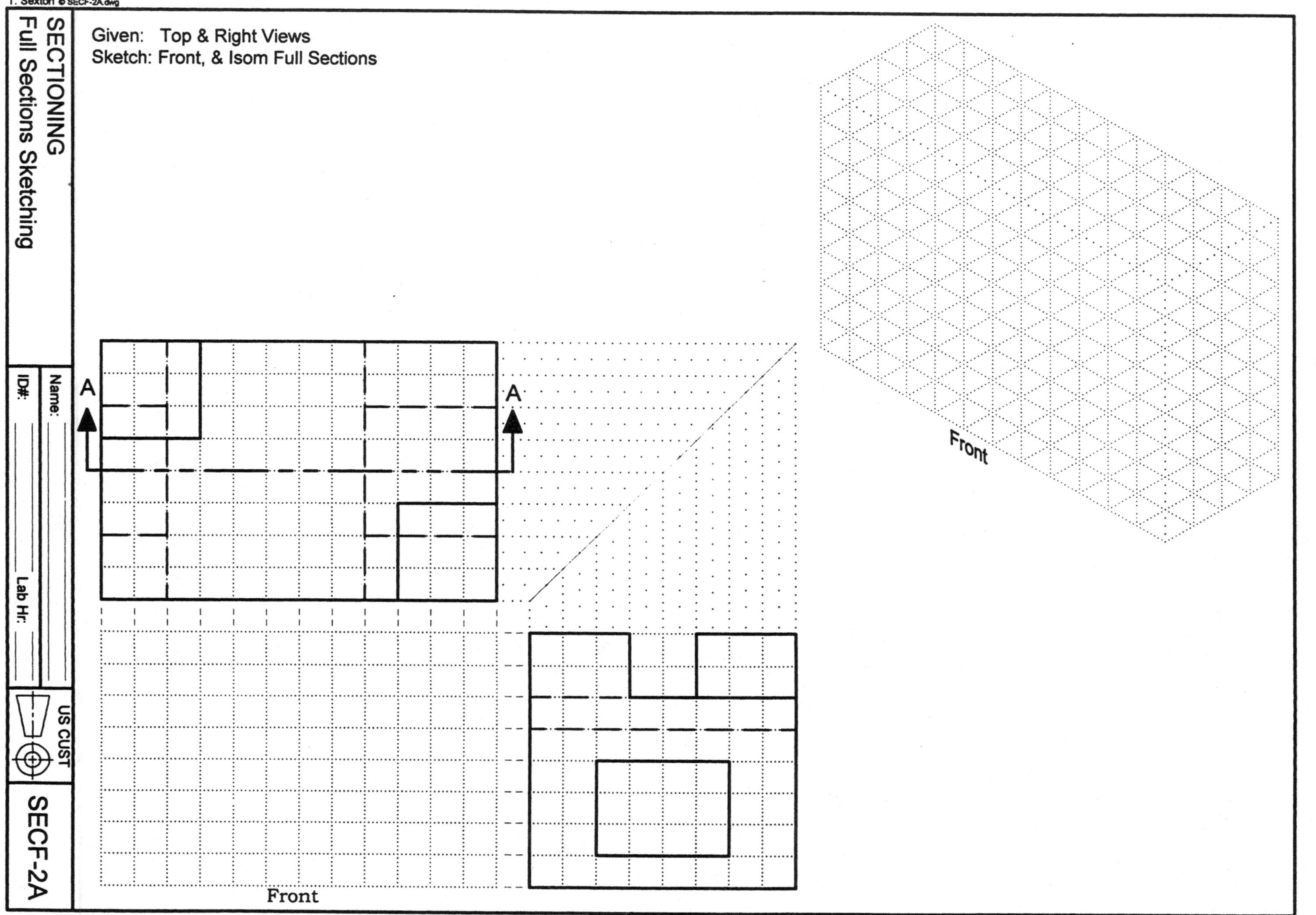
T. Sexton © SECF-2A.dwg
Given: Top & Right Views
Sketch: Front, & Isom Full Sections
A
A
Front
Front
SECTIONING
Full Sections Sketching
Name:
ID#:
Lab Hr:
US CUST
SECF-2A

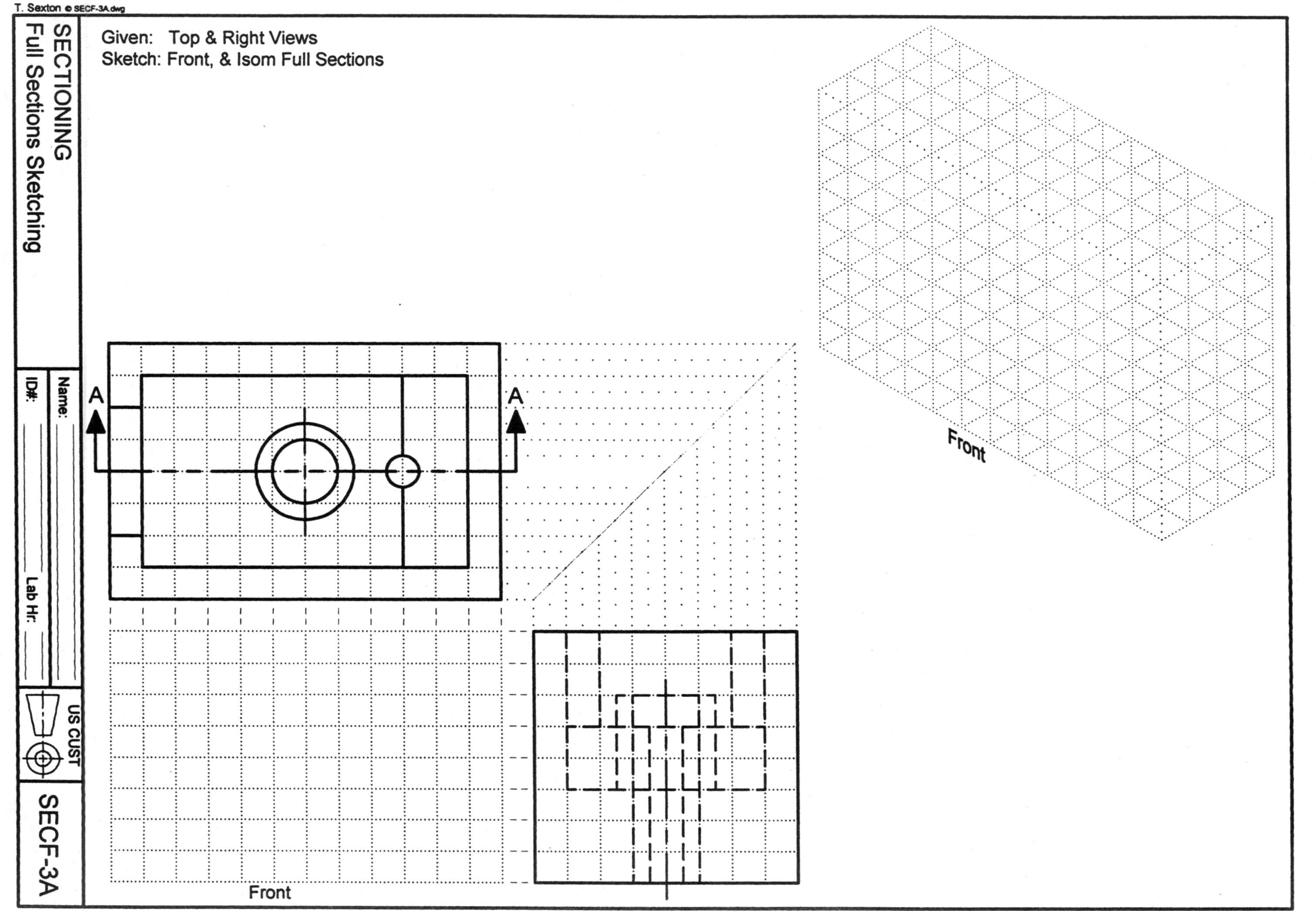
Given: Top & Right Views
Sketch: Front, & Isom Full Sections
A
A
Front
Front
T. Sexton © SECF-3A.dwg
SECTIONING
Full Sections Sketching
Name:
ID#:
Lab Hr:
US CUST
SECF-3A

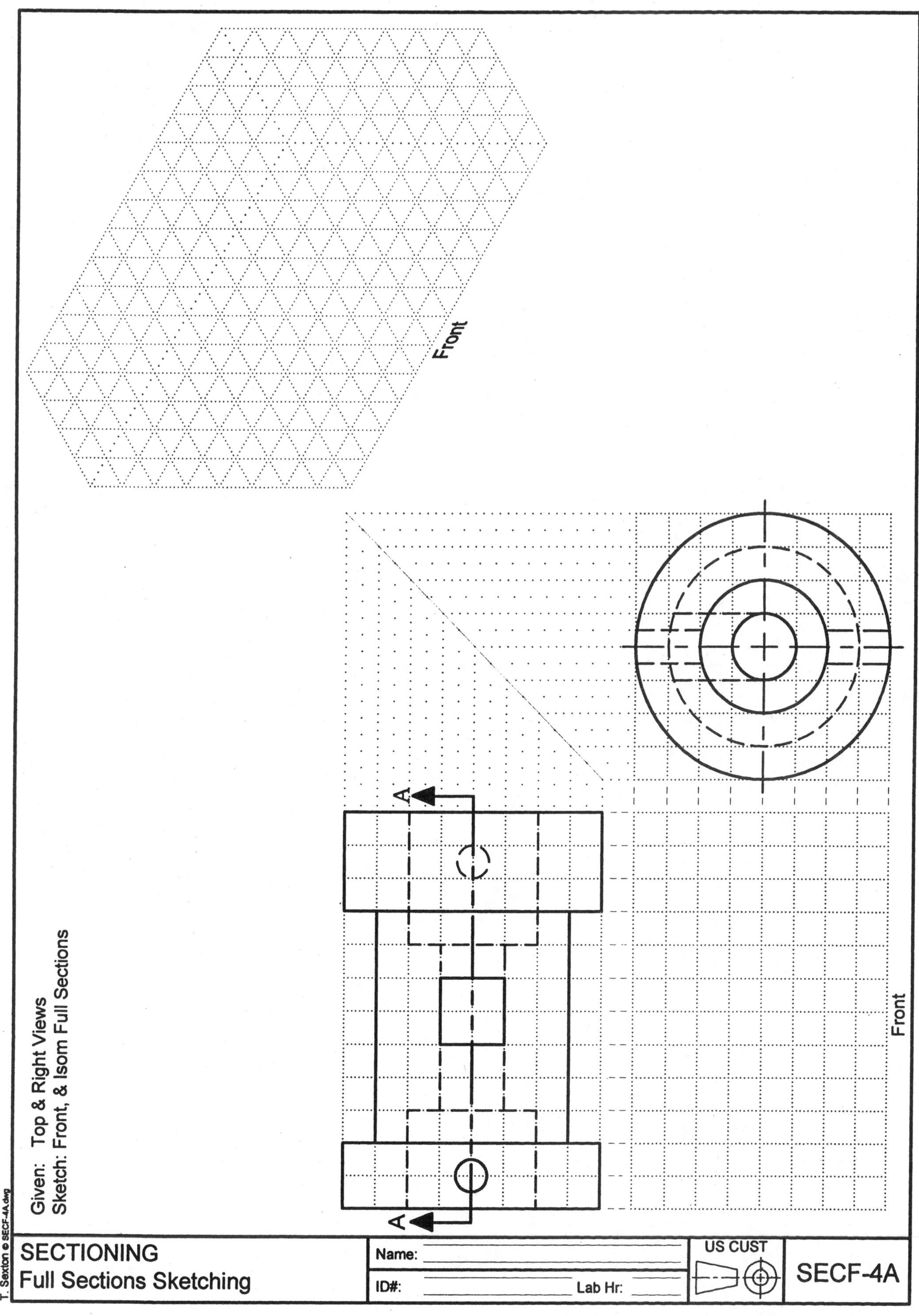
Given: Top & Right Views
Sketch: Front, & Isom Full Sections
Front
A
A
Front
T. Sexton © SECF-4A.dwg
SECTIONING
Full Sections Sketching
Name:
ID#:
Lab Hr:
US CUST
SECF-4A

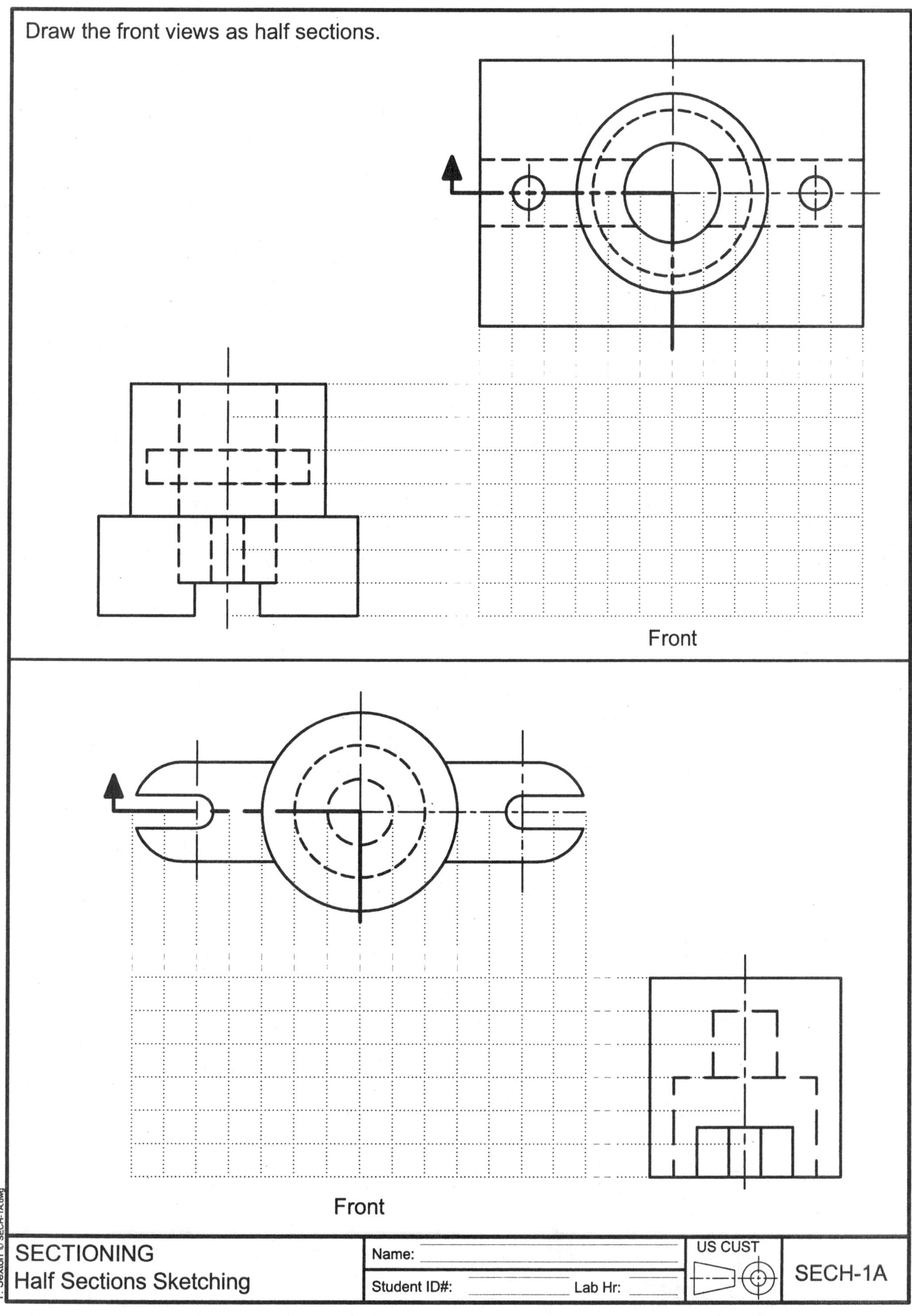
Draw the front views as half sections.
Front
Front
SECTIONING
Half Sections Sketching
Name:
Student ID#:
Lab Hr:
US CUST
SECH-1A
T. Sexton © SECH-1A.dwg

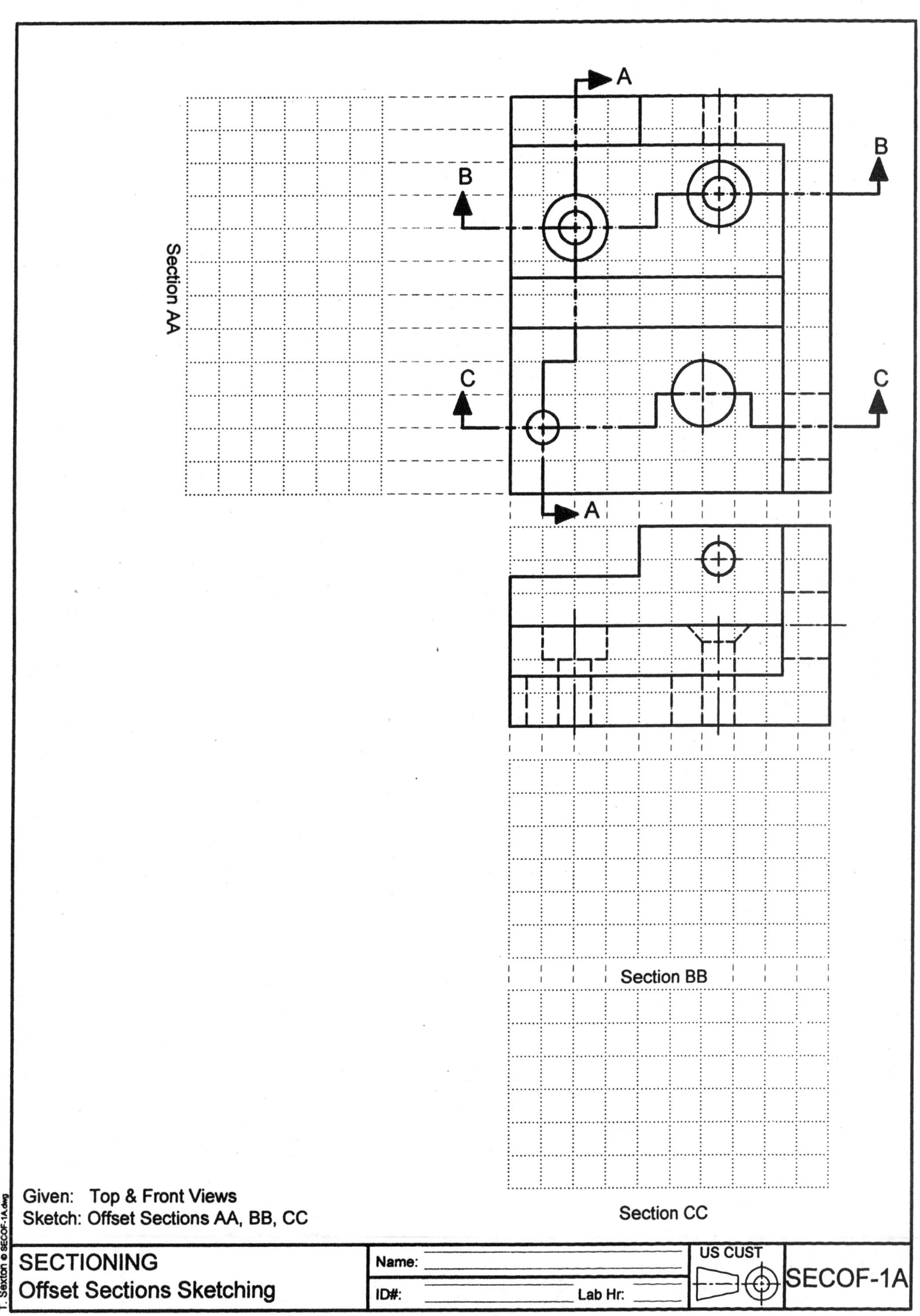

Given: Top & Front Views
Sketch: Offset Sections AA, BB, CC

SECTIONING
Offset Sections Sketching

Name:
ID#:
Lab Hr:

US CUST

SECOF-1A

T. Sexton © SECOF-1A.dwg

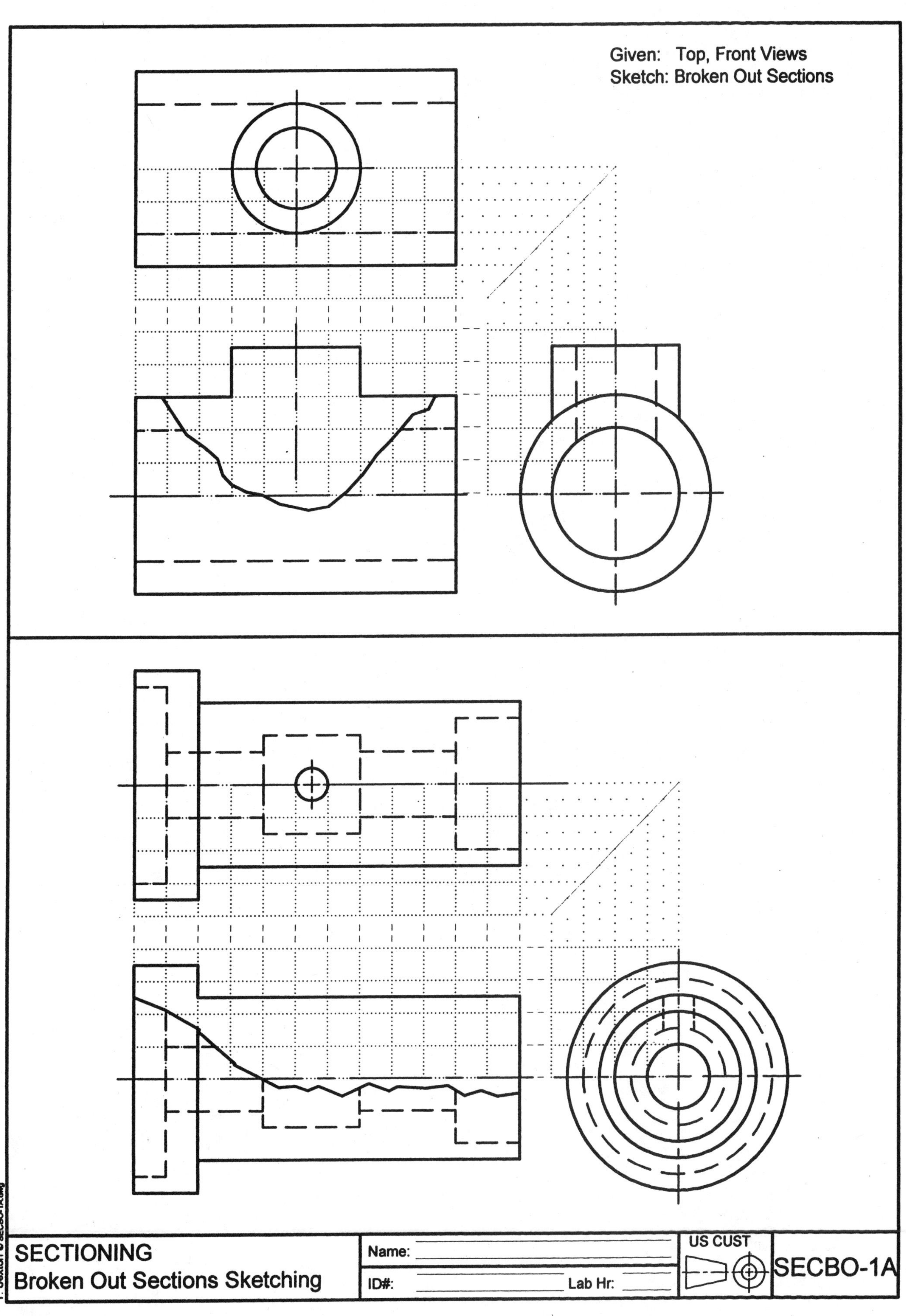
Given: Top, Front Views
Sketch: Broken Out Sections
SECTIONING
Broken Out Sections Sketching
Name:
ID#:
Lab Hr:
US CUST
SECBO-1A
T. Sexton © SECBO-1A.dwg

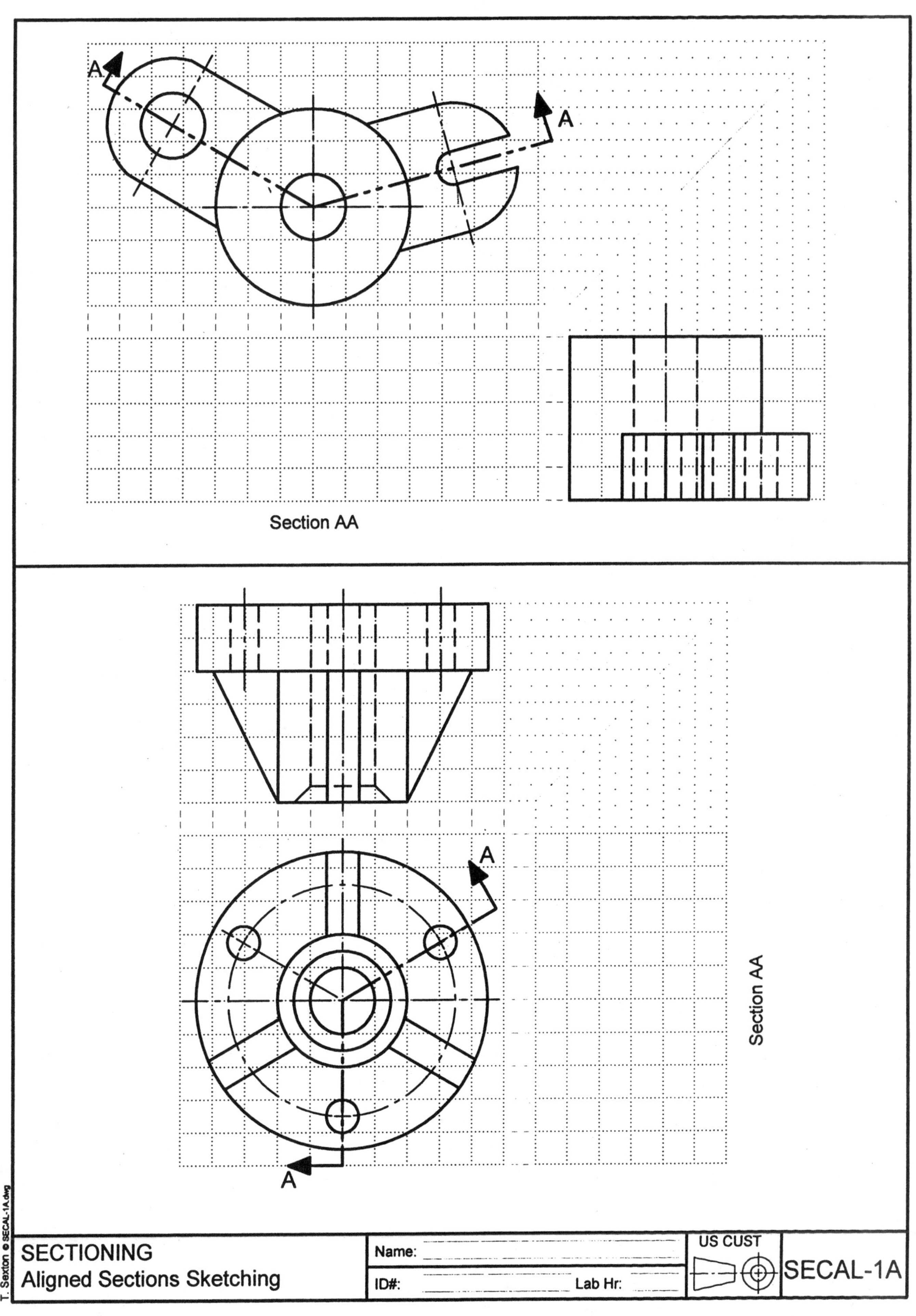

A
A
Section AA
A
A
Section AA
SECTIONING
Aligned Sections Sketching
Name:
ID#:
Lab Hr:
US CUST
SECAL-1A
T. Sexton © SECAL-1A.dwg

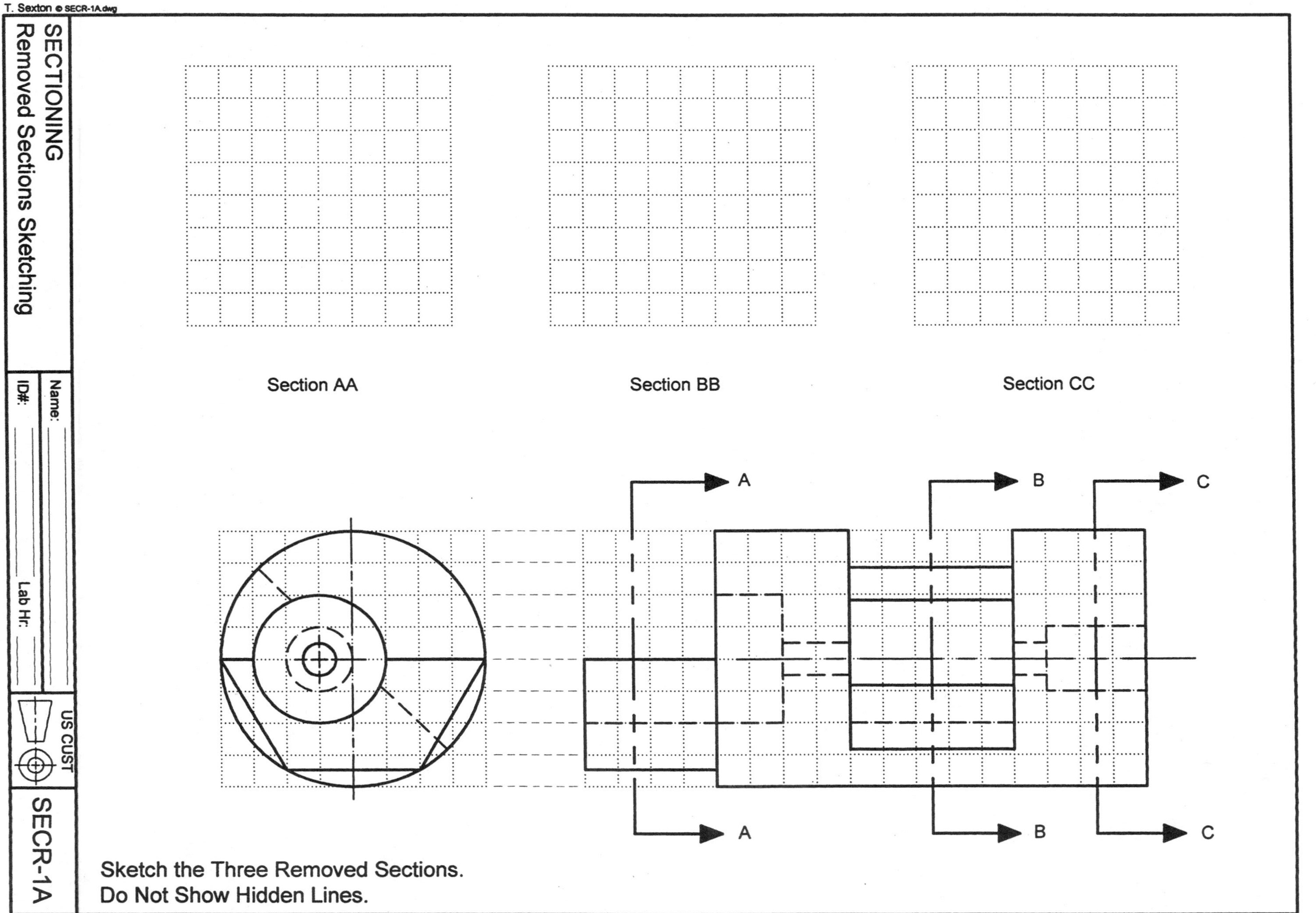
Section AA
Section BB
Section CC
A
B
C
A
B
C
Sketch the Three Removed Sections.
Do Not Show Hidden Lines.
T. Sexton © SECR-1A.dwg
SECTIONING
Removed Sections Sketching
Name:
ID#:
Lab Hr:
US CUST
SECR-1A

Chapter 8

Measurement & Scale

The majority of linear measurement is done full or actual size. Short measurements are typically made using units of inches or millimeters. Longer measurements are made using units of feet and inches or meters. But when you want to draw or sketch an object that is too large for your paper, the drawing must be proportionately reduced in size. If you are working on a part that is too small to see small details, then the part must be proportionately enlarged. This chapter discusses the method and equipment necessary to increase or decrease the size of an object or building when drawing or sketching on paper.

Full Size Metric Measurement:

Machine part drawings using SI (units, commonly referred to as METRIC, are based on the millimeter. Figure M&S 1 shows a full size metric scale with 10mm, 35mm, and 47mm marked off.

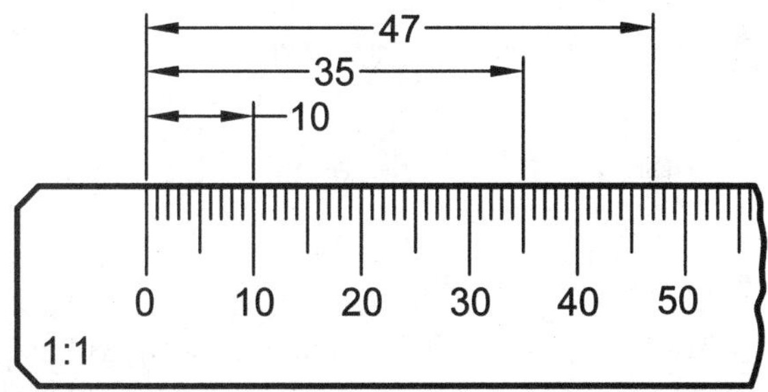

Figure M&S 1: Full size metric scale.

Architectural drawings using the metric system are based on the meter. Lengths less than one meter long are noted in decimal format such as 2.5 (two and one half meters). Metric drawings do not label units of measure with either mm for millimeter or m for meter.

Full Size US Customary:

Machine part drawings using US Customary units, commonly referred to as INCHES, are based on the inch. Inexpensive rulers and tape measurers divide the inch into 16 parts. Figure M&S 2 shows a full size scale (ruler) with the inch divided into 16 parts or every 1/16th of an inch. Figure M&S 2 also shows the measurements 1.25, 1.63, 2.12 inches marked off. Even though the inch is divided into fractional units, dimensions are displayed using the decimal inch format. Work requiring more accuracy than 1/16 of an inch must use a scale that divides the inch into 32 or even 64 parts. Inch based machine drawings do not label units of measure with either the abbreviation (in) for inches or (") for the inch mark.

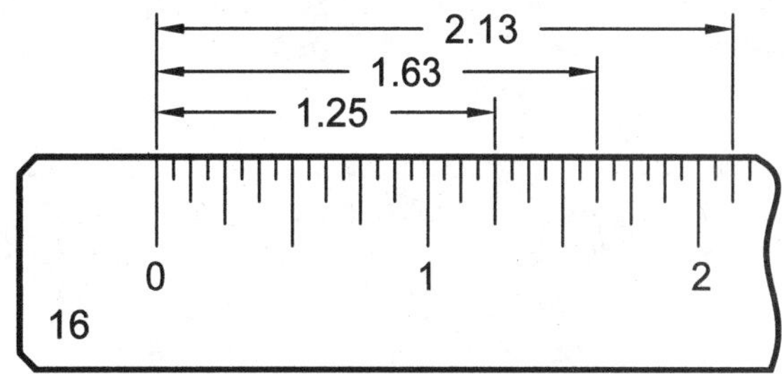

Figure M&S 2: Full size inch scale.

Architectural drawings using US Customary units use one foot as the base unit. The foot is divided into 12 inches and then into fractions of an inch. Architectural drawings display units of measure using the foot mark (') and the inch mark ("). The feet are separated from the inches by a dash. For example, two feet five and one half inches would be displayed as 2'–5 1/2". If the measurement is less than one foot, display just the inches, for example, 6 1/4". If there are no inches, a zero is displayed, for example, 10' – 0".

Furniture and cabinet drawings use the inch as their base unit. On these types of drawings the unit of measure is labeled using the inch mark ("). The foot mark (') is used only if the project is very large and using feet is part of the dimensioning format.

Scale:

A scale is simply a ratio with the format # = # for inch based scales and # : # for metric based scales. The # on the left is the measurement on the drawing while the number on the right is the actual measurement as illustrated in Figure M&S 3. For example, 1 = 2 (half scale) means that one inch measured on the drawing is actually representing 2" on the actual part. This means the object is being reduced by a factor of one half. The architectural scale 1/4" = 1'–0" (ratio of 1:48) means that 1/4" measured on a floor plan drawing actually represents 1'–0" on the actual building.

Left Side – Right Side	
1:2	
1=2	
Measurement On Drawing	Measurement Actual Part.

Figure M&S 3: Specifying scale on a drawing using half size is used as an example. The number on the left is what is measured on the drawing while the number on the right is the measurement on the actual part or building.

> *You should **NEVER** measure a drawing! Read its dimensions because drawings can be printed at any scale or the drawing may be drawn one way and dimensioned another.*

Scales:

When constructing drawings using drafting instruments, you lay off linear measurements using a scale. Scales have markings that make it easy to reduce or enlarge the drawing in relationship to the actual size of the object. Physically scales look like a ruler and can be made of wood, metal or plastic. They can be flat or triangular in cross section. They are usually 6 or 12 inches long. Figure M&S 4 shows a 6" flat and a 12" triangular scale.

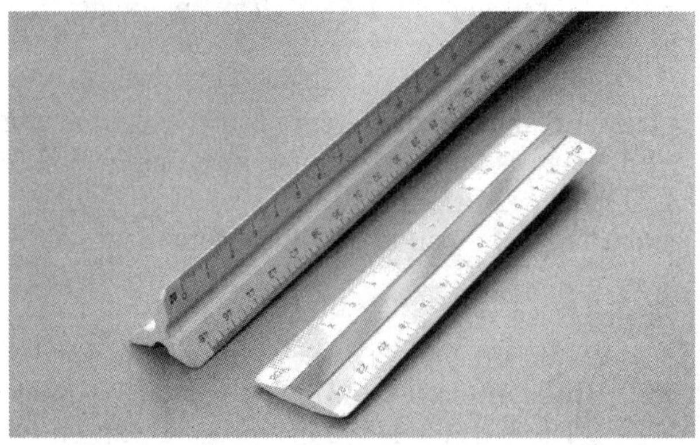

Figure 4: 6" flat and 12" triangular scales.

The four basic types of scales are:

1. civil engineering scale which is sometimes called an engineering scale
2. mechanical engineering scale
3. architectural scale
4. metric scale

Civil Engineering Scale:

The civil engineering or engineering scale is based on the inch, with the inch being divided into a multiple of ten: 10, 20, 30, 40, 50, or 60. When using the civil scale, measurements are displayed in decimal format such as 105.6'. Using the decimal feet format, as opposed to fractions, makes it easier to run calculations on surveying data. Figure M&S 5 illustrates two civil engineering scales. In M&S 5A the inch is divided into ten parts and in M&S 5B the inch is divided into twenty parts. Note that the civil engineering scale has every inch marked off. This is called a chain divided scale.

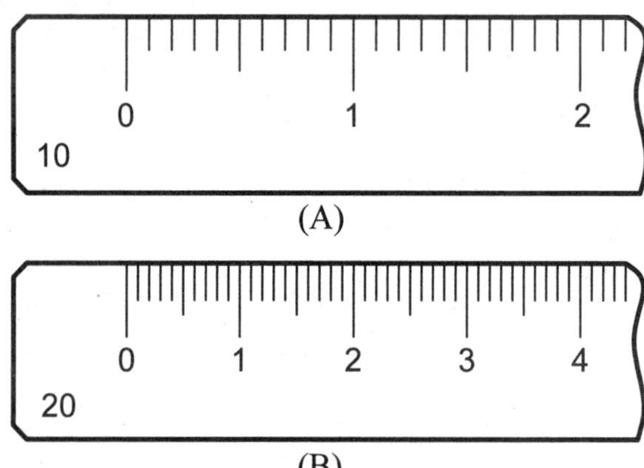

Figure M&S 5: Civil engineering scales of base A) 10 and base B) 20.

A civil scale can be used to measure off any multiple of the base. For example, on the 30 scale in Figure M&S 6 the 30 can represent 3.0 feet, 30 inches, 300 feet, 30 meters, or 300 meters. The 6.9 can represent 6.9 inches, 69 feet, 690 feet, 69 meters, or 690 meters.

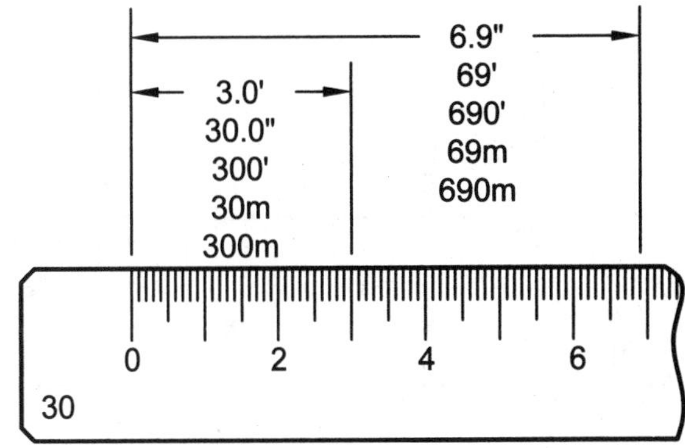

Figure M&S 6: A measurement on the civil engineering scale can represent any multiple of the base.

The civil engineering scale can also be used to draw inch based machine parts. Figure M&S 7 illustrates a base 10 civil scale with full scale measurements laid out in decimal inches.

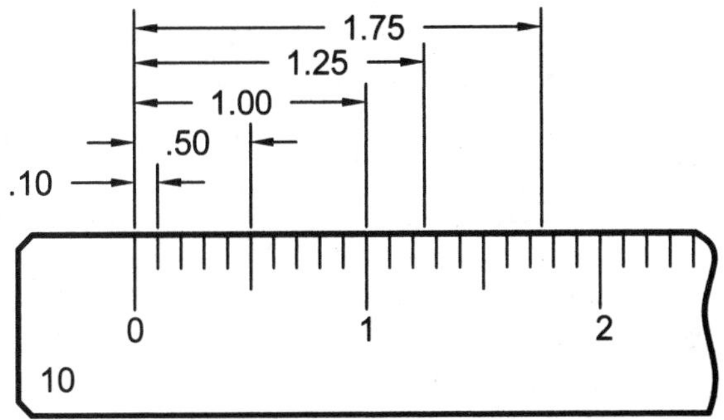

Figure M&S 7: Full scale measurements on a base 10 civil scale. All measurements are in the decimal inch format.

Mechanical Engineering Scales:

The mechanical engineering scale is used when constructing drawings using fractional inches such as 1 1/2" or 8 3/4". Since almost all machine drawings are done using decimal fractions, the mechanical scale is seldom used. Figure M&S 8A illustrates increments on a full (1 = 1) mechanical scale. M&S 8B illustrates increments on a half (1 = 2) mechanical scale. Note that full inches are measured to the right while fractions of an inch are measured to the left. Figure M&S 9 shows 2.25, 2.50, 3.75 inches laid out on a half (1 = 2) scale. All fractional values are converted to their decimal equivalents.

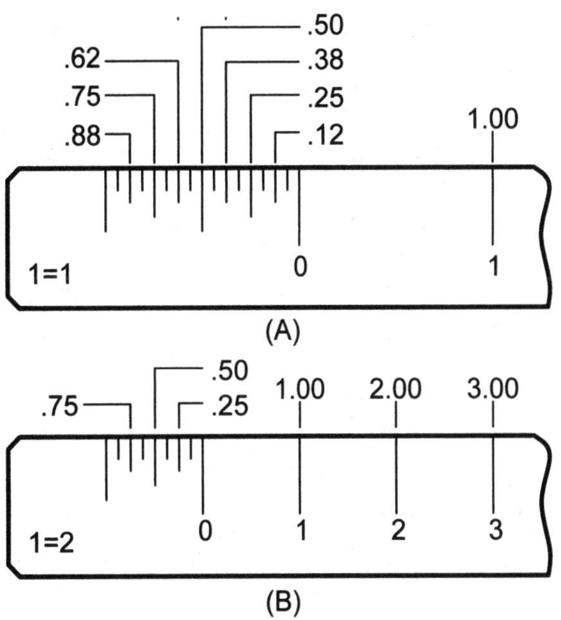

Figure M&S 8: Mechanical scales: A) 1 = 1 scale increments, and B) 1 = 2 scale increments.

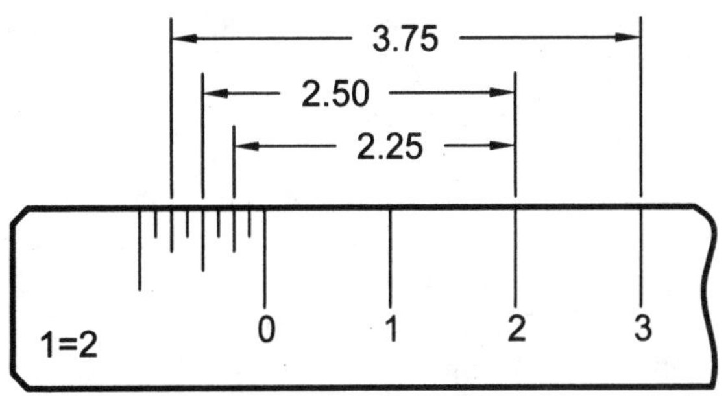

Figure M&S 9: Measurements marked off on a 1 = 2 mechanical scale.

Architectural Scales:

The base unit on the architectural scale is one foot. The most common inch based architectural scales are: 1/8" = 1'-0", 1/4" = 1'-0", 3/8" = 1'-0", 1/2" = 1'-0", 3/4" = 1'-0", 1" = 1'-0", and 3" = 1'-0". On architectural scales, only a single foot is divided into fractional parts and it is located at one end of the scale. This is called an open divided scale. Figure M&S 10 illustrates the 3/4" = 1'-0" architectural scale. To the left of the zero, 3/4 of an inch is divided into 12 inches. The smallest mark is equal to one inch. To the right of the zero only full foot units are marked off. Read only the lower numbers 0, 1, and 2 as the upper numbers 28, 26, and 24 are unit markings of the 3/8" = 1'-0" scale starting from the other end of the scale. Triangular scales have two scales along each edge: one reading from right to left and one from left to right. To increase accuracy read to the left and right of the zero at the same time. Do not separately lay out full feet and then the inches. It is not as accurate as lying them out together.

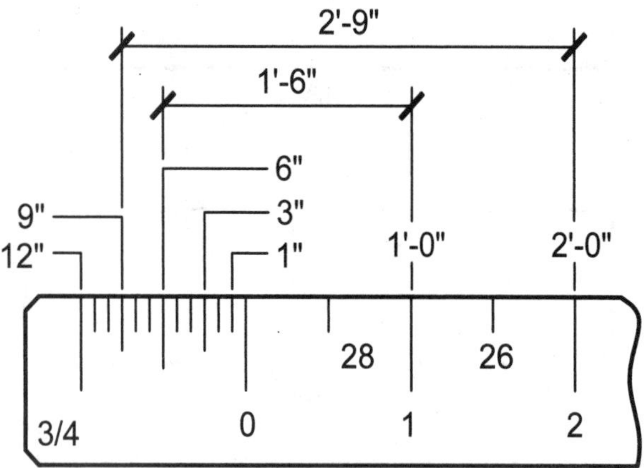

Figure M&S 10: 3/4″ = 1′-0″ scale with full feet reading to the right of the zero and inches reading to the left.

Figure M&S 11 illustrates the 1/4" = 1'-0" scale with 3'-0", 4'-6", and 6'-10" marked off and the 1/8" = 1'-0" scale with 7'-0", 10'-4", and 14'-10" marked off. Note how with 10'-4" and 14'-10" you need to read to both sides of the zero. Since 1/4" is small, it is divided into only six parts with each mark representing 2 inches. The 1/4" = 1'-0" scale reads from right to left using only the lower numbers. The upper numbers are for the 1/8" = 1'-0" scale which reads from left to right on the same edge.

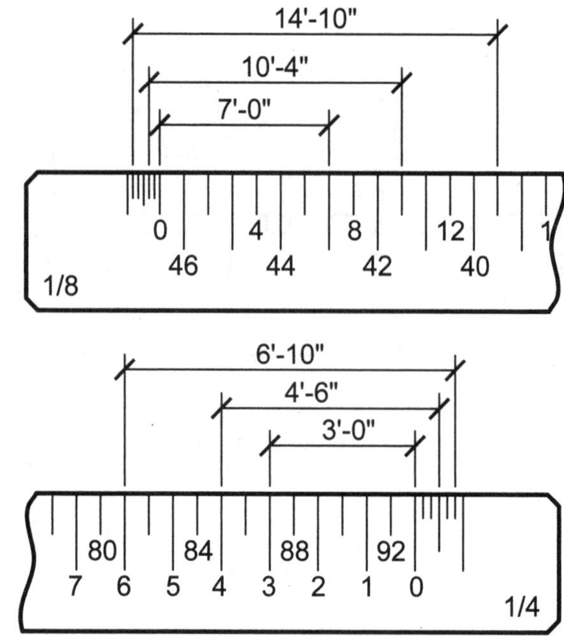

Figure M&S 11: 1/4″ = 1′-0″ scale with 3′-0″, 4′-6″, and 6′-10″ marked off and the 1/8" = 1'-0" scale with 7'-0", 10'-4", and 14'-10" marked off.

Metric Scales:

Metric machine drawings use the millimeter as their base unit. Figure M&S 12A illustrates the 1 : 2 scale with 40, 70 and 94 mm marked off. Figure M&S 12B illustrates the 1 : 5 scale with 125, 210, and 275 mm marked off. Note that all metric scales have millimeter markings along their entire edge. This is called a chain divided scale.

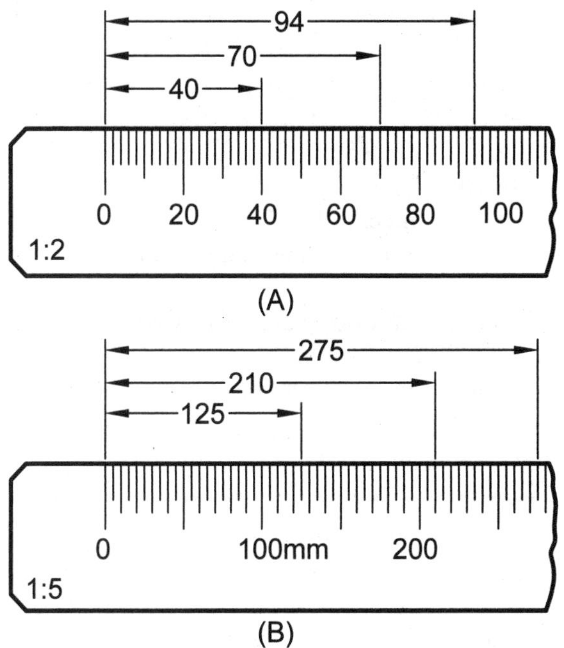

Figure M&S 12: A) metric measurements on a 1 : 2 scale, B) metric measurements on a 1 : 5 scale.

Metric architectural scales 1 : 50 and 1 : 100 are used for floor plans, with the meter as their base unit. A measurement less than one meter is designated in decimal format. For example 1.5 is equal to one and one half meters or 1500mm. Figure M&S 13A illustrates the 1 : 50 floor plan scale with 1.5m, 1.75m, and 2.25m marked off and M&S 13B illustrates the 1 : 100 floor plan scale with 2m, 3.5m, and 4.6m marked off. Metric architectural scales 1 : 20 or 1 : 25 are used to draw details such as wall sections.

Inch based and metric based architectural scales are similar:

Inches		Metric	
1/8″=1′-0″ (1:96)	vs.	1:100	Floor
1/4″=1′-0″ (1:48)	vs.	1:50	Floor
1/2″=1′-0″ (1:24)	vs.	1:25	Section
3/4″=1′-0″ (1:16)	vs.	1:20	Detail

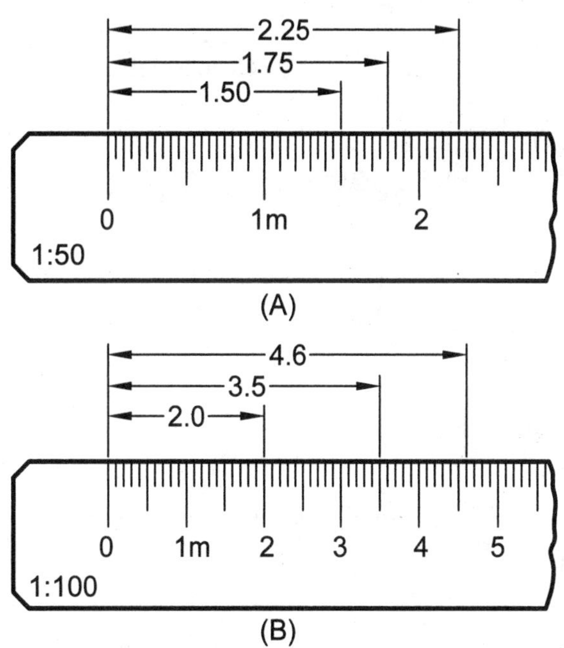

Figure M&S 13: A) Floor plan scale 1:50 with 1.5m, 1.75m, and 2.25 marked off, and B) floor plan scale 1:100 with 2m, 3.5 m and 4.6m marked off.

Bar Scales:

A bar (or graphic) scale is not a physical scale but a graphic way of expressing the scale of a drawing. The bar scale is used extensively on maps and related drawings. A graphic scale is simply a rectangle divided into scaled units. Figure M&S 14 illustrates three graphic scales: 1/4" = 1'-0", 1" = 200', and 1" = 400'. Since the bar scale is graphic, it will change proportionally with the scale of the drawing when it is printed/plotted or photocopied. Printing/plotting and photocopying can produce drawings at any scale. A bar scale can give you only approximate distances.

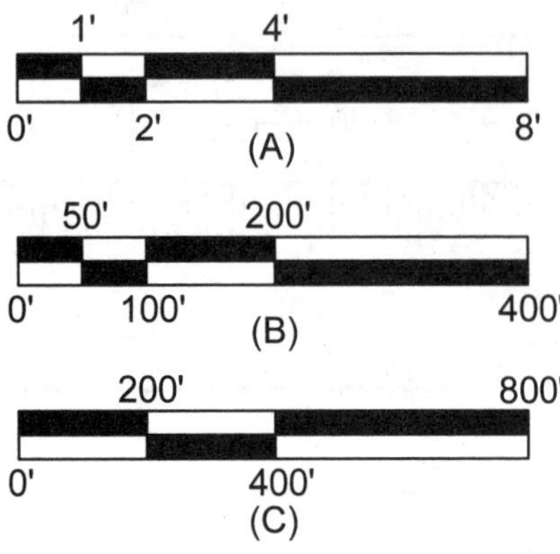

Figure M&S 14: Sample bar scales A) 1/4″ = 1′-0″, B) 1″ = 200′, C) 1″ = 400′

> ***Always*** *read the dimensions on a drawing and NEVER measure the drawing!*

Table M&S 1 lists the scales used with both metric and inch based drawings.

Table M&S 1

Scale	Ratio
Inch Architectural:	
Commercial floor plans and elevations	
1/8" = 1'-0"	1 : 96
Residential floor plans and elevations	
1/4" = 1'-0"	1 : 48
Building and wall sections	
1/2" = 1'-0"	1 : 24
3/4" = 1'-0"	1 : 16
1" = 1'-0"	1 : 12
Large Details	
1 1/2" = 1'-0"	1 : 8
3" = 1'-0"	1 : 4
Inch Civil Engineering:	
1" = 10'	1 : 120
1" = 20'	1 : 240
1" = 30'	1 : 360
1" = 40'	1 : 480
1" = 50'	1 : 720
1" = 60'	1 : 600
1" = 100'	1 : 1200
1" = 200'	1 : 2400
Inch Machine:	
1 = 1	
Reducing	Enlarging
1 = 2	2 = 1
1 = 4	4 = 1
Metric Machine:	
1 : 1	
Reducing	Enlarging
1 : 2	2 : 1
1 : 5	5 : 1
1 : 10	
Metric Architectural:	
Residential floor plans and elevations	
1 : 50	
Commercial floor plans and elevations	
1 : 100	
Building and wall sections	
1 : 20	
1 : 25	

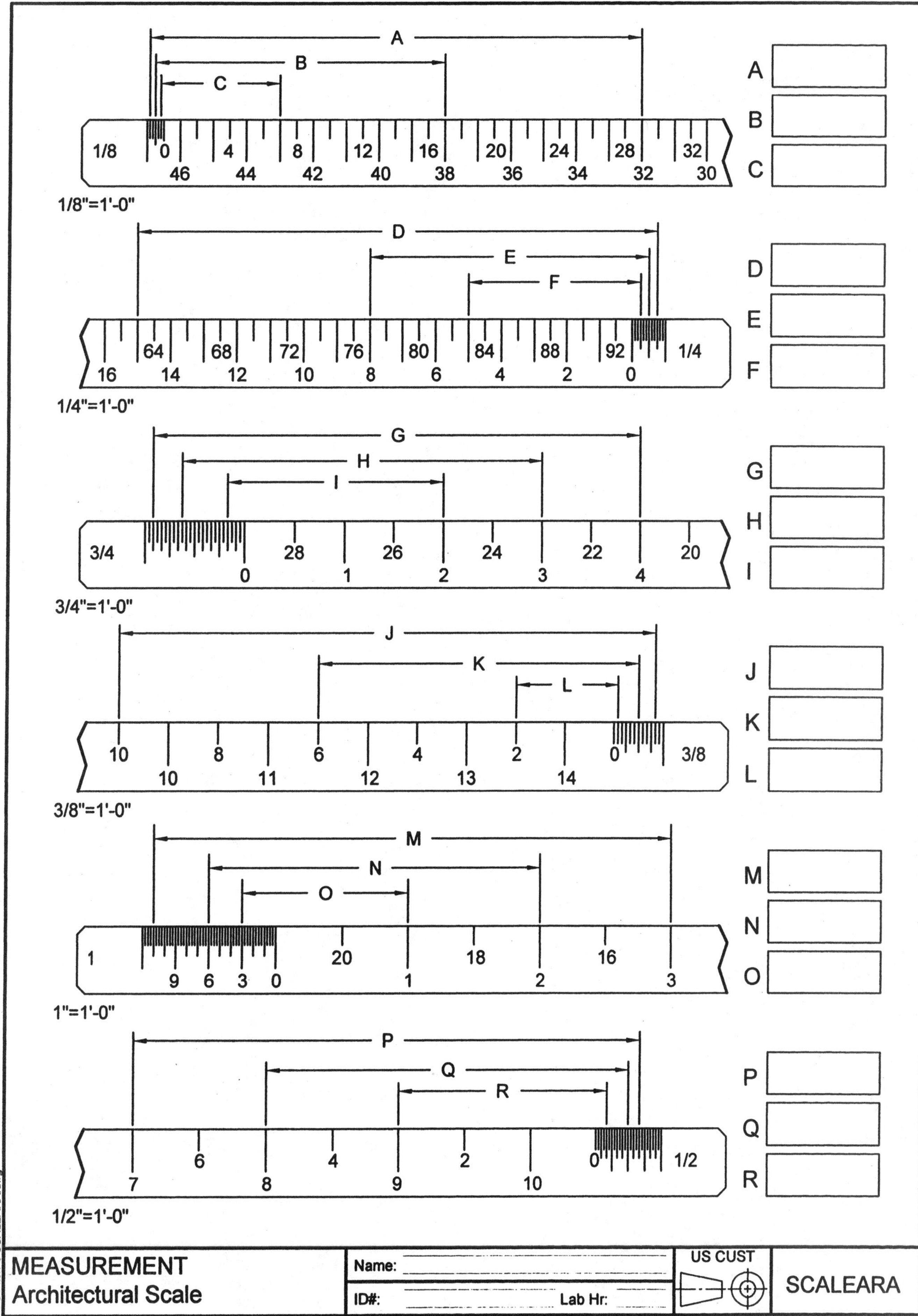

A
B
C
1/8
0 4 8 12 16 20 24 28 32
46 44 42 40 38 36 34 32 30
1/8"=1'-0"
D
E
F
64 68 72 76 80 84 88 92
16 14 12 10 8 6 4 2 0
1/4
1/4"=1'-0"
G
H
I
3/4
28 26 24 22 20
0 1 2 3 4
3/4"=1'-0"
J
K
L
10 8 6 4 2 0
10 11 12 13 14
3/8
3/8"=1'-0"
M
N
O
1
20 18 16
9 6 3 0 1 2 3
1"=1'-0"
P
Q
R
6 4 2 0
7 8 9 10
1/2
1/2"=1'-0"
T. Sexton ⊕ SCALEARA.dwg
MEASUREMENT
Architectural Scale
Name:
ID#:
Lab Hr:
US CUST
SCALEARA

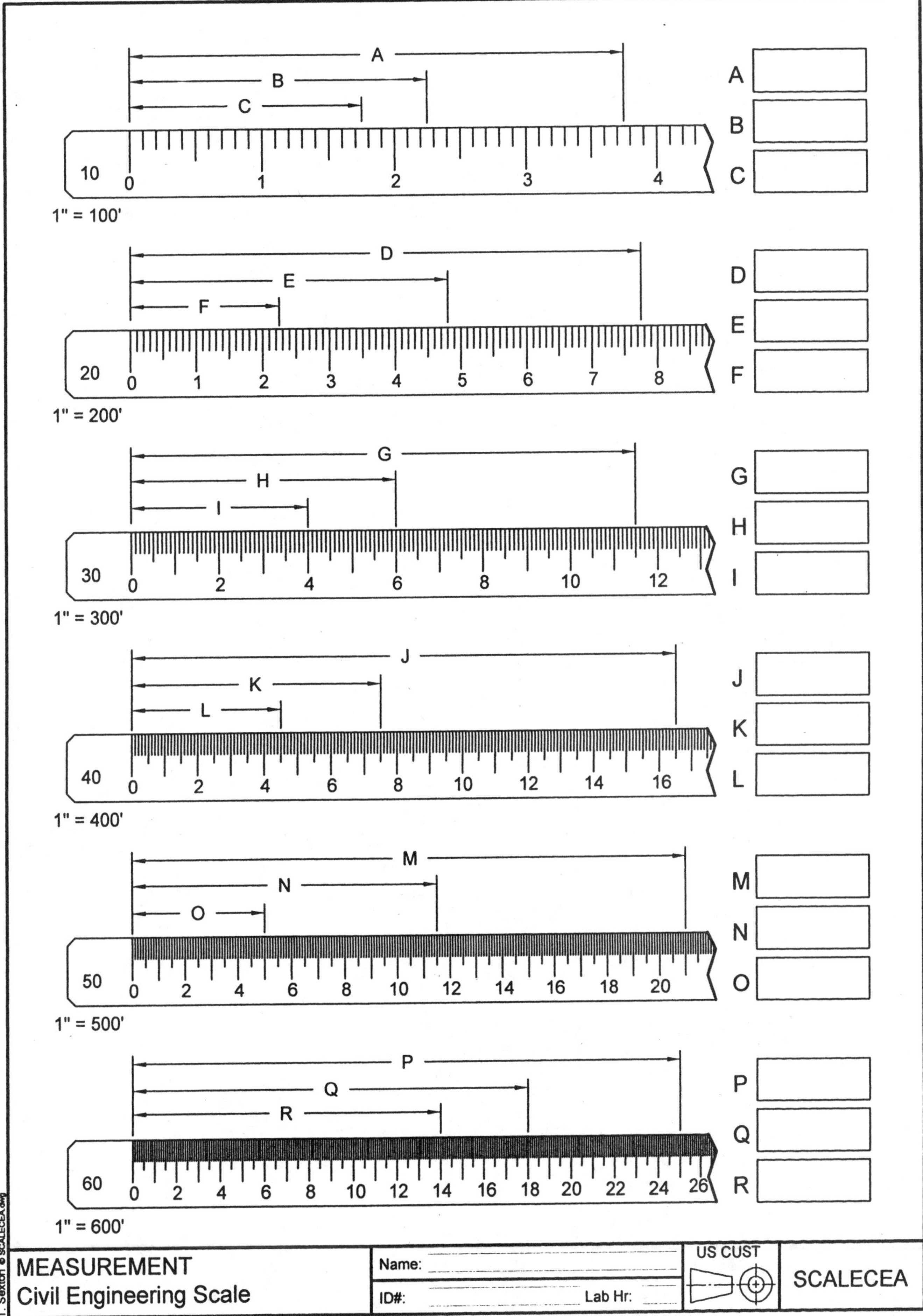

A
B
C
10
0 1 2 3 4
1" = 100'
D
E
F
20
0 1 2 3 4 5 6 7 8
1" = 200'
G
H
I
30
0 2 4 6 8 10 12
1" = 300'
J
K
L
40
0 2 4 6 8 10 12 14 16
1" = 400'
M
N
O
50
0 2 4 6 8 10 12 14 16 18 20
1" = 500'
P
Q
R
60
0 2 4 6 8 10 12 14 16 18 20 22 24 26
1" = 600'
MEASUREMENT
Civil Engineering Scale
Name:
ID#:
Lab Hr:
US CUST
SCALECEA
T. Sexton © SCALECEA.dwg

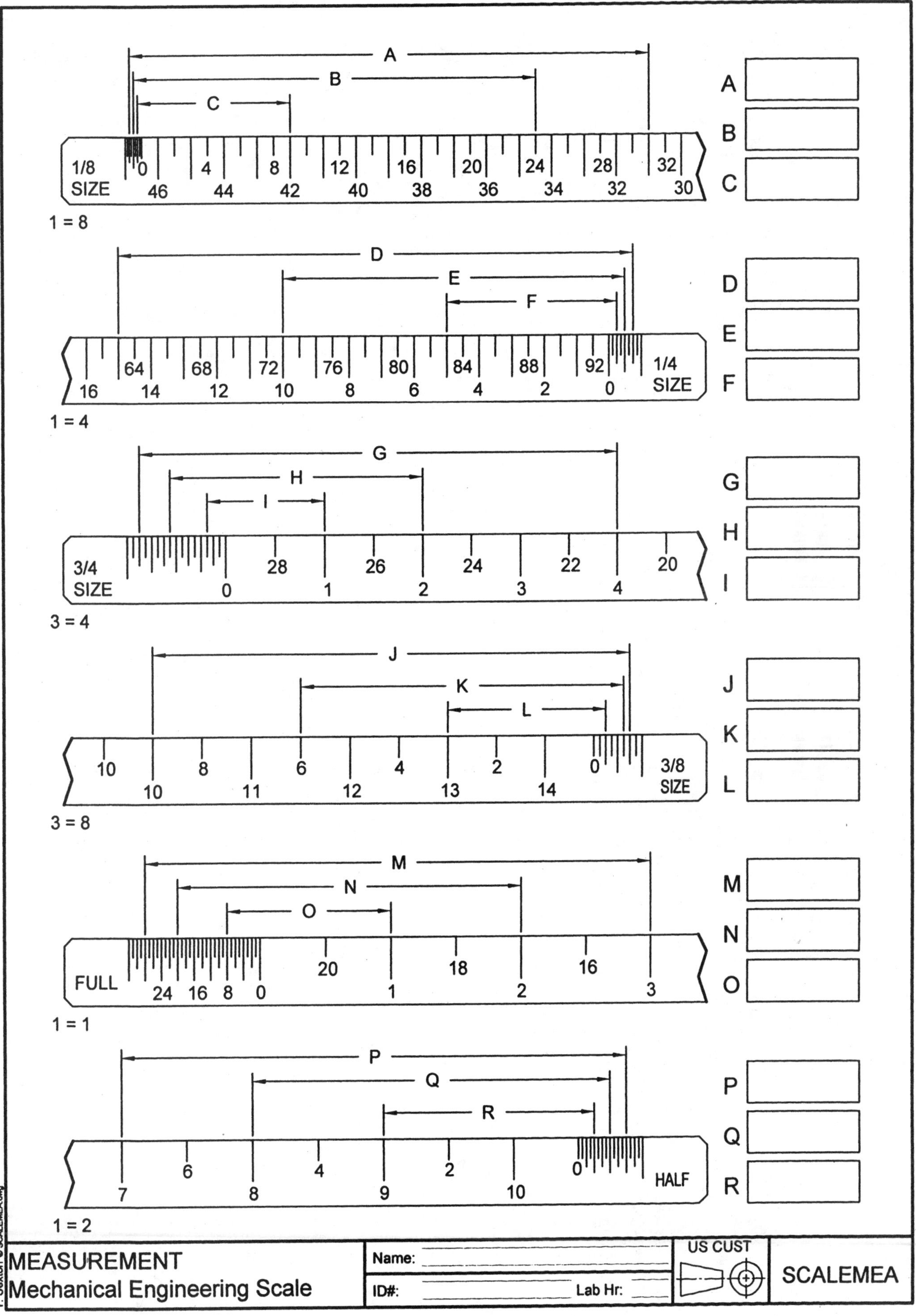
A
B
C
1/8
SIZE
0 4 8 12 16 20 24 28 32
46 44 42 40 38 36 34 32 30
1 = 8
D
E
F
64 68 72 76 80 84 88 92
16 14 12 10 8 6 4 2 0
1/4
SIZE
1 = 4
G
H
I
3/4
SIZE
28 26 24 22 20
0 1 2 3 4
3 = 4
J
K
L
10 8 6 4 2 0
10 11 12 13 14
3/8
SIZE
3 = 8
M
N
O
FULL
24 16 8 0
20 18 16
1 2 3
1 = 1
P
Q
R
6 4 2 0
7 8 9 10
HALF
1 = 2
A B C D E F G H I J K L M N O P Q R
T. Sexton © SCALEMEA.dwg
MEASUREMENT
Mechanical Engineering Scale
Name:
ID#:
Lab Hr:
US CUST
SCALEMEA

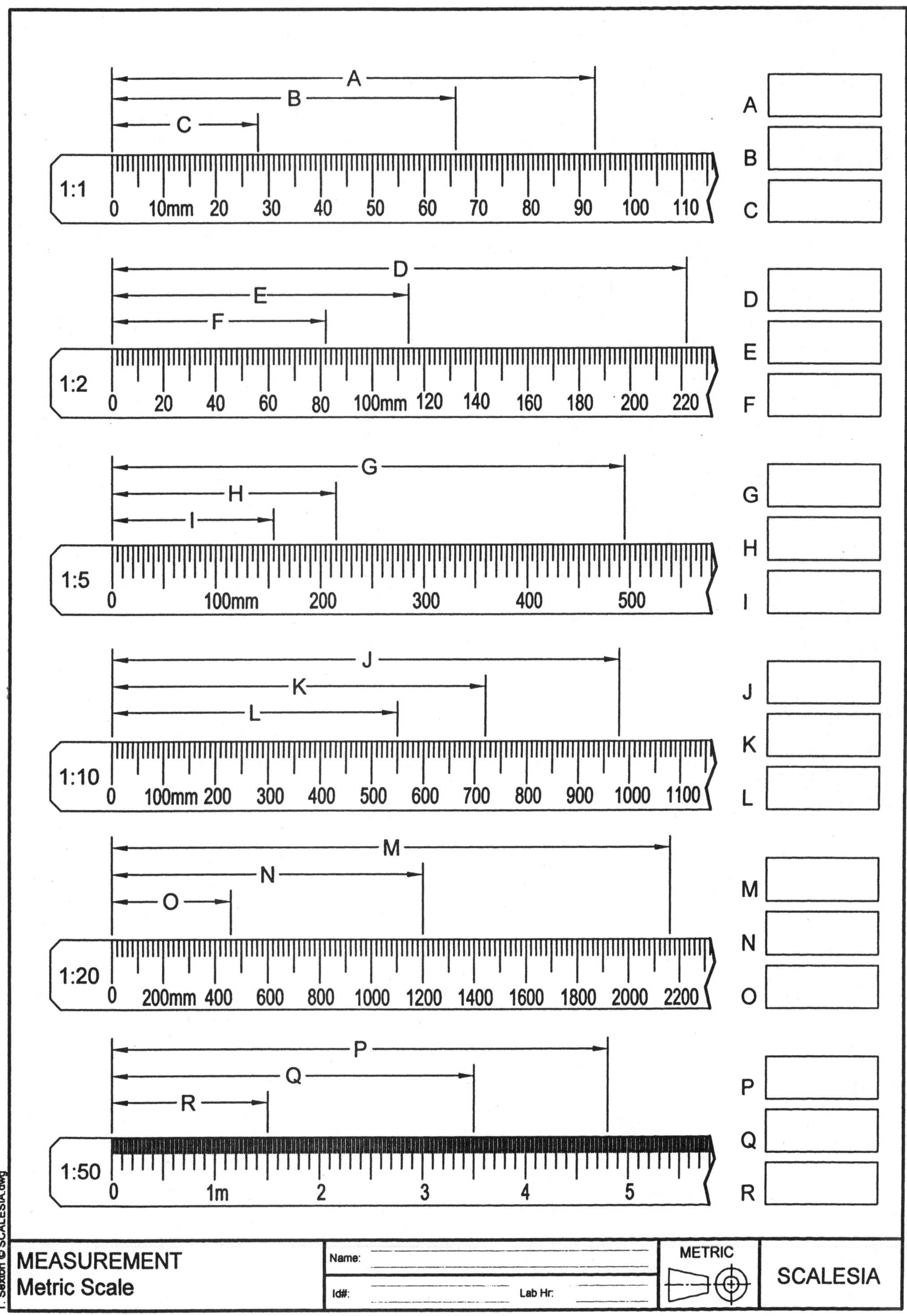

1:1
0 10mm 20 30 40 50 60 70 80 90 100 110
1:2
0 20 40 60 80 100mm 120 140 160 180 200 220
1:5
0 100mm 200 300 400 500
1:10
0 100mm 200 300 400 500 600 700 800 900 1000 1100
1:20
0 200mm 400 600 800 1000 1200 1400 1600 1800 2000 2200
1:50
0 1m 2 3 4 5
A B C D E F G H I J K L M N O P Q R
MEASUREMENT
Metric Scale
Name:
Id#:
Lab Hr:
METRIC
SCALESIA
T. Sexton © SCALESIA.dwg

Chapter 9

Dimensioning

Why Dimension?

Dimensions are numeric values that specify size and location of geometric features. Dimensioning also includes symbols and text that help describe geometric characteristics or a special operation performed on the object. A drawing without dimensions would be of little value because the drawing would only describe the object's shape not its exact size or the location of geometric features. This chapter is based on the American National Standards Institute (ANSI) standard Y14.5 - 1994 *Dimensioning and Tolerancing*.

Dimensioning Line Types, Terminology, and Symbols:

Dimension Lines are two thin solid lines terminating in arrowheads with a dimension placed between as illustrated in Figure DIM 1A and Figure DIM 2. An architectural dimension is placed above a continuous dimension lines as illustrated in the two examples in Figure DIM 1A. Unlike machine drawings that use only arrowheads as terminators, architectural drawings can have several different types of terminators such as dots or tick marks as illustrated in Figure DIM 1B and DIM C.

Extension Lines are a pair of thin solid lines extending from the geometric feature on the object to slightly beyond the dimension line as illustrated in Figure DIM 1 and Figure DIM 2. A centerline can also be used as an extension line as illustrated in Figure DIM 2. Leave a small gap between the object and an extension line. The gap should be approximately 1/2 the lettering height (LH) as illustrated in Figure DIM 2. An extension line should extend beyond the last arrowhead to a length equal to the height of the lettering (H) as illustrated in Figure DIM 2.

Center Line is a thin long-dash-long line used to locate the center of a geometric feature or indicate symmetry of a geometric feature as illustrated in Figure DIM 1 and Figure DIM 2. When a hole appears as a circle as when looking down into a hole, two centerlines are used starting with a "plus sign" at the center of the circle as illustrated in Figure DIM 1B. When looking at the hole from the side, a single centerline is used such as in Figure DIM 1C.

Leader Line is a thin solid line beginning with a short horizontal line then extending on an angled until it touches the object. The angled line terminates in an arrowhead as illustrated in Figure DIM 1 and Figure DIM 2. Imagine it as a finger pointing at something. The arrowhead should always touch what is being noted and not float in the air.

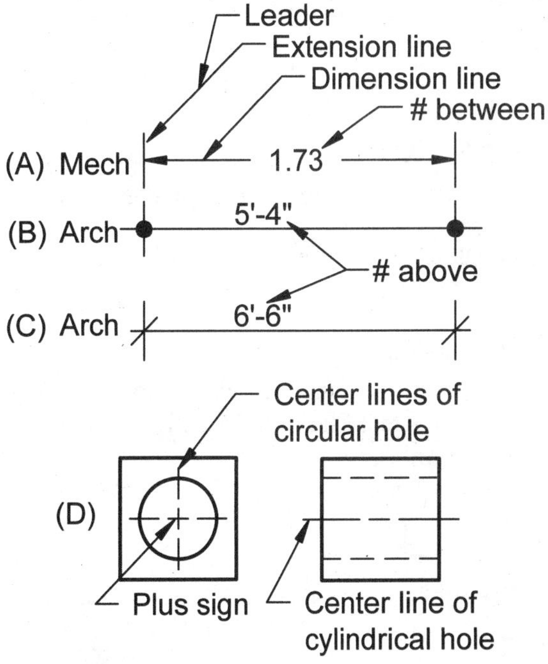

Figure DIM 1: Dimensioning line types and terminology.

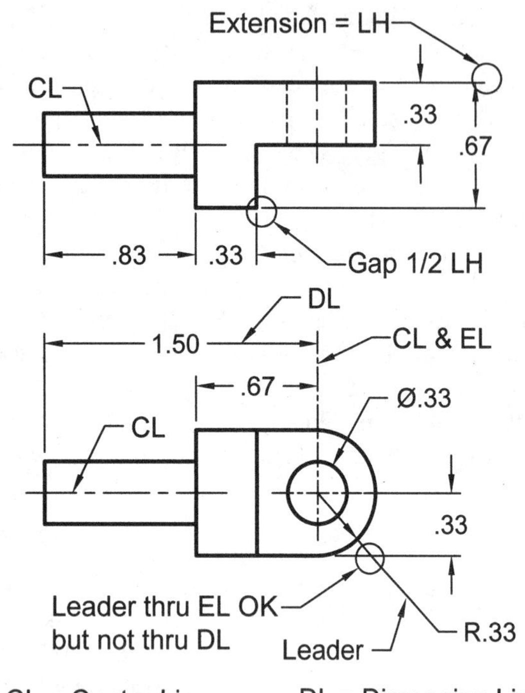

Figure DIM 2: Application of dimensioning line types and terminology.

Special Symbols:

To minimize language on a drawing special symbols are used. Figure DIM 3 illustrates the most common symbols.

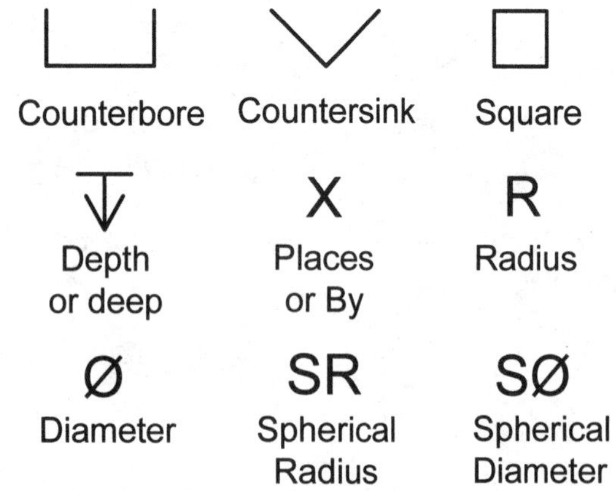

Figure DIM 3: Special symbols used on detail drawings.

Units of Measure:

Machine drawings use two types of dimensioning systems: 1) US Customary commonly referred to as METRIC with its basic unit being the millimeter. In Architectural/Engineering/Construction (AEC) drawings, the basic unit for US Customary is the foot and for SI it is the meter.

On machine drawings inches are normally displayed using the decimal format. For example, three inches = 3.00, three and three quarters inches = 3.75, and a fraction of an inch such as one quarter = .25.

For fractional inches there is ***NO*** leading zero before the decimal point, for example, three eighths of an inch = .38. A portion of a millimeter uses the leading zero, for example, one half of a millimeter = 0.5.

On machine drawings the following are ***NOT*** displayed: 1) the abbreviation in, 2) the inch mark " for inches, nor 3) the abbreviation mm for millimeters. On AEC drawings units are displayed, for example, four feet six inches = 4'-6", eight and one half inches = 8 1/2" (just inches, no feet), and ten feet zero inches = 10'-0" (zero inches should be displayed). Don't forget the dash that separates the feet from the inches. Table DIM 1 summarizes and gives examples of how different types of units are displayed. The number of decimal places used depends on the degree of accuracy required and the abilities of the manufacturing equipment. Each additional decimal point can increase the cost of the product. To give some "ballpark" idea of what adding decimal places means, a sheet of paper is approximately .003 inches thick and 0.1 millimeters is approximately equal to .004 inches.

Table DIM 1

Units of Measure Summary		
UNIT	**EXAMPLE**	**NOTES**
Machine Drawings		
millimeter	44 35.3 0.6	No unit abbreviations Leading zero required
inches	8.62 40.50 .625	No unit abbreviations Decimal fractions No leading zero
Degrees	68° 26.5° 68°40'32"	Degree symbol used Decimal of a degree Degree/Minutes/Seconds
AEC Drawings		
Feet & Inches	20' - 6 1/2" 8' – 0" 4 1/2"	Feet (') & inch (") symbols shown Zero inches shown Zero feet omitted
Feet & decimal fractions of a foot	11.62'	Used on plot plans & maps (') mark shown
Meters	10.500	Meters & portions of a meter to 3 decimal places No m for meter
Degrees	35°15'45"	Degrees/Minutes/Seconds

Dimensioning Standards:

There are two different sets of rules for dimensioning drawings called *standards*. The United States follows standards developed by the American National Standards Institute (ANSI). ANSI standards cover both inch and metric drawings. Most countries outside the United States follow standards developed by the International Standards Organization (ISO). ISO standards cover drawings done in millimeters. In addition to using one of these two standards, most companies have "in house" standards developed for their particular industry or product. This text follows ANSI standards.

Even when following standards, there is often more than one correct way to dimension a drawing. For example the depth of an object could be dimensioned in the top or right view. On rare occasions, when a drawing becomes very complex, standards may need to be violated to insure clarity.

Dimensioning Strategy:

To help insure maximum clarity, dimensions and detailing notes must be positioned in a neat and organized manner. Misinterpretation of dimensions or notes can result in very costly manufacturing or construction mistakes.

The two most important factors that need to be understood in order to properly dimension a drawing are: 1) how does the part function and 2) how does the part mate with other parts?

A very practical way to determine if a drawing is dimensioned well is if you can use the drawing to make the product to specifications without asking any questions.

There is no single strategy for attacking a dimensioning problem. But if you have a consistent strategy, you are less likely to overlook required dimensions. The following strategy is suggested:

1. Dimension all the height dimensions first.
2. Dimension all the width dimensions second.
3. Dimension all the depth dimensions third.
   (1, 2, and 3 can be completed in an order.)
4. Dimension holes, arcs, and solid cylinders.
5. Apply the remaining notes such as counter bores, keyways, and thread specs.
6. Increase clarity by repositioning dimensions and notes as required.
7. Review your work.
8. Have someone "proofread" your dimensioning for completeness and clarity.
9. Based on your review, make any additions and/or corrections.

Unidirectional vs. Aligned:

Machine drawings follow the *unidirectional* and notes read from the bottom of the page as illustrated in Figure DIM 4A. AEC drawings use the *aligned system* of dimensioning in which notes and horizontal dimensions read from the bottom of the page while vertical dimensions read from the right side of the page as illustrated in Figure DIM 4B.

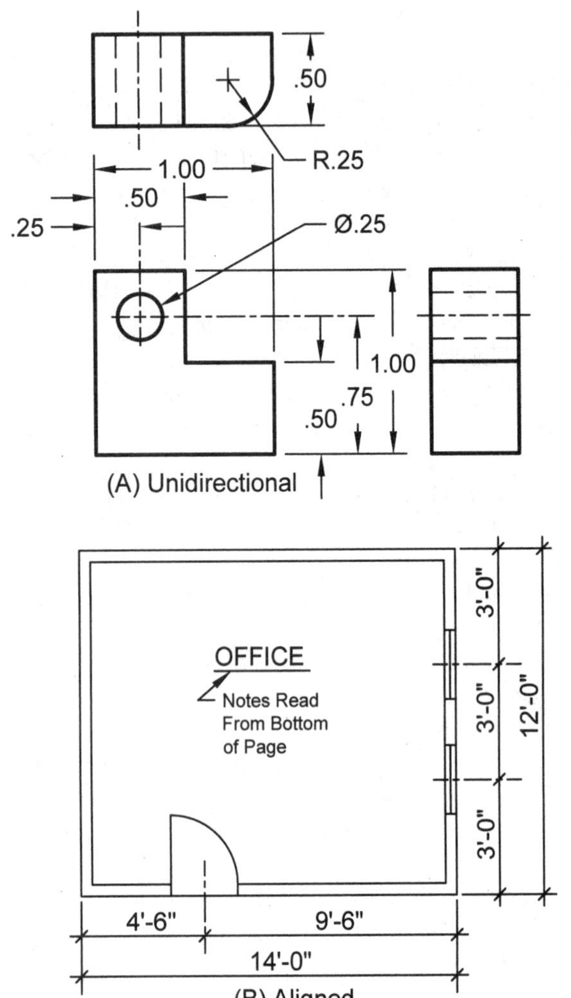

Figure: DIM 4: A) Unidirectional system for machine drawings vs. B) aligned system for AEC drawings.

Dimensioning Techniques:

ANSI standards provide rules for dimensioning drawings. These rules should be followed as closely as possible. The following list of general dimensioning techniques and illustrations follow the ANSI standards.

1. Place dimensions between views that share dimensions when possible as illustrated in Figure DIM 5.
2. Avoid dimensioning to hidden lines as illustrated in Figure DIM 6.

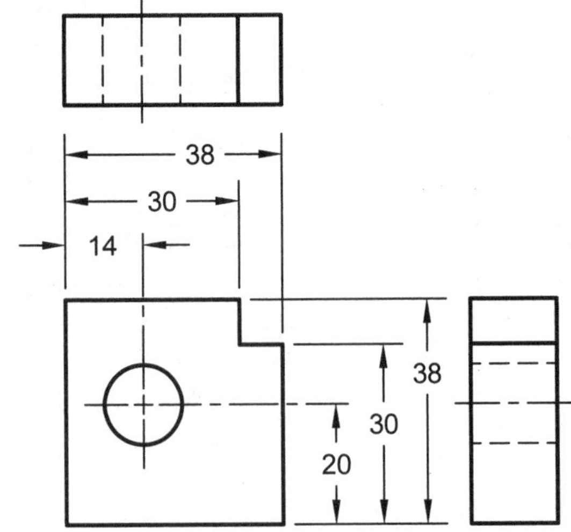

Figure DIM 5: Place dimensions between the views that share dimensions

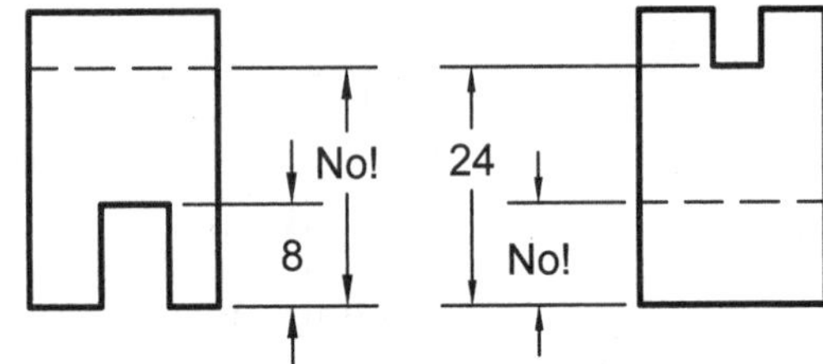

Figure DIM 6: Avoid dimensioning to hidden lines.

3. Dimension in the "profile" view as illustrated in Figure DIM 7 and Figure DIM 8. If you look at a person face-to-face, you can't tell how long their nose is until they turn 90° to show you their profile. By dimensioning in the profile view, the dimension is close to the geometric feature it is describing. A rule of thumb: if you need to look at another view to tell what is being dimensioned, it's in the wrong view – so move it.

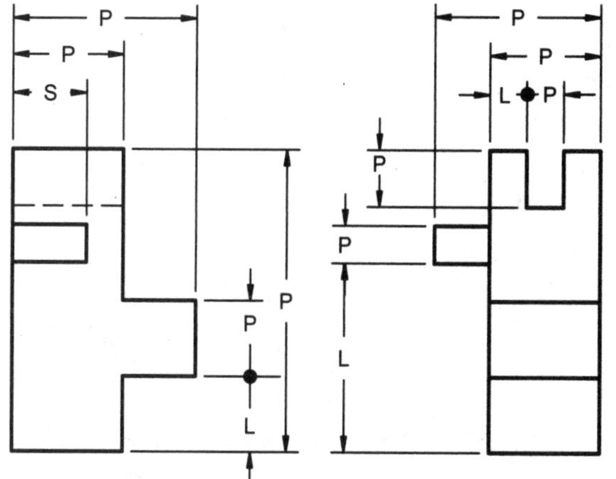

Figure DIM 7: Dimension in the "Profile" (P) view that shows the part's geometric characteristic (S)=Size (L)=Location.

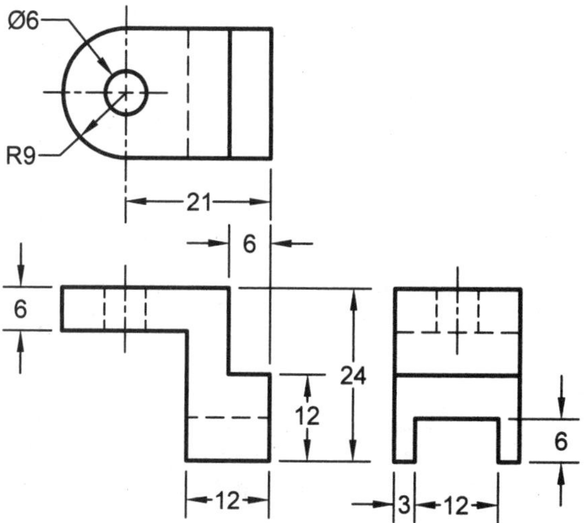

Figure DIM 8: Each dimension is placed in the profile view.

4. Keep dimensions lined up in well organized groups so the drawing appears clear and uncluttered as illustrated in Figure DIM 9.
5. Keep dimensions off the views. Placing dimensions on a view distracts from its shape description as illustrated in Figure DIM 10.

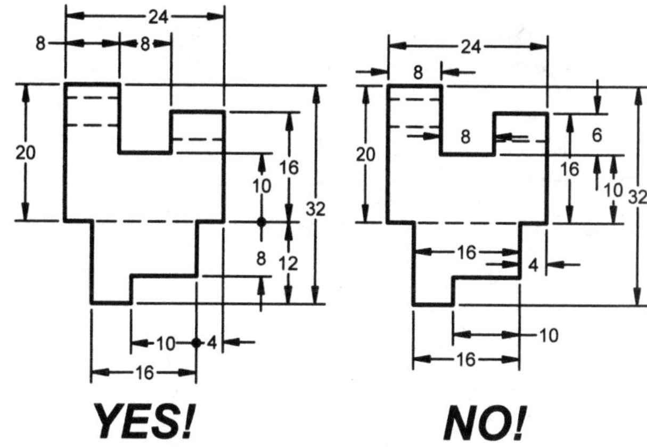

Figure DIM 9: Keep dimensions lined up and in well organized groupings

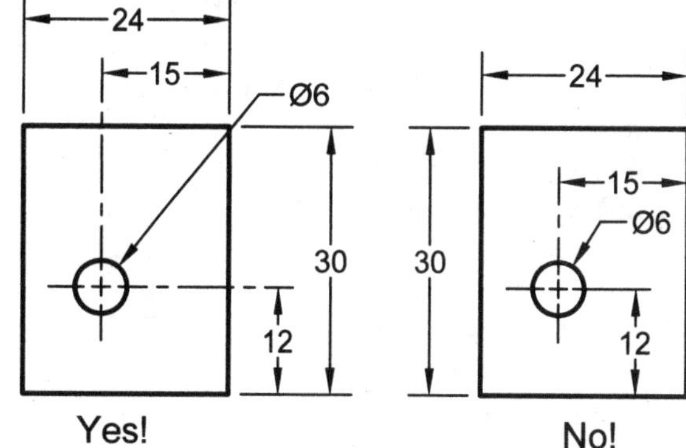

Figure DIM 10: Keep dimensions off view.

6. The dimension line closest to the object should be located a MINIMUM of three times the lettering height (3xLH) away from the drawing as illustrated in Figure DIM 11. The distance between successive dimensioning lines should be a MINIMUM of two times the lettering height (2xLH) as illustrated Figure DIM 11. The extension line extends beyond the last dimension line (or arrowhead) to a length equal to the lettering height (LH) as illustrated in Figure DIM11.
7. The size of the gap between the object and an extension line is about one half the lettering height (1/2xLH) as illustrated in Figure DIM 11.
8. When an extension line must be drawn on the object to dimension an interior detail there is no gap where it crosses the visible line as illustrated in Figure DIM 11. When two

extension lines cross there is no gap as illustrated in Figure DIM 11.

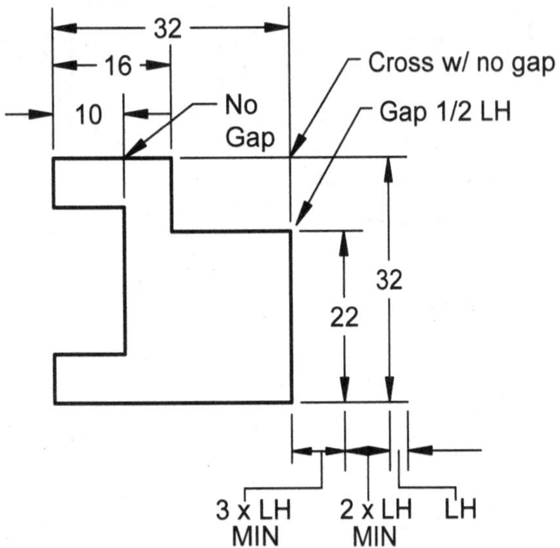

Figure DIM 11: Extension Line Gap/ No Gaps and Dimension Line placement.

9. Leaders may cross extension lines but not dimension lines as illustrated in Figure DIM 12.

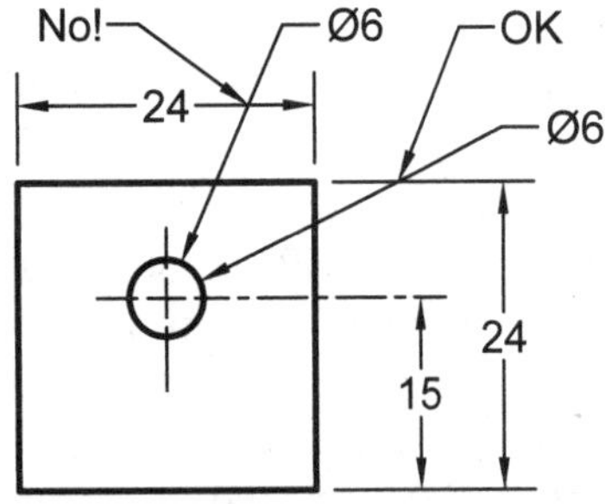

Figure DIM 12: Leaders may cross an extension line but not a dimension line.

10. Dimension the shortest features first then work out to the overall dimensions. This avoids having extension lines pass through dimension lines as illustrated in Figure DIM 13.
11. Never repeat a dimension as illustrated in Figure DIM 14.

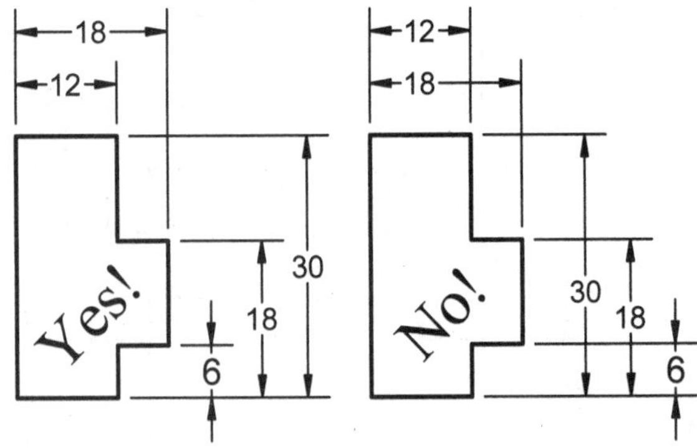

Figure DIM 13: Dimension from shortest to longest dimension to avoid having an extension line pass through a dimension line.

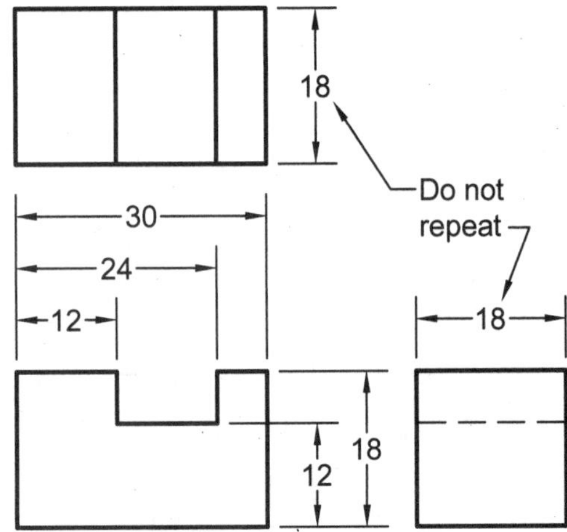

Figure DIM 14: Never repeat a dimension.

12. Avoid chain dimensioning that result in an accumulation of tolerances. If a dimension does not give a maximum and minimum value or a +/- (plus or minus) value, it defaults to the drawing's general tolerance (+/- value) labeled in or near the title block. In Figure DIM 15A the tolerance on the chain of dimensions adds up to +/- 3 units. The overall dimension is 36 with a tolerance of +/- 3 units. But the desired value is 36 +/- 1 units. So leave out the least important sub-line dimension as illustrated in Figure DIM 15B or use baseline dimensioning which references dimensions back to an datum surface as illustrated in Figure DIM 15C.
13. Stagger the location of stacked dimensions as illustrated in Figure DIM 16 because it makes the dimensions easier to read.

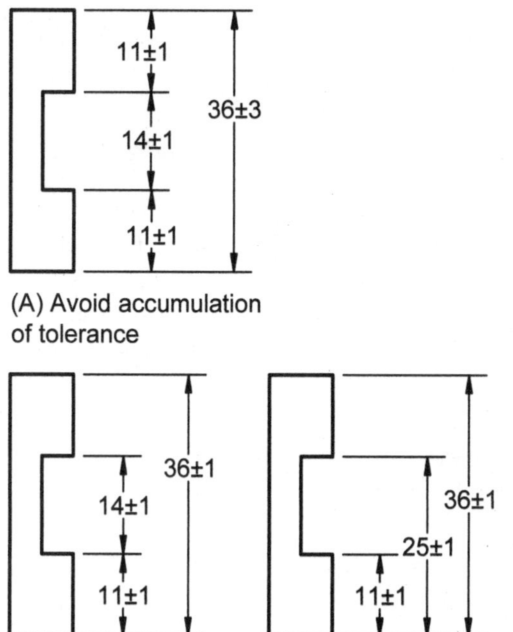

(A) Avoid accumulation of tolerance

(B) Correct leaving out the least important dimension

(C) Correct using baseline dimensioning

Figure DIM 15: (A) Tolerances accumulated to +/- 3. To avoid the accumulation of tolerances, (B) eliminate the least important sub-line dimension, or (C) dimension from a baseline.

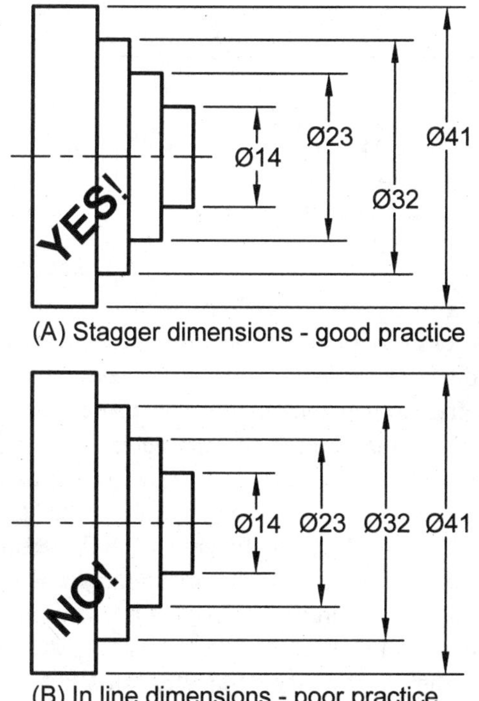

(A) Stagger dimensions - good practice

(B) In line dimensions - poor practice

Figure DIM 16: Stagger the location of dimensions to make it easier to read.

Linear Dimensions:

For machine drawings, Figures DIM 17 and DIM 18 illustrate acceptable formats for linear dimensions. Figure DIM 17 is for metric dimensions Figure DIM 18 is for inch dimensions. The number of decimals places shown in Figure DIM 18 depends on the degree of accuracy required. Figure DIM 19 illustrates a drawing with properly formatted metric linear dimensions.

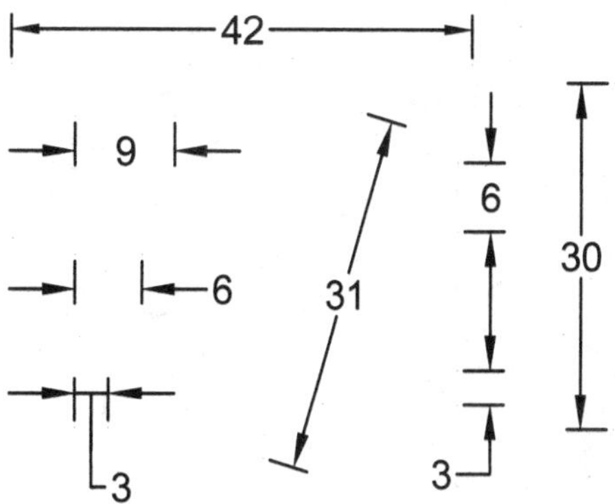

Figure DIM 17: Linear dimension formats in millimeters.

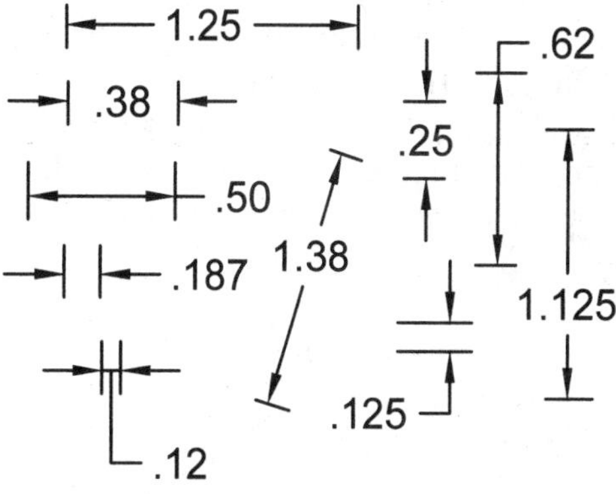

Figure DIM 18: Linear dimension formats in inches.

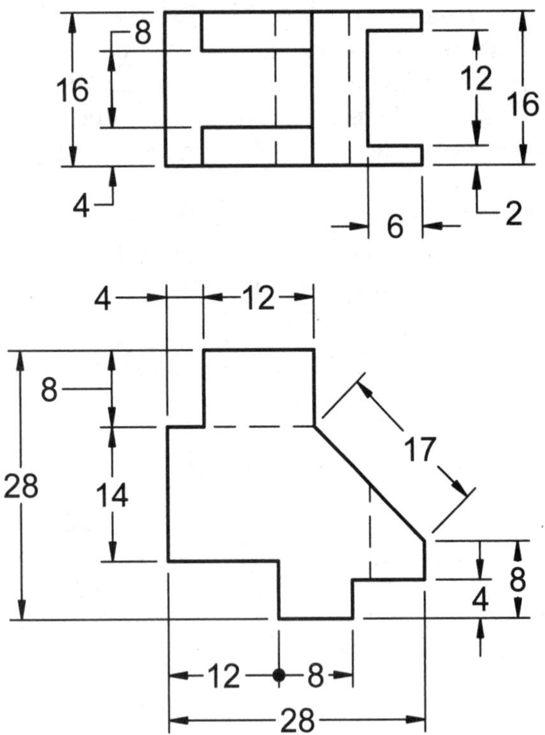

Figure 19: Applied metric linear dimensions.

Oblique Extension Lines:

Dimension lines normally run perpendicular to extension lines. However, when space is extremely limited, oblique extension lines can be used as illustrated in Figure DIM 20. Note how the dimension lines are still drawn vertically.

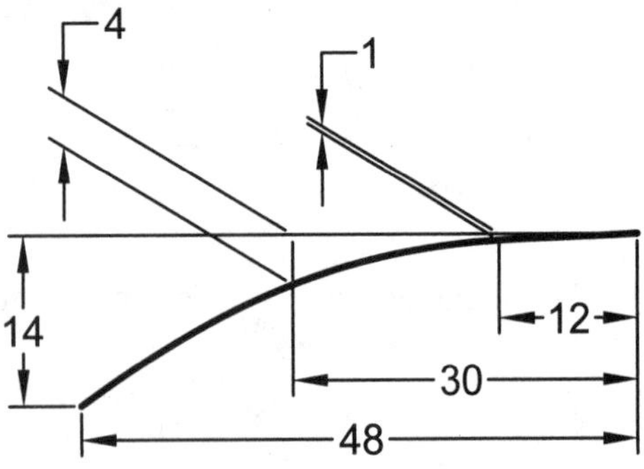

Figure DIM 20: Oblique extension lines.

Angular Dimensions:

Angles can be dimensioned using:

1) two linear measurements, which is used when a high degree of accuracy is required, as illustrated in Figure DIM 21A;
2) a linear measurement and an angle as illustrated in Figure DIM 21B; or with just an angle as illustrated in Figure DIM 21C, DIM 21D, DIM 21E, and DIM 21F.

Angular dimensions can be expressed in either degrees plus decimals of a degree as in 35.5°, which makes for ease of calculations, or in degrees° minutes' seconds" as in 35°30'10". When only minutes are expressed, the minutes are proceeded by 0° as in 0°45'. When only seconds are expressed the seconds are proceeded by 0°0' as in 0°0'35".

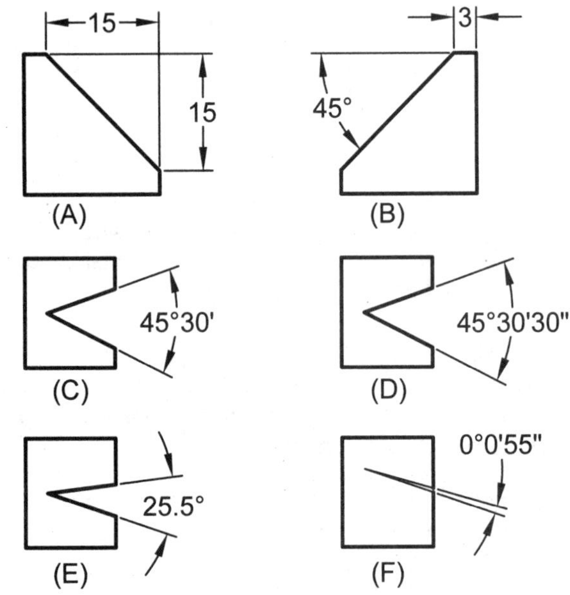

Figure DIM 21: Angular Measurements: A) linear only, B) linear and angle, C) degrees/minutes, D) degrees/minutes/seconds, E) decimal degree, F) seconds only.

Holes and Cylindrical Features:

A circle can represent either a hole ("a void cylinder") or a solid cylinder. In both cases the diametral dimension should be preceded by the symbol ∅ that represents diameter. It is best to dimension a hole in the view showing it as a circle which means you are looking into or through the hole. Figure DIM 22A illustrates the dimensioning of a small diameter hole. Figure DIM 22B illustrates the dimensioning of a medium sized diameter hole. Figure DIM 22C illustrates the dimensioning of a large diameter hole. Figure DIM 22D illustrates the dimensioning of multiple holes with the same diameter.

As illustrated in Figure DIM 22, large and medium circles can have the arrowheads inside the circle pointing away from the circle's center. With smaller circles the arrow should point toward the center of the circle. The leader should be radial to the circle's center which means that if the leader is extended, it will pass through the circle's center. Since holes appear as circles, dimension them as a diameter (∅), never as a radius (R).

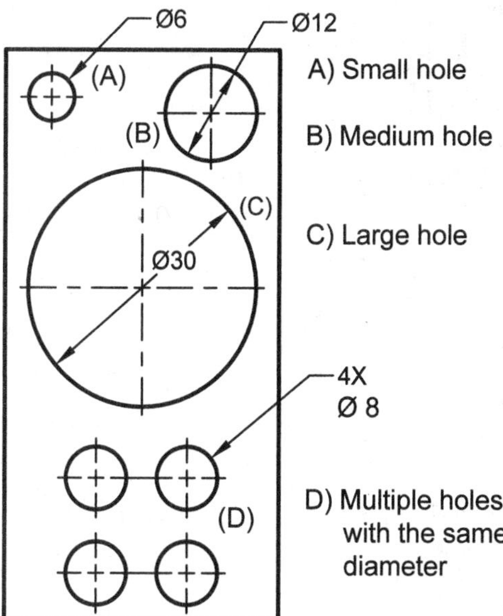

Figure DIM 22: Dimensioning: A) small diameter hole, B) medium diameter hole, C) large diameter hole, D) multiple holes of the same diameter.

Figure DIM 23A illustrates a hole drilled completely through the material. Only the diameter is labeled. Figure DIM 23B illustrates a hole drilled to a specific depth. The hole's diameter is written first followed by the hole's depth. The hole's depth is measured to the end of the cylindrical portion of the hole. Figure DIM 23C illustrates a counterbored hole while Figure DIM 23E shows a photo of a counterbored tool. The syntax for a counterbored hole is: the diameter of the small hole, the diameter of the large hole, and the depth of the larger hole. Counterbored holes are used with a variety of fasteners, e.g., hex head bolts. The larger diameter hole allows the head of the fasteners to sit below the surface of the part. Figure DIM 23D illustrates a countersunk hole while Figure DIM 23F shows a photo of a countersink tool. The syntax for a countersunk hole is: the diameter of the small hole, the diameter of the large hole, and then the included angle (usually 82° or 90°) of the cone shaped portion. Countersunk holes are used with flat head fasteners.

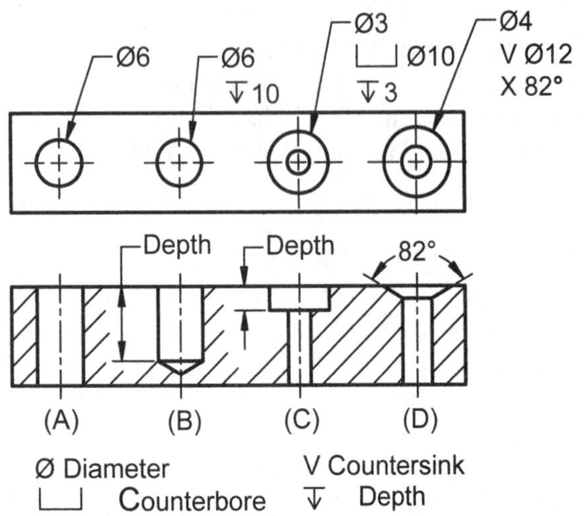

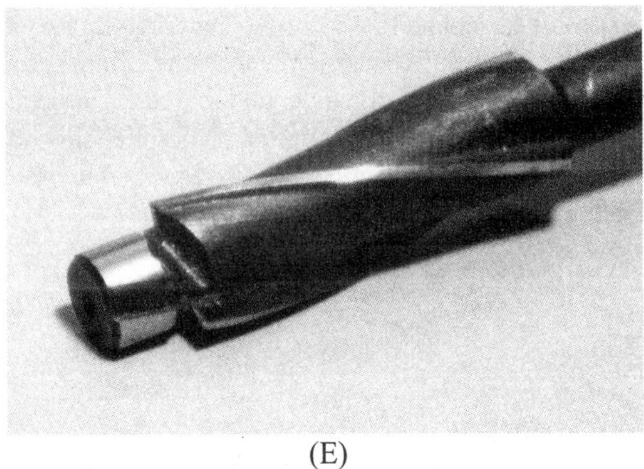

(E)

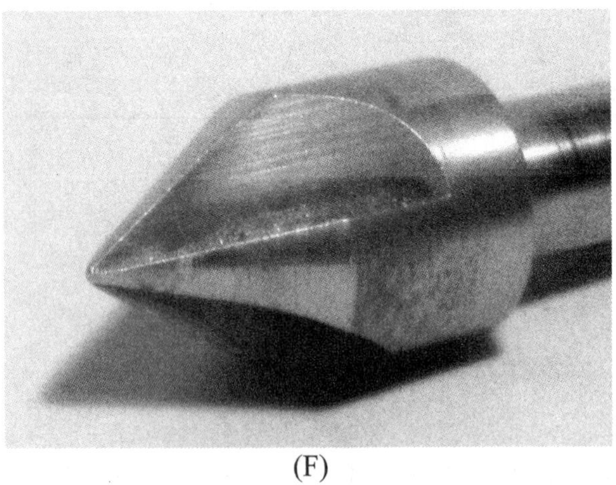

(F)

Figure DIM 23: Dimensioning techniques for: A) drilled through hole, B) hole drilled to a specific depth, C) counterbored hole, D) countersink hole, photo of a counterbore tool, and F) photo of a twist drill.

Unlike holes, solid cylinders are dimensioned in the view where they appear as a rectangle. Figures DIM 24A and DIM 24B illustrate the dimensioning of solid cylinders in the profile view. Figure DIM 25 illustrates the application of dimensioning of both holes and solid cylinders according to specifications.

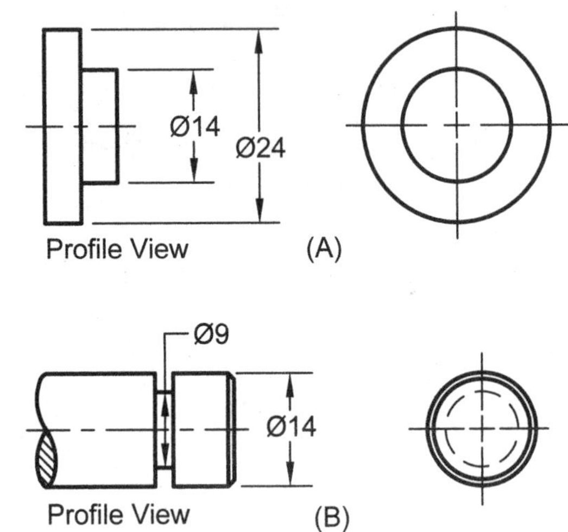

Figure DIM 24: Solid cylinders dimensioned in their profile views.

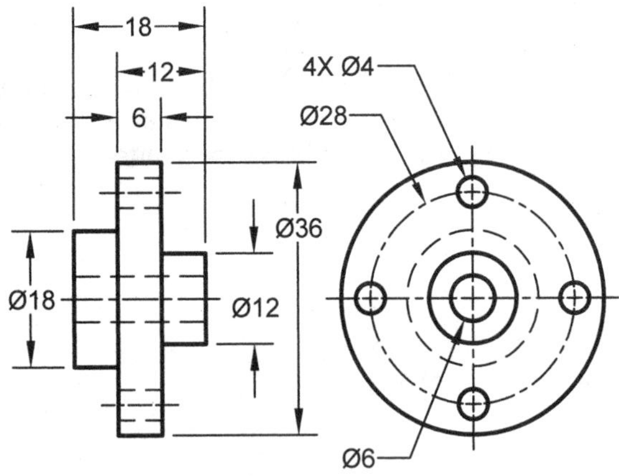

Figure DIM 25: Applications of the rules for dimensioning holes and solid cylinders.

Arcs and Radii:

An arc is a segment of a circle. The method of dimensioning an arc depends on its size and the location of its center. Figure DIM 26A illustrates two large radii arcs where the radius dimensions fits between the arc's center and the arc. Figure DIM 26B illustrates arcs where the center marks are drawn but the dimension will not fit between the arc's center and the arc. Figure DIM 26C illustrates arcs with very small radii where the location of the centers are not important. Figure DIM 26D illustrates the dimensioning of a very large radius arc using a "false center" because the true center is off the view or off the page. Except for small radii, the tail of the arrowhead should start at the arc's center with the arrowhead pointing away from the arc's center.

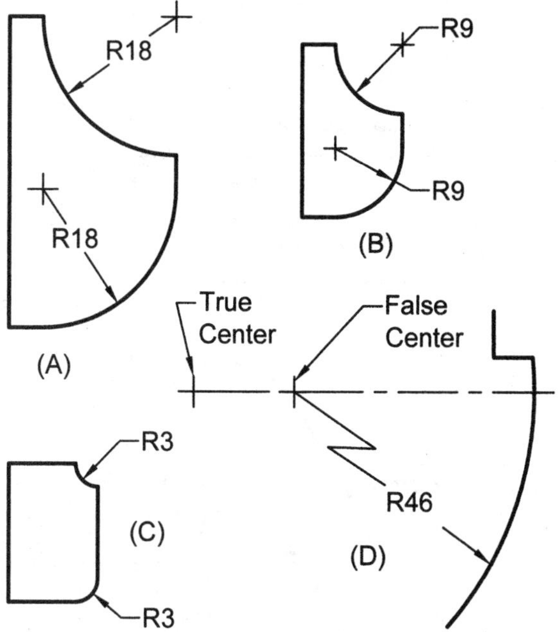

Figure DIM 26: Dimensioning Radii.

Spheres and Rounded Cylinder Ends:

A spherical feature is dimensioned using a leader line accompanied by abbreviations S for spherical and ∅ for diameter. This results in S ∅ or "spherical diameter" as illustrated in Figures DIM 27A and DIM 27B. A rounded cylindrical end can be dimensioned in one of two ways depending on whether it has a full radius as in Figure DIM 27C or a partial radius as in Figure DIM 27C. Both the full and partial radius use SR or "spherical radius".

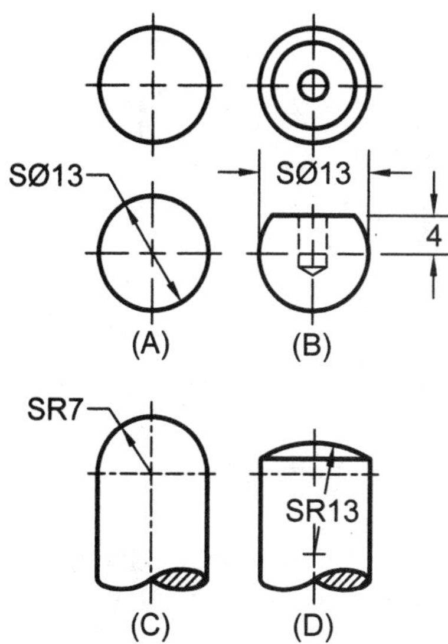

Figure DIM 27: Dimensioning techniques for A) & B) spherical shapes; and C) & D) rounded ends of a cylinder.

Rectangular Stock with Rounded Ends:

Rectangular stock with rounded ends can be dimensioned using one of two methods: 1) for rectangular stock with fully rounded ends, two overall dimensions are noted and the radii of the ends are noted with the abbreviation 2X R with no numerical value noted is illustrated in Figure DIM 28A; 2) for rectangular stock with partially rounded ends, two overall dimensions and the value of the radii are noted 2X R9 as illustrated in Figure DIM 28B.

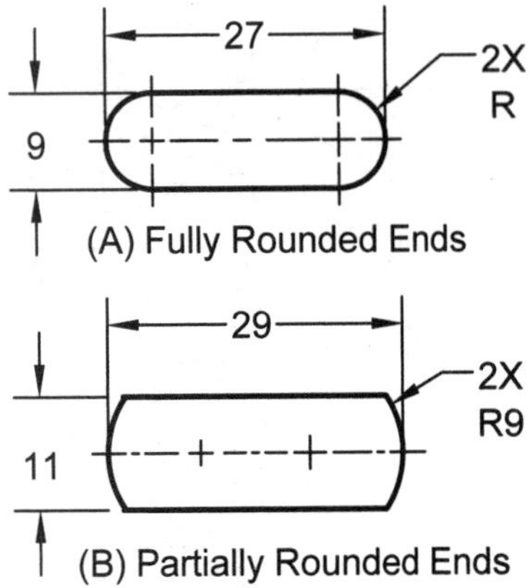

Figure DIM 28: Dimensioning rectangular stock with A) fully rounded ends and B) partially rounded ends.

Slotted Holes:

Slotted holes can be dimensioned using one of several different methods illustrated in Figures DIM 29A, DIM 29B, DIM 29C, and DIM 29D. In Figures DIM 29A, DIM 29B, and DIM 29C the radius R is noted but no numerical value is given. In Figures DIM 29D the radii value are specified.

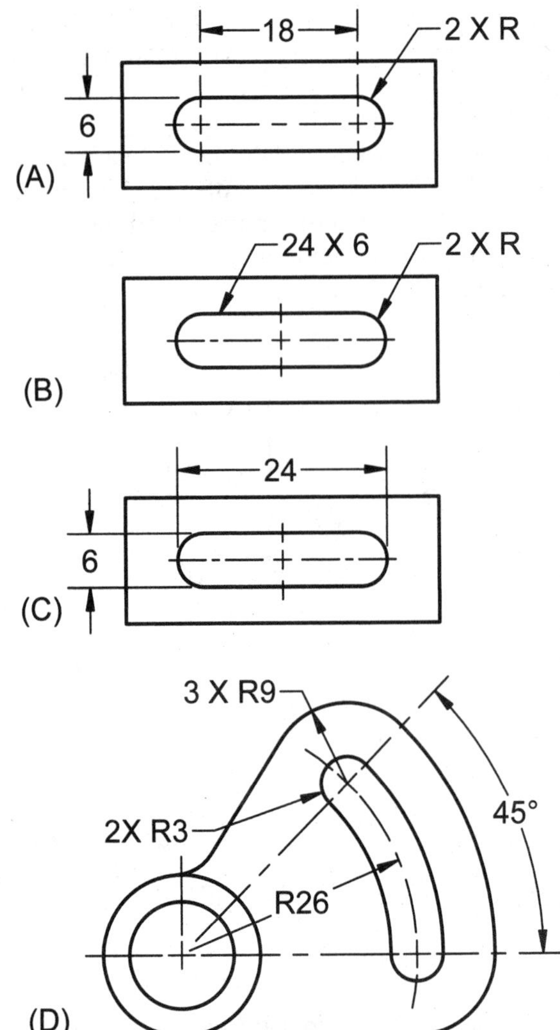

Figure DIM 29: Dimensioning techniques for a slotted hole – note that no numerical value is given for the radius (R) in A, B, and C.

Location Dimensions:

Location dimensions position a geometric feature but do not give the features size. Rectangular features are located using dimensions to their sides as in Figure DIM 30A while cylindrical features are located using dimensioning to their centerlines as in Figure DIM 30B.

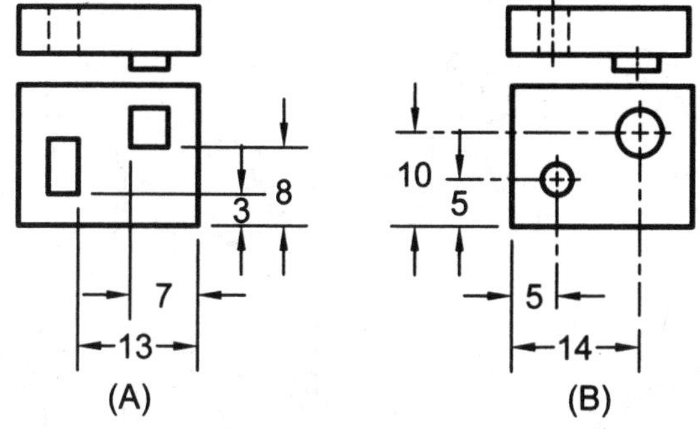

Figure DIM 30: Location dimensions A) for rectangular features, and B) for cylindrical features.

Holes Located About a Center Point:

Location dimensions for holes that are distributed along an arc or circle can be dimensioned in a variety of ways depending on the manner in which the holes are distributed. Figure DIM 31A illustrates five equally spaced small holes distributed along a circular center line called a "circle of centers" or "bolt circle" as the holes commonly have bolts passing through them in order to fasten the part to a mating part. Figure DIM 31B uses (x, y) coordinates measured from the primary centerlines. The diameter of the bolt circle is not dimensioned because it is not needed to locate the small holes. However the bolt circle's diameter can be labeled as a reference using either Ø22 REF or (Ø22). A reference dimension is noted for convenience but is not used in any calculations, e.g., programming a computer controlled machine (CAM) to locate the center of a hole for drilling purposes.

In Figure DIM 31A the holes are equally spaced around the center point of a bolt circle. The bolt circle's diameter is specified and a note specifies the number of holes and that they are equally spaced. Holes that are not equally spaced, as illustrated in Figure DIM 32A, are located by giving the bolt circle's diameter and positioning the holes on the bolt circle using angular measurements in degrees. Holes positioned along an arc, as illustrated in Figure DIM 32B, are located by specifying the arc's radius and locating the holes using angular measurement in degrees.

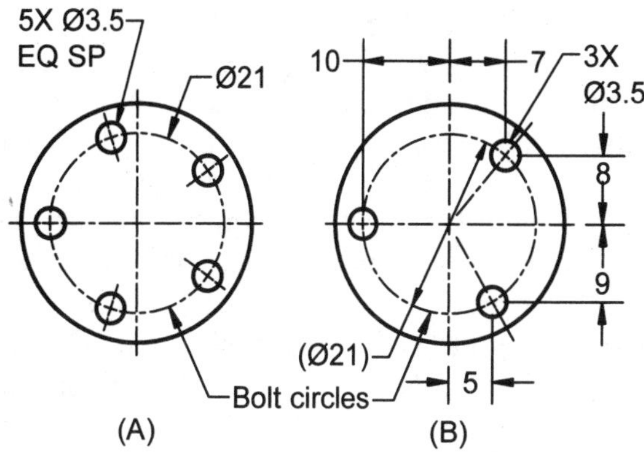

Figure DIM 31: A) circles located using a bolt circle diameter and an equal spacing note, B) holes located using coordinates and a reference diameter.

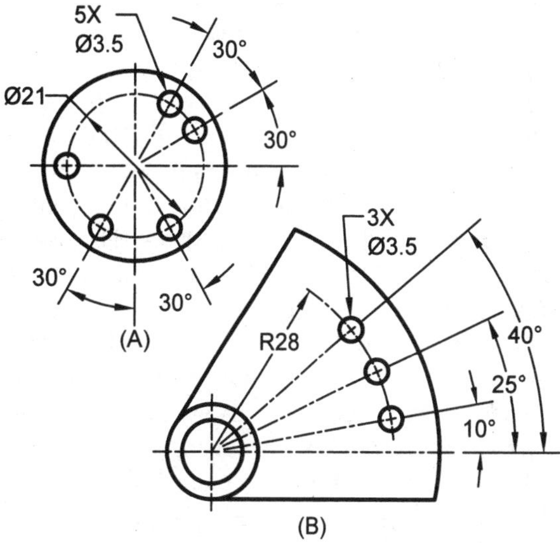

Figure DIM 32: A) Holes located using a bolt circle diameter and angular measurements in degrees, B) holes located along an arc using the arc's radius and angular measurement in degrees.

Specifying Machining Operations:

In the past it was common for a drawing to specify a specific machining operation such as "REAM to .38" DIAM". Now the preferred method is to specify only the desired holes diameter. This avoids unnecessary language and leaves the method of machining up to the machinist.

Understanding a few basic machining operations is very helpful in placing dimensions and notes on a drawing. It also helps tremendously when trying to visualize the geometry formed by different machining operations.

Drilling and Reaming:

A *drilling* operation uses several different types of drills. The most common for metal and plastic is the twist drill illustrated in Figure DIM 33B. Figure DIM 33A shows a hole drilled partially through the material and a hole drilled completely through the material. When the hole is not drilled completely through the stock it leaves an 118° cone shape at the bottom of the hole. Angles other than 118° are used but 118° is the most common. The diameter and the hole's depth are specified on the drawing as illustrated on the left side of Figure DIM 33A. When specifying the depth of a hole, the cone shape portion is not included. If a hole is drilled completely through, only its diameter is specified as illustrated on the right side of Figure DIM 33A.

A reaming operation uses a cylindrical bit with cutting flutes spaced around its diameter as illustrated in Figure DIM 33B. Reaming is a two-step operation: 1) the hole is drilled slightly smaller than the desired diameter (e.g., a few thousandths of an inch), and 2) a spinning ream is fed down the hole removing a very small amount of material leaving the hole the desired diameter. The shallow flutes on a ream results in a more massive cross section compared to a twist drill. The additional mass in combination with the ream removing only a small amount of material makes for a more rigid tool and a more precise hole. As with a drilled hole, the size of a reamed hole is specified by noting the hole's diameter, or by noting the hole's diameter and depth if the hole is reamed to a specified depth.

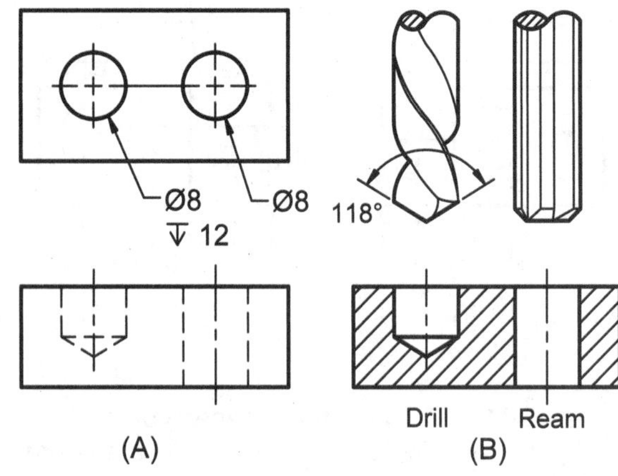

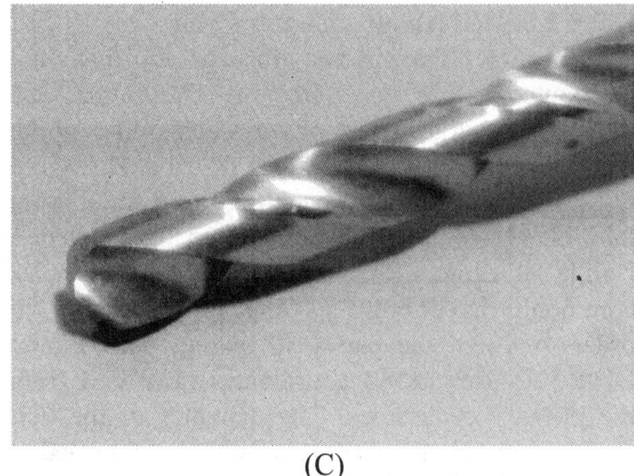

(C)

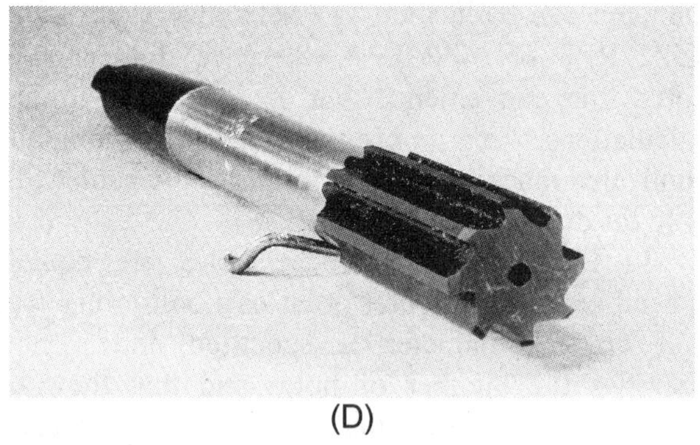

(D)

Figure DIM 33: A) dimensioning notes for a drilled hole at a specified depth or a drilled through hole, B) illustrations of a twist drill and a reaming tool, C) photo of a twist drill, and D) photo of a ream.

Counter Bore and Counter Sinks:

A countersink hole, illustrated in Figure DIM 34A, is used to accept flathead fasteners as illustrated in Figure DIM 34C. A countersink hole is formed in a two step operation: 1) drill the hole, and then 2) guide a countersink tool, illustrated in Figure DIM 34B and a photo of a countersink tool in Figure DIM 34D, concentrically into the drilled hole forming an 82° or 90° cone shape to a specific depth or until the diameter of the cone is cut to a specific diameter as measured on the surface.

A counterbored hole, as illustrated in Figure DIM 34A, is used to accept a variety of fasteners such as the hex head or round head cap screws as illustrated in Figure DIM 34C. A counterbored hole is formed in a two step operation: 1) drill the smaller hole using a twist drill illustrated in Figure 34B and a photo of the actual tool in Figure 34C, and then 2) guide the counterbore tool, as illustrated in Figure DIM 34B and an actual counterbore tool DIM 34E, concentrically into the smaller diameter drilled hole to cut the larger diameter hole to a specified depth as illustrated in Figure DIM 34B. Note that the counterbore tool has a small cylinder on its end which guides the tool into the smaller hole drilled in the first step. This ensures that the smaller and larger holes are concentric.

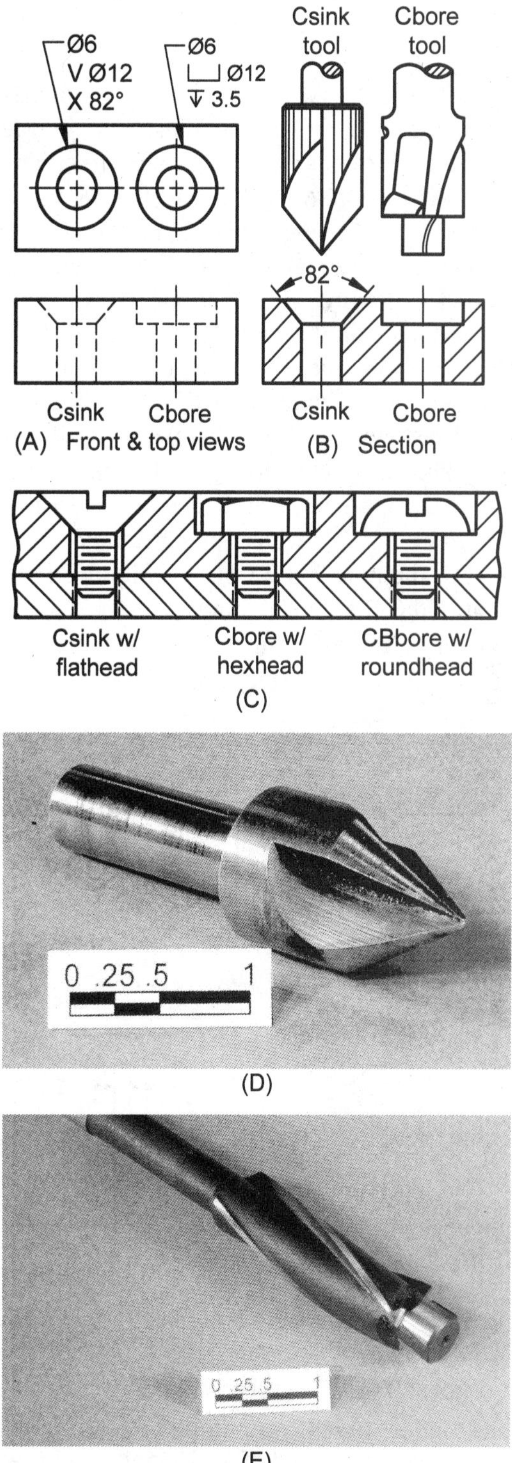

Figure DIM 34: A) top and front views of a countersink (left) and counterbored (right) holes with dimensioning notes, B) countersink (left) and counterbore (right) tools and a section cut through to expose the shape of the holes, C) countersink and counterbored holes with various fasteners inserted, D) countersink tool, E) counterbore tool.

Chamfers:

A *chamfer* is formed by removing material from the edge of two 90° intersecting planes to form an inclined surface as illustrated in Figure 34A. This type of chamfer can be labeled either 6 X 45° CHAM or 6 X 6 CHAM. A chamfer can also be formed by removing material at an angle around the entire diameter at the end of a cylinder as illustrated in Figure DIM 35B and DIM 35C. The chamfered end of the cylinders in DIM 35B and DIM 35C dulls the sharp edge and aids when assembling the cylinder into a hole. The ends of threaded fasteners are also chamfered for the same reason. The most common chamfer angle is 45°. Figure DIM 35B illustrates a cylinder's 45° chamfer labeled with either the note 2 X 45° CHAM or the note 2 X 2 CHAM. Figure DIM 35C illustrates the method of dimensioning a chamfer that is cut at an angle other than a 45°. Specify its angle (30° in this example) and its depth (3 in this example).

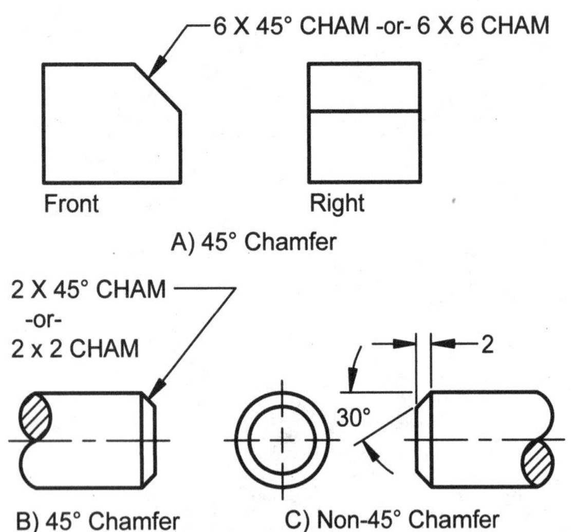

Figure DIM 35: Labeling a chamfer: A) a chamfer at the intersection of two intersecting planes labeled with a note, B) a 45° chamfer around the end of a cylinder labeled with two different notes, C) a 30° chamfer around the end of a cylinder with an angle and depth dimension.

Keyways and Keyseats

A *key* is a mechanical link between a shaft and a hub that transmits the torque from the shaft to the hub as illustrated with pulley and shaft in Figure DIM 36. Half the height of the key fits into a groove in the shaft called a *keyseat*. The other half of the key fits into a groove in the hub called a *keyway* as illustrated in Figures DIM 36 and DIM 37. Figure DIM 36 shows how the keyseat and the keyway are dimensioned for a standard rectangular shaped key. Figure DIM 37 shows how to dimension a keyseat for a woodruff key which has a radial bottom. Figures DIM 36 and DIM 37 are the preferred methods of dimensioning but sometimes a keyseat or keyway will simply be labeled with its width and depth such as KEYSEAT 6WD X 6DP where WD = width and DP = depth. Sometimes instead of dimensioning a keyseat for a woodruff key, the actual woodruff key number will be called out such as WOODRUFF KEY #608. The first digit represents the keys thickness in thirty-seconds of an inch (6/32 = .18" thick) and the final two digits represent the keys diameter in one-eighth of an inch (8/8 = 1" in diameter).

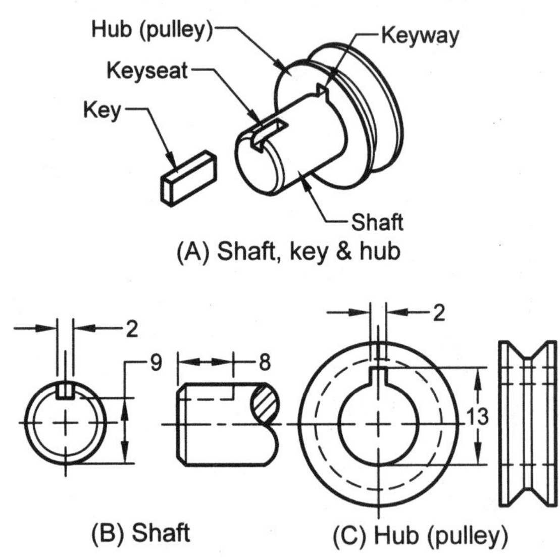

Figure DIM 36: Dimensioning of a keyseat and keyway for a standard rectangular shaped key.

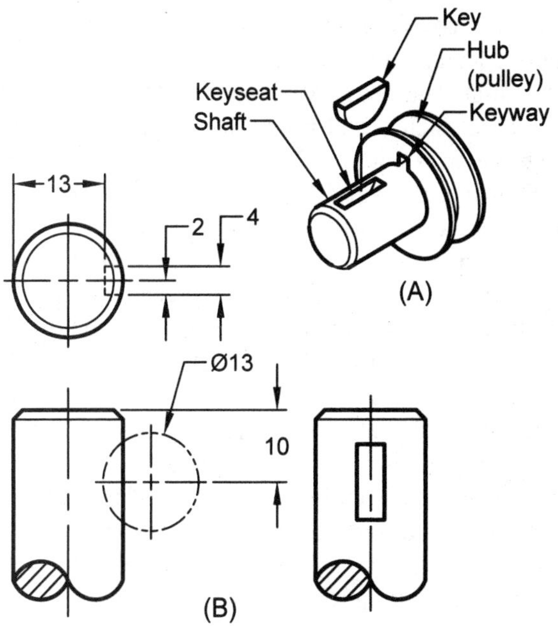

Figure DIM 37: Dimensioning a keyseat for a woodruff key.

Necks and Undercuts:

A *neck* is a groove cut into the outside of a cylinder where the cylinder changes from a small to a larger diameter as illustrated in Figure DIM 38. A neck allows the shoulder of the large cylinder or fastener to sit flush in a hole as illustrated in Figure DIM 38E. Without the neck a small radius is left by the cutting tool that prevents the fastener from seating flush. Figures DIM 38A, and DIM 38B show how to dimension a neck using a note while Figure DIM 38C uses dimensions. A neck at the transition from threads to a larger diameter allows the thread cutting tool to easily be removed as it approaches the larger diameter as illustrated in Figure DIM 38D. This type of neck is referred to as a *thread relief*.

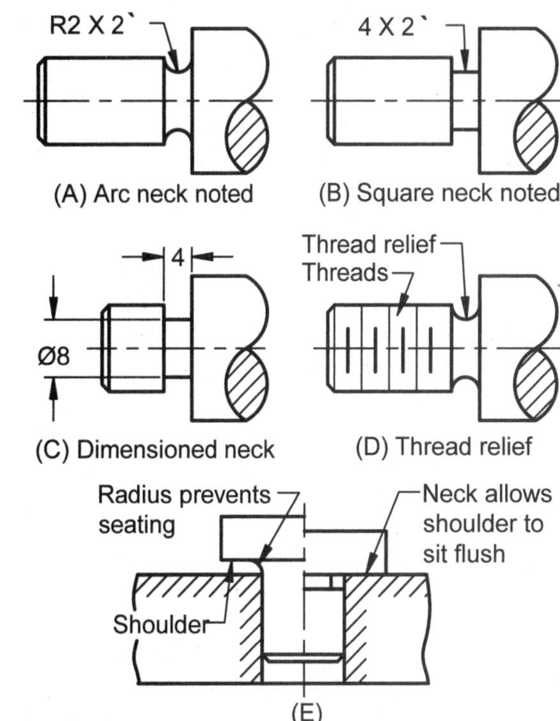

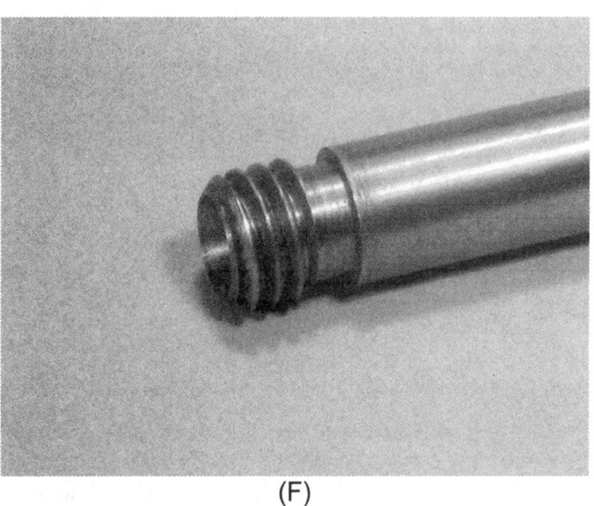

(F)

Figure DIM 38: A) an arced neck labeled with a note, B) a square neck labeled with a note, C) a neck dimensioned instead of using a note, D) a neck used with threads is commonly called a thread relief, E) the neck allows a shoulder to sit flush, F) illustrates of a thread relief.

An *undercut* is a groove cut into the inside face of a cylinder as illustrated in Figure DIM 39. An example use of an undercut is to provide a groove for an "O" ring which is a rubber seal in the shape of a torus (doughnut). Or a snap ring retaining clip which looks like a horse shoe with the opening that allows it to be pinched to make its outside diameter smaller or spread apart to make its inside diameter smaller. It is used to prevent something such as a bearing from sliding along the interior axis of the cylinder.

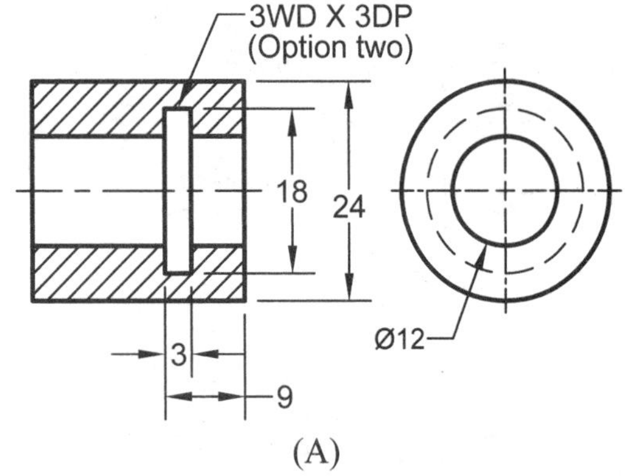

(A)

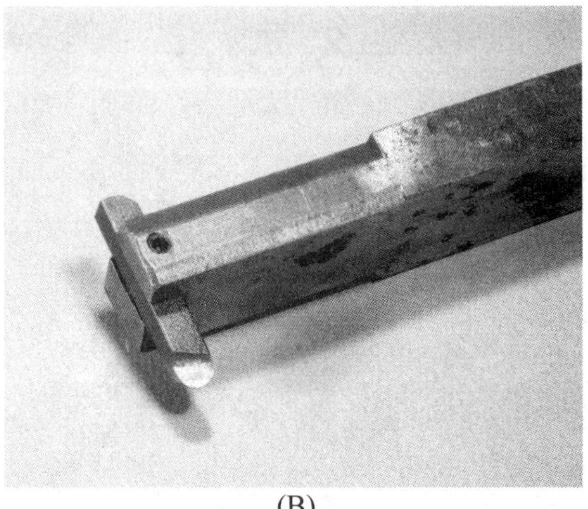

(B)

Figure DIM 39: A) an undercut can be dimensioned or optionally labeled with the note 3WD X 3DP, B) illustrates a boring bar used to cut an undercut.

Knurling:

Knurling results in putting a straight or diamond pattern on a cylinder as illustrated in Figure DIM 40A and DIM 40B. Knurling does not cut into the cylinder but displaces material under high pressure. Since no cutting takes place, the diameter of the knurled section is slightly larger than the cylinder's diameter. Knurling provides a better gripping surface and can also be used to enlarge a cylinder's diameter so the cylinder can be press fit into a hole. A press fit simply means that the cylinder has a slightly larger diameter than the hole so the cylinder is permanently fixed in the hole. On a drawing you don't actually draw the knurled pattern; you simply note it as in Figure DIM 40C or DIM 40D. The note "96DP straight" in Figure DIM 40C is an inch based note. DP stands for dimetral pitch which is equal to N/D where N is the number of grooves on the circumference of the cylinder and D is the diameter of the cylinder. Thus a DP of 96 would have 96 teeth on a one inch diameter cylinder. The most common dimetral pitches are 64DP, 96DP, 128DP, and 160DP (DP96 is preferred). In Figure DIM 40D the note "P 0.8 DIAMOND" is a metric note. P stands for pitch which is the distance between the grooves. Pitch in this example is 0.8mm. Figure DIM 41A illustrates a diamond knurl while Figure DIM 41B illustrates the tool used to form the diamond pattern.

If the knurling is used to provide a press fit, then the diameter after knurling needs to be labeled by appending the following note to the callouts in Figure DIM 40A and DIM 40B: "Ø# MIN AFTER KNURLING."

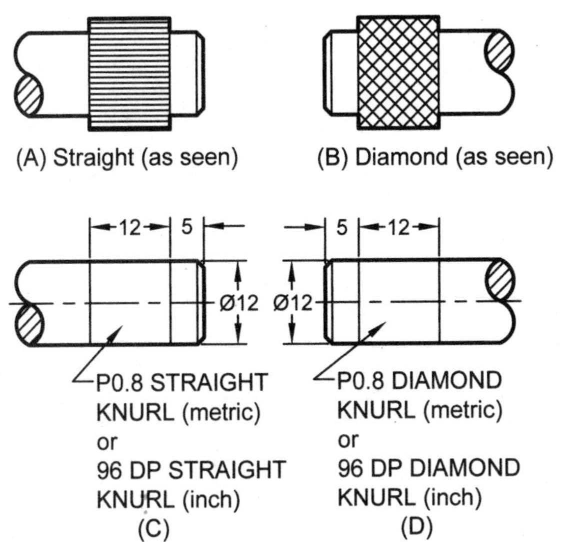

Figure DIM 40: A) straight knurls as seen, B) diamond knurls as seen, C) note for a straight knurl, D) note for a diamond knurl.

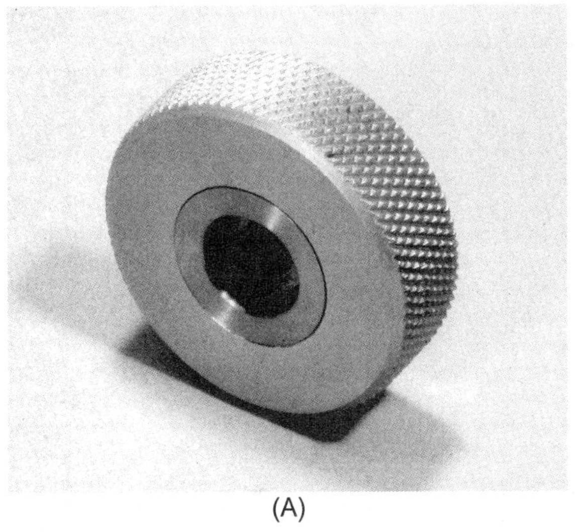

(A)

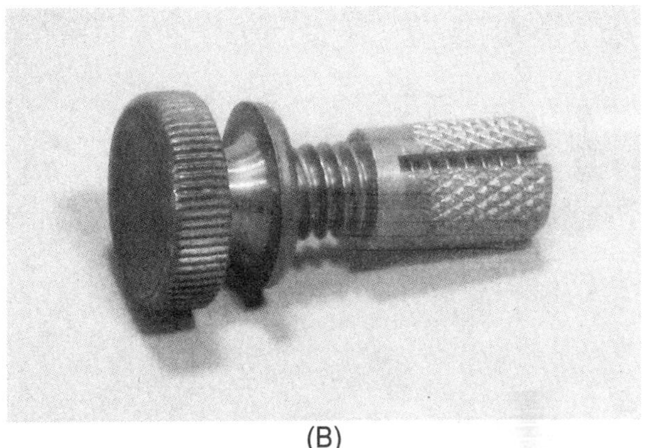

(B)

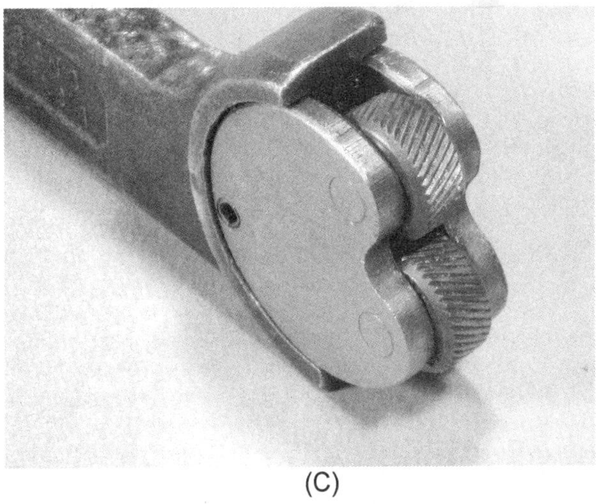

(C)

Figure DIM 41: A) sample of a diamond knurl on an adjusting wheel, B) diamond and straight patterns C) a diamond knurling tool.

NOTES:

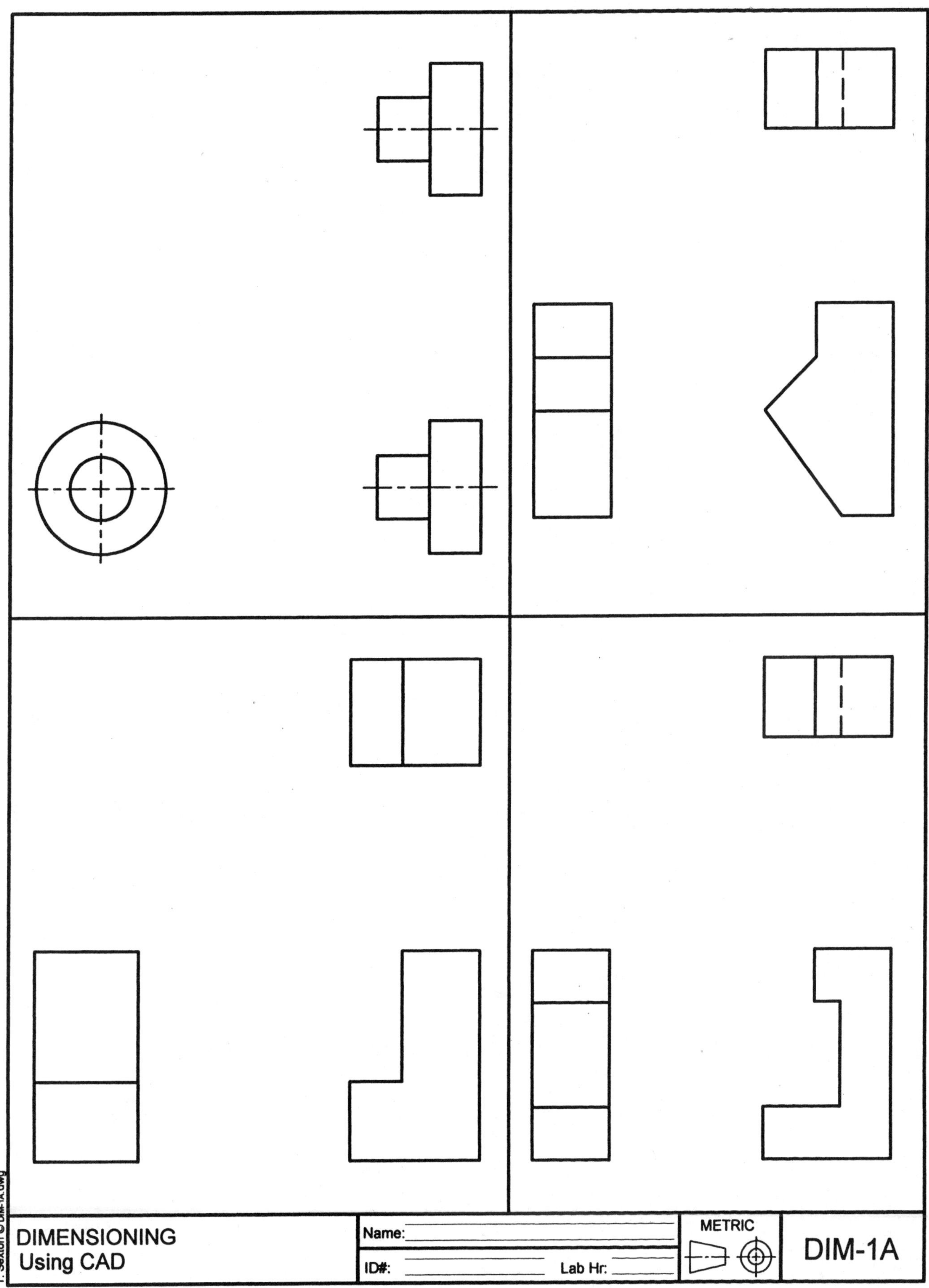
DIMENSIONING
Using CAD
Name:
ID#:
Lab Hr:
METRIC
DIM-1A
T. Sexton © DIM-1A.dwg

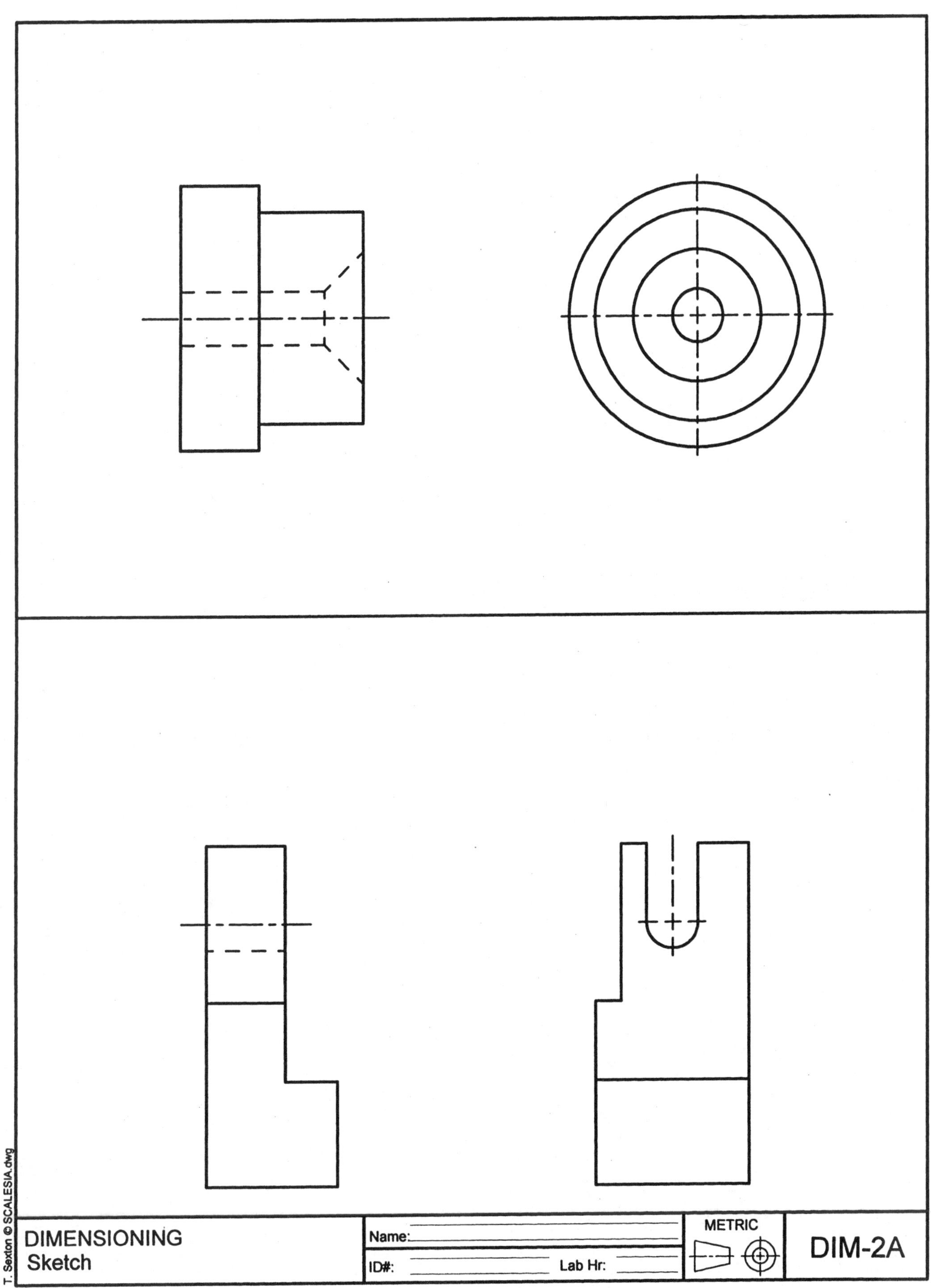
DIMENSIONING
Sketch
Name:
ID#:
Lab Hr:
METRIC
DIM-2A
T. Sexton © SCALESIA.dwg

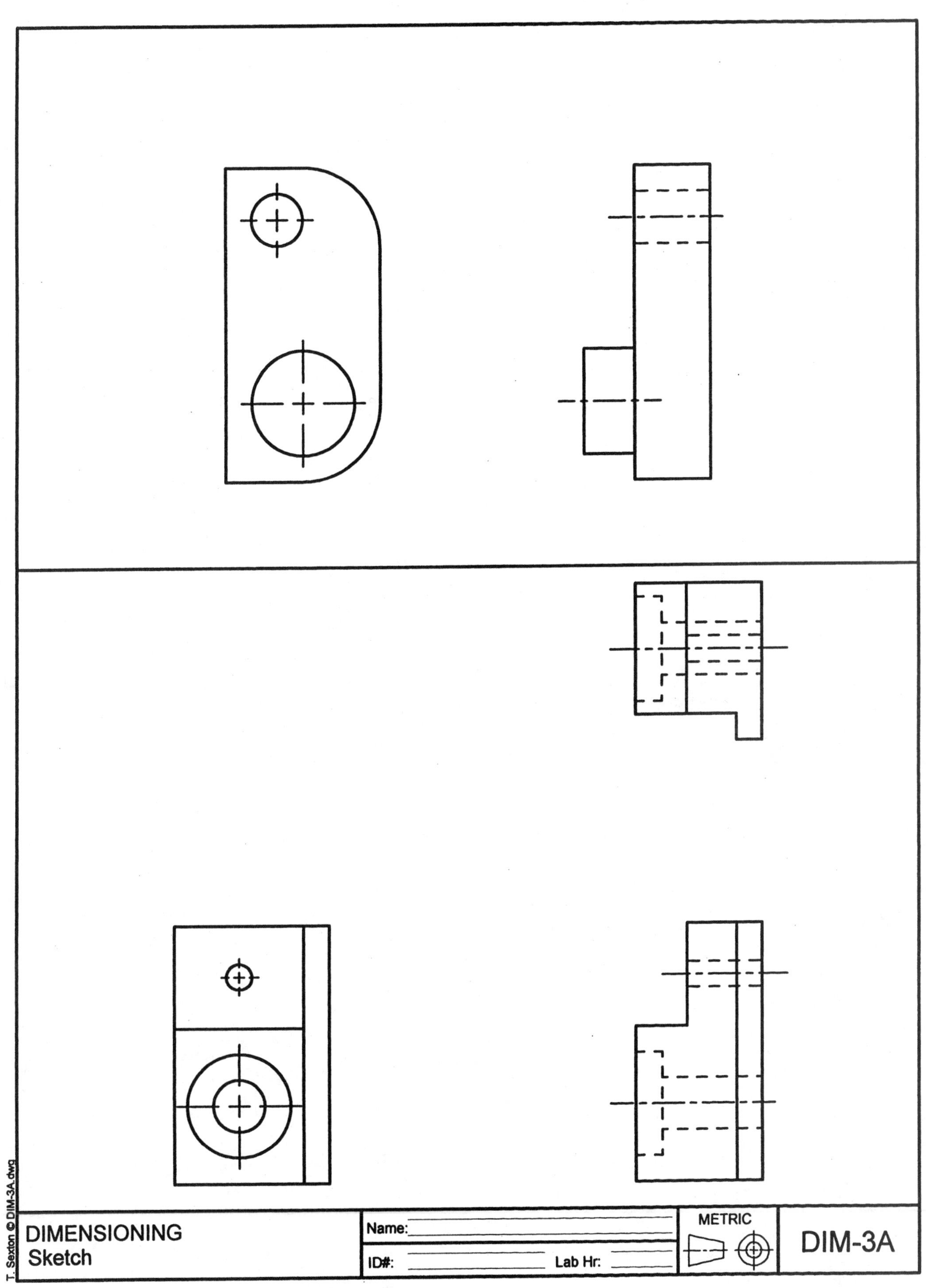

T. Sexton © DIM-3A.dwg
DIMENSIONING
Sketch
Name:
ID#:
Lab Hr:
METRIC
DIM-3A

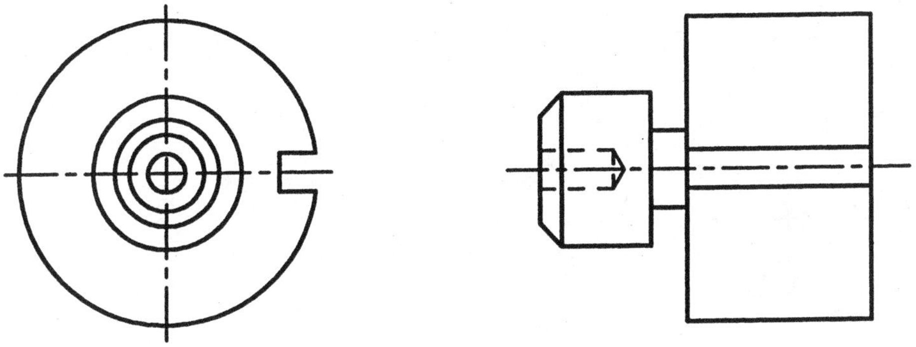

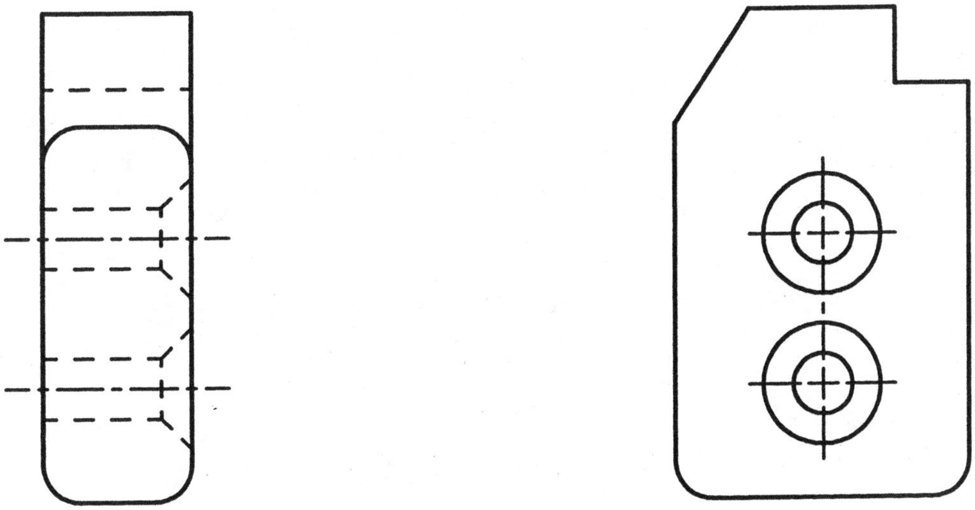

DIMENSIONING
Sketch

Name:

ID#: Lab Hr:

METRIC

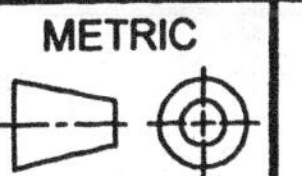

DIM-4A

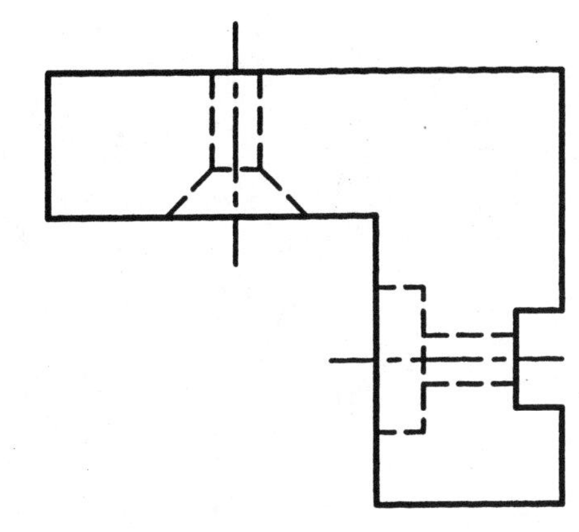

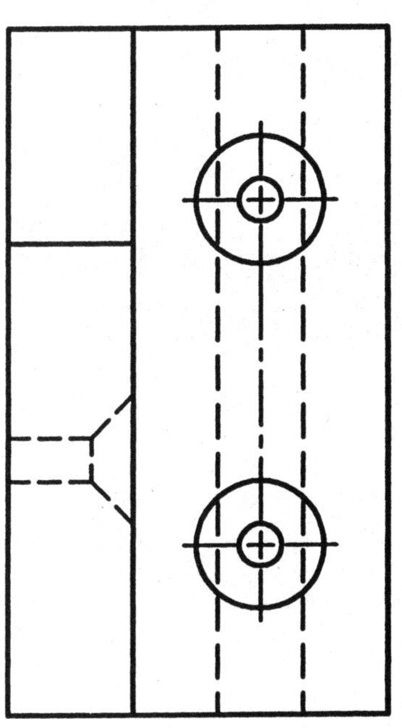

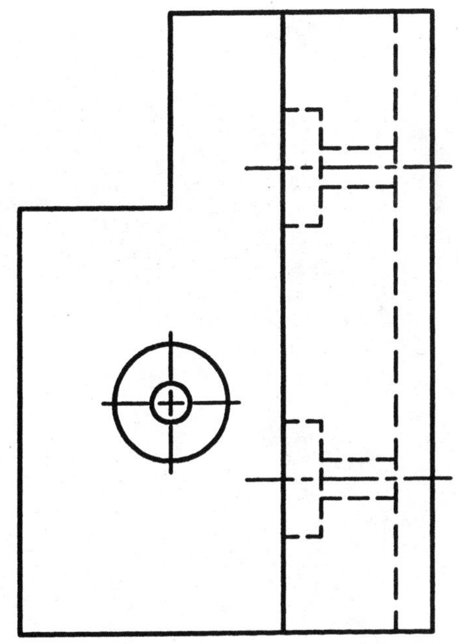

DIMENSIONING Using CAD	Name:________ ID#: ________ Lab Hr: ______	METRIC 	DIM-5A

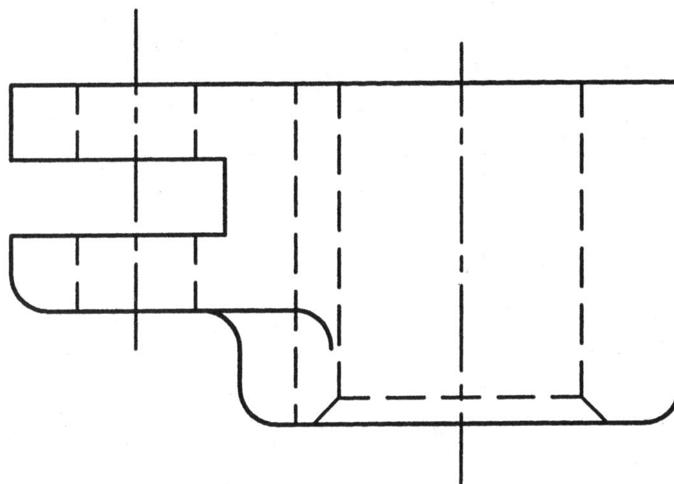

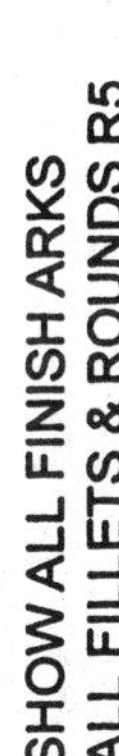

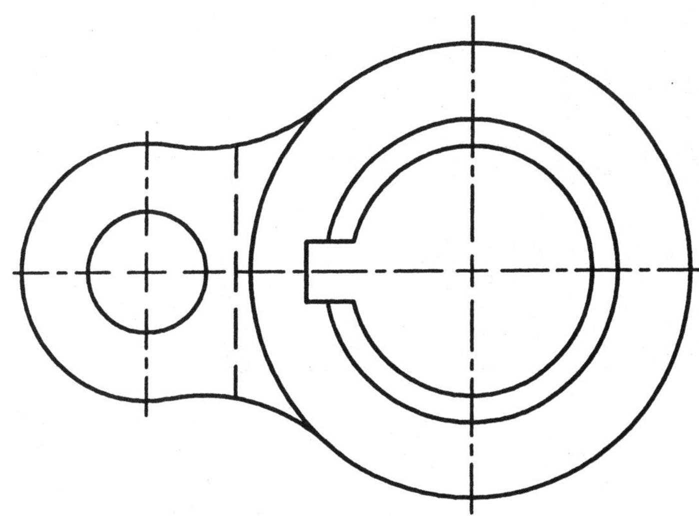

DIMENSIONING Using CAD	Name: Answer Key ID#: Lab Hr:	METRIC 	DIM-6A

Chapter 10

General Tolerance

Introduction:

A project requires one hundred cylindrical dowels 30mm long. A sample of the dowels is measured and shows a variation in the 30mm length. The question is how much variation in length is acceptable to meet the requirements of the project. To keep costs down the widest variation meeting the requirements of the project is desirable. It is determined that a length between 79 to 81mm will work well. The range of 2mm (81-79 = 2mm) is called the tolerance. Tolerance is the permissible range a size or location dimension may vary on an *individual part*. A tolerance should be as large as possible to reduce manufacturing costs.

What Determines the Tolerance?

Several factors must be considered when determining a tolerance value. Most important of these is the personal and institutional knowledge based on the experience of past projects. The process used to manufacture a product is another factor, for example, die casting which injects molten metal into a smooth metal mold allows for a greater degree of dimensional accuracy than pouring the molten metal into a sand casting mold. Within a manufacturing area such as metal machining, different processes are limited to a specific numeric tolerance ranges. For example, the process of lapping and honing can hold an extremely tight tolerance range followed in descending precision by grinding, broaching, reaming, turning, milling, and drilling. Another factor is the ability of a specific machine to produce the level of tolerance desired. Consider the following: is the initial quality of the machine a "Ford" or a "Royals Royce"; is the ambient temperature extremely high or low, is there large fluctuations in temperature; is the machine located adjacent to a machine or part transportation route which can cause the machine to vibrate; or is proper and timely maintenance on the machine being preformed which can prevent critical parts from wearing out prematurely.

Specifying Tolerance:

Any dimension that does not have a maximum and minimum value specified on the drawing is assigned the drawing's general tolerance.

General tolerances are normally noted in or near the title block as illustrated in Figure GNT 1. In Figure GNT 1 the title block assigns tolerances based on the number of decimal places 0.X ± 0.3 or 0.XX ± 0.05 and angular dimensions receive a ± .5°.

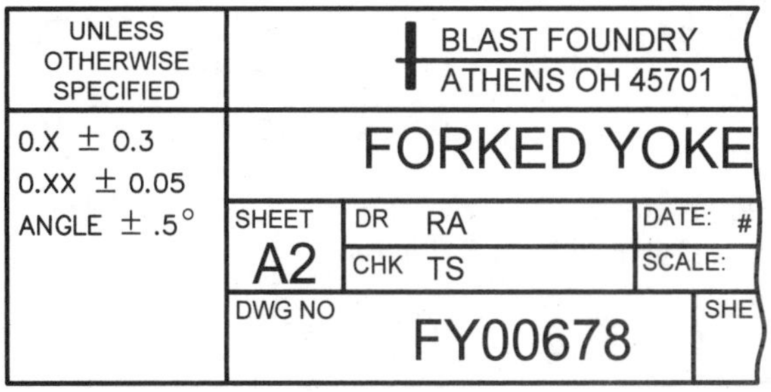

Figure GNT 1: General tolerance placed near the title block.

There are three formats for specifying a size or location tolerance: limit dimensions, unidirectional or bidirectional tolerance.

The ***limit dimension*** format specifies the maximum value over the minimum value or in the case of a single line note and leader the format is the minimum–maximum value as illustrated in Figure GNT 2A.

The ***unidirectional format*** specifies a plus and minus tolerance value with either the plus or minus value being zero as illustrated in Figure GNT 2B.

The ***bidirectional format*** uses the plus and minus format with either a single numeric value for both the plus and minus value or different values for both the plus and minus values as illustrated in Figure GNT 2C.

Note that the + and – values use full height lettering. When space is tight the tolerance value may be displayed as illustrated in Figure GNT 3.

Maximum on top
25.40
25.00
Ø25.00 - 25.04
Minimum first
(A) Limit Format

25.00 +0.04 -0.00
Ø25.00 +0.04 -0.00
Full height lettering
(B) Unidirectional

25.02 ± 0.02
(or)
25.02 +0.01 -0.02
Ø25.02 ± 0.02
(C) Bidirectional

Figure GNT 2: Tolerance formats A) limit, B) unidirectional, and C) bilateral.

32.00 ± 0.02
32.00 +0.01 -0.02
(A) Metric

1.250 ± .002
1.250 +.002 -.001
(B) Inches

Figure GNT 3: Bidirectional tolerance when there is limited space in A) metric and B) inches.

Mating Parts

In Figure GNT 4 the T-bolt fits into the T-slot and must be able to slide along the slot freely yet must not chatter excessively. The tolerance of the T-slot is .7512"–.7500" = .0012" and the tolerance on the T-bolt is .7484" – .7476" = .0008". Note that the T-bolt is always smaller than the width of the T-slot. As illustrated in Figure GNT 4, the minimum clearance between the T-slot and the T-bolt is calculated by taking the smallest T-slot value minus the largest T-bolt value or .7500" – .7484" = .0016". The maximum clearance is calculated using the largest T-slot value minus the smallest T-bolt value or .7512" – .7476" = .0036". The tightest fit (or minimum clearance) between the T-slot and T-bolt is called the allowance. Allowance is the minimum clearance or maximum interference between two mating parts. An interference fit can be illustrated by a shaft always having a diameter larger than the hole it is going into. An interference value is noted as negative allowance.

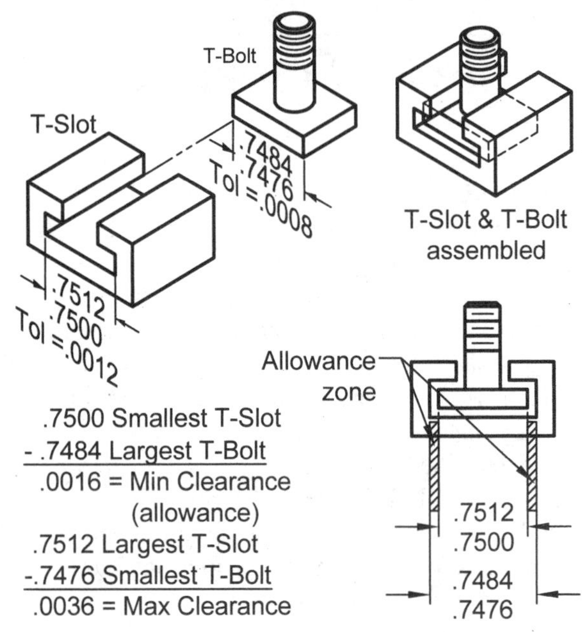

Figure GNT 4: Calculating minimum clearance (allowance) and maximum clearance for mating parts.

ANSI Inch Based System of Limits & Fits

The American National Standards Institute tables B4.1 1967 (R1979) and the *Machinery's Handbook* supply the maximum and minimum tolerance values for mating holes and shafts. The tables are categorized as either basic hole or basic shaft. The basic hole system is used when standard machine tools such as drills, reams, broaches, or end mills are used to create the hole while the shafts are machined to establish the correct fit between the hole and shaft. The basic hole system uses the smallest hole diameter as the basic size for calculating the tolerance and allowances between the hole and the shaft. The basic shaft system is used in industries, such as the textile industry, which use standard bar stock for the shaft and then varies the diameter of the mating part's hole to establish the correct fit between the hole and shaft. The largest shaft is used as the basic size for calculating tolerances and allowance between the shaft and hole.

ANSI Inched Based Standard Fits		
Running Fit	RC	Running & Sliding
Location Clearance Fits	LC	Clearance Location
	LT	Transition Location
	LN	Interference Location
Force Fit	FN	Forced & Shrink

Figure GNT 5: Summary table of the types of fits in the inch based ANSI system.

The ANSI inch based tolerance tables are further categorized into three types of fits: running, location, and force. Figure GNT 5 is a table summarizing the types of fits in the ANSI inch based system. Running fits (RC) allow the shaft to rotate in the hole while taking into consideration factors such as lubrication requirements, the shafts rotation speed, and ambient temperature. Location fits (LC, LT, and LN) are used only to determine the location of mating parts. There are three types of location fits: location clearance (LC), location transition (LT), and location interference (LN). Force fits (FN), also referred to a press or shrink fits, are interference fits meaning the shaft is always larger than the hole.

Figure GNT 6 illustrates an excerpt from the ANSI inch based basic hole tolerance and fit tables and the method of calculating the tolerance limits of the hole and shaft. Calculations are based on a basic hole size of Ø1.2500 inches and the running clearance fit RC 3. Column one labeled "Normal Size Range" lists the range for the basic hole's diameter. In our example the basic hole diameter of Ø1.250" falls between 1.19 and 1.97. Column two labeled "Limits of Clearance" lists the minimum clearance (or allowance) and the maximum clearance in thousandths of an inch. Column three labeled "Hole H7" lists the limit values for the hole in thousandths of an inch. Therefore the maximum hole size is equal to Ø1.250" + .001" = Ø1.251". The minimum hole size is equal to Ø1.250" + 0 = Ø1.250". The hole ranges in size from Ø1.250" to Ø1.251". The fourth column labeled "Shaft f7" lists the limit values of the shaft in thousandths of an inch. Therefore the maximum size of the shaft is equal to Ø1.250" – .001" = Ø1.2490". The minimum size of the shaft is equal to Ø1.2500" – .0016" = Ø1.2484". The size of the shaft ranges from Ø1.2490 to Ø1.2484.

The "Limits of Clearance" figures can be checked by calculating the tightest fit by subtracting the largest shaft from the smallest hole Ø1.2500" – Ø1.2490" = .0010" as illustrated in Figure GNT 6. The maximum clearance is found by subtracting the smallest shaft from the largest hole Ø1.2510" – Ø1.2484" = .0026" as illustrated in Figure GNT 6. Remember a negative fit value (shaft always larger than the hole) is an interference fit.

Nominal Size Range Inches	Limits of Clearance	Class RC 3 Standard Limits	
Over - To		Hole H7	Shaft f6
1.19 - 1.97	1.0 2.6	+1.0 0	-1.0 -1.6
All numbers in thousands of an inch			

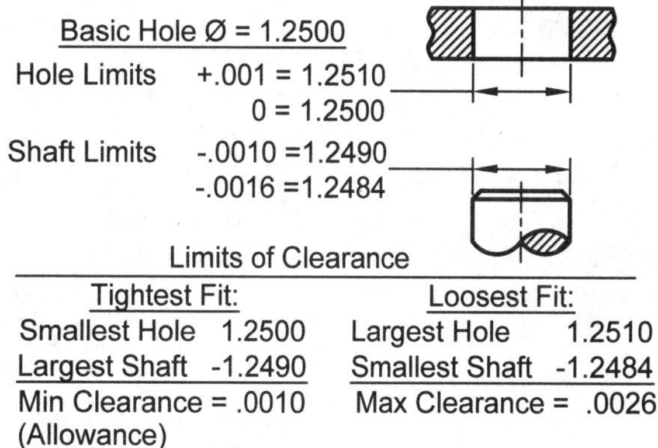

Figure GNT 6: Exert from the ANSI inch based system of limits of limits of fits with calculations for a running fit RC3 and the basic hole size of Ø1.250 inches.

ISO System of Limits & Fits:

The American National Standards Institute tables B4.2 1978 (R1984) *Preferred Metric Limits and Fits* and the *Machinery's Handbook* supply the maximum and minimum tolerance values for mating holes and shafts. The basic concepts discussed with the ANSI inch based system also apply to metric limits and fits. However, the metric based system has some unique terminology and concepts. Refer to Figure GNT 7 when discussing the following list of terms.

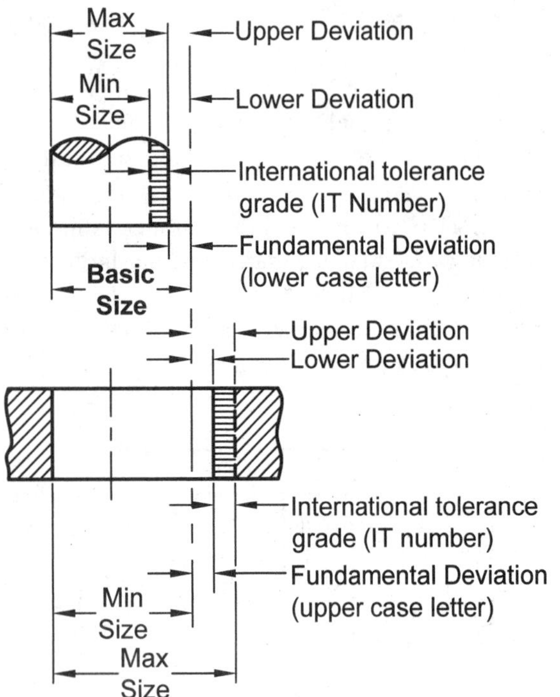

Figure GNT 7: Illustrations of ISO limits and fits terminology.

Upper deviation is the difference between the basic size and the maximum size of the part.

Lower deviation is the difference between the basic size and the minimum size of the part.

Fundamental deviation is the deviation closest to the basic size. In the Figure GNT 9, the uppercase letter "H" in the designation ∅25HB/f7 is the fundamental deviation of the hole while the lowercase letter "f" is the fundamental deviation of the shaft. The sum of the fundamental deviation of the hole plus the fundamental deviation of the shaft corresponds to the allowance in the ANSI inch based system.

International tolerance (IT) grade is a system of tabularized tolerances found in the ANSI standards and the *Machinery's Handbook* that vary based on the basic size. There are eighteen grades ranging from IT01, IT0, IT1... to...IT16. The lower the IT number, the tighter the tolerance. The IT grades in Figure GNT 8 ranging from 01 to about 8 are used for measuring instruments, 8 to 16 for materials, 5 to 11 for defining fits, and 12 to 16 for large tolerances.

Tolerance grade (or zone) is the tolerance and its position relative to the basic size. Tolerance grade is the combination of the fundamental deviation plus the international tolerance grade. In Figure GNT 9B the tolerance grade for the hole is H8 and the tolerance grade for the shaft is f7.

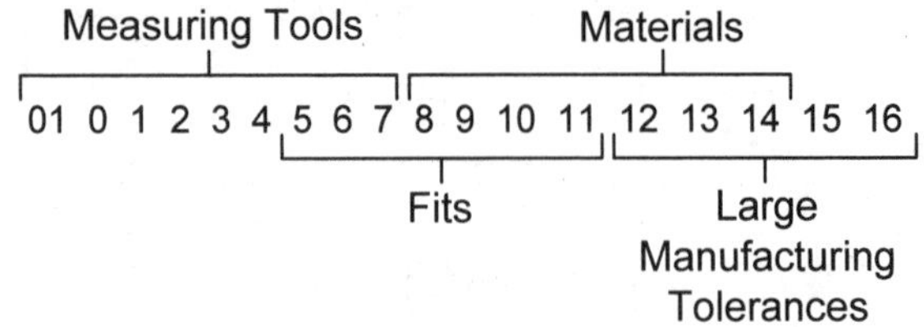

Figure GNT 8: International tolerance (IT) grade applications.

Metric Fit Symbols:

The metric tolerance symbols ∅25H8/f7 in Figure GNT 9A represents the close running fit H8/f7 for a hole-basic ∅25mm hole. ∅25 is the hole-basic diameter, H8 the hole's tolerance grade, f7 the shaft's tolerance grade, and the H8f7 together represents the fit between the hole and shaft. Figure GNT 9B illustrates the symbol used to specify the fit of the hole only. Figure 9C illustrates the symbol used to specify the fit of the shaft only. In Figure GNT 9B tolerance grade designation H8 for the hole breaks down into "H" the fundamental deviation of the hole, plus "8" the international tolerance (IT) grade of the hole. In Figure 9C the tolerance designation f7 for the shaft breaks down into "f", the fundamental deviation of the shaft, plus the "7", the international tolerance grade of the shaft.

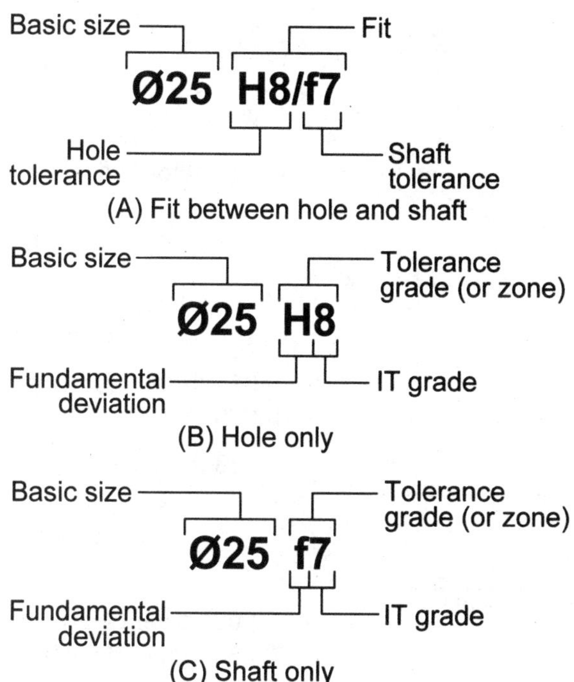

Figure GNT 9: Hole-basic size ∅25 and symbols used to specify fits A) the H8/f7 represents the composite fit between the hole and shaft, B) H8 represents the fit on the hole, and C) f7 the fit on the shaft.

Figure GNT 10 illustrates three methods for displaying a hole-basic fit for a 25mm diameter hole. Figure GNT 10A is the preferred method which simply labels the limits of the hole and shaft using the ISO symbols ∅25H8 for the hole and ∅25f7 for the shaft. Figure GNT 10B labels the diameters using the ISO symbols ∅25H8 and ∅25f7 and adds the limit values in parentheses meaning they are not required but are used as a reference only. Figure GNT 10C labels the diameters using the limit values and adds the ISO symbols in parentheses meaning they are not required but are used as a reference only.

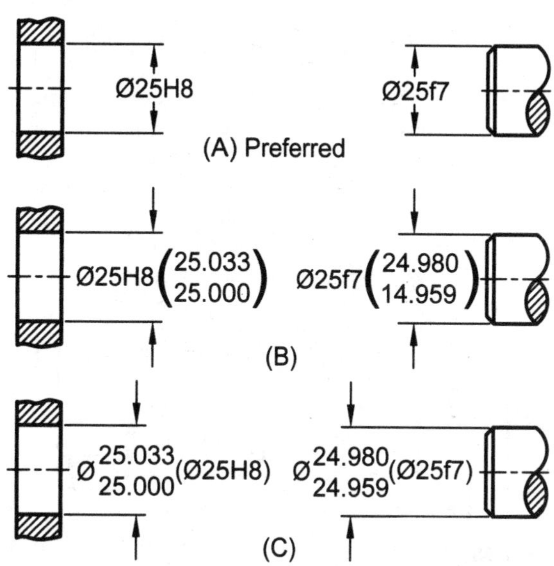

Figure GNT 10: Three methods for specifying ISO tolerances on a hole or shaft. Note that method (A) is preferred.

Figure GNT 11 represents an excerpt from the preferred ANSI metric tolerance and fit tables for a ∅25 basic hole metric close running fit H8/f7 and all the calculations associated with such fits. The tolerance and fit table supplies the maximum and minimum values for the hole and shaft as well as the maximum and minimum clearances. Figure GNT 12 illustrates the calculations from Figure GNT 11. The table in Figure GNT 13 summarizes the preferred ISO clearance, transition, and force fits for both hole-basic and shaft-basic systems of fits.

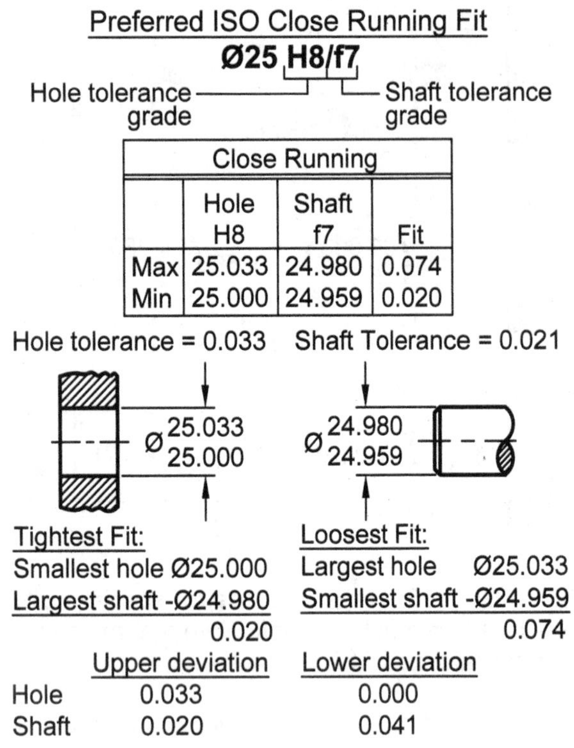

Figure GNT 11: Calculations based on the ANSI metric preferred tolerance and fit tables using a basic-hole size of ∅25mm and a H8f7 fit.

Preferred ISO Close Running Fit
Ø25 H8/f7
Ø24.980 (max shaft Ø)
Ø24.959 (min shaft Ø)
0.021 (shaft tolerance)
0.020 (upper deviation)
0.041 (lower deviation)
0.074 (max clearance)
0.020 (min clearance)
0.033 (hole tolerance)
0.000 (lower deviation)
0.033 (upper deviation)
Ø25.000 (min hole Ø)
Ø25.033 (max hole Ø)

Figure GNT 12: Graphic illustration of the calculations in Figure GNT 11 for a preferred ISO basic-hole fit ∅25H8/f7.

ISO Prefered Fits			
Type of Fit	Hole Basic	Shaft Basic	Description
Clearance Fit	H11/c11	C11/h11	Loose Running
	H9/d9	D9/h9	Free Running
	H8/f7	F8/h7	Close Running
	H7/g6	G7/h6	Sliding
Transition Fit	H7/h6	H7/h6	Location clearance
	H7/k6	K7/h6	Location Transition
	H7/n6	N7/h6	Location Transition
Interference Fit	H7/p6	P7/h6	Location Interference
	H7/s6	S7/h6	Medium Drive
	H7/u6	U7/h6	Force Fit

Figure GNT 13: List of preferred hole-basic and shaft-basic fits.

Terminology Review:

Actual size – is a part's size obtained by actually measuring the part.

Allowance – is the minimum clearance or maximum interference between two mating parts.

Basic hole system – when standard machine tools such as drills, reams, broaches, or mills are used to create holes the smallest hole diameter is used as the basic size for calculating the tolerances and allowance between the hole and shaft.

Basic shaft system – in industries using standard cylindrical bar stock, e.g., textile industry, the largest shaft diameter is used as the basic size for calculating tolerances and allowances for the shaft and the hole it fits into.

Basic size – (sometimes called nominal size) is the theoretical dimension from which size limit values are determined by applying plus-and-minus tolerances.

Bilateral tolerance – allows variation from the basic size in both the positive and negative directions, e.g., 12.0 + 0.2 / -0.2 –or- 12.0 +0.2 / -0.4.

Clearance fit – always has some dimensional clearance between mating parts, e.g., a shaft has a diameter always smaller than the hole diameter it mates with.

Interference fit – results when the internal member of mating parts is always larger than the external part. This requires the internal part to be forced or press fit into the external part.

Limit dimensions – are tolerance values expressed by noting the dimension's maximum and minimum values, e.g., 12.02/11.98 or .501/.499

Transition fit – has either a clearance fit or interference fit depending on the dimensional limits of the mating parts, e.g., the diameter of a shaft can be either slightly smaller or larger than the hole it mates with.

Tolerance – is the difference between the maximum and minimum allowable sizes on an individual part.

Unilateral tolerance – allows variation from the basic size in either the positive or negative direction but not both, e.g., 12.0 + 0.2 / -0.0.

NOTES:

Chapter 11

Geometric Dimensioning and Tolerancing (GD&T)

Introduction:

Chapter 8 Dimensioning discussed the basic concepts for describing the size and location of a part's geometry. But when a manufactured part demands extremely accurate geometry, the part must be defined using geometric dimensioning and tolerancing symbology. The American National Standards Institute standard *Y14.5-1994 Dimensioning and Tolerancing* governs the application of geometric dimensioning and tolerancing (GD&T) symbology. GD&T symbols are adjectives that further describe the location of a feature or refine the geometric shape of the feature.

The two most important questions to ask when applying GD&T symbols are:

1. How is the part used, and
2. How does it mate to another part?

Maximum material condition (MMC) should be taken quite literally as it describes a feature when it contains the maximum amount of material (steel, cast iron, plastic, etc.). For example, the dowel pin in figure GDT 1A is at MMC when its diameter is at its maximum value of ∅12.2mm. The hole in Figure GDT 1A is at MMC when its diameter is at its minimum value of ∅11.8mm.

Least material condition (LMC) describes a feature when it contains the least amount of material. For example, the dowel pin in Figure GDT 1A is at LMC when its diameter is at its minimum value of ∅11.8mm. The hole in Figure GDT 1A is at LMC when it is at maximum value of ∅12.2mm.

Regardless of feature size (RFS) describes the condition when a tolerance is applied to a feature regardless of how large or small the actual feature measures.

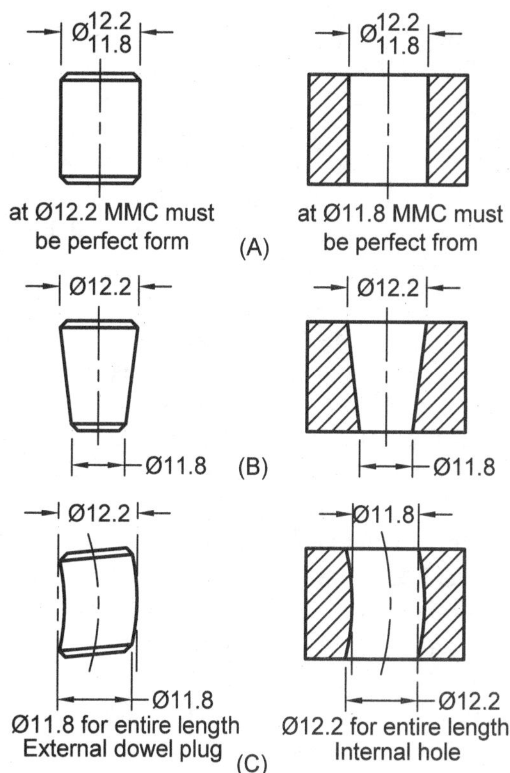

Figure GDT 1: When only size tolerance is specified, Rule # 1 allows variation in form and size within the limits of the feature's stated size limits.

Rule # 1 Envelope Principle

> *"Where only a tolerance of size is specified, the limits of size of an individual feature prescribe the extent of which variations in its geometric form, as well as size, are allowed."* ANSI Y14.5 1994

Figure GDT 1 illustrates the concept behind Rule #1. The only descriptor called out for the external dowel pin in Figure GDT 1A is its limits of size ∅12.2mm to ∅11.8mm. If the diameter of the dowel pin is at its maximum size, or MMC, as it is in Figure GDT 1A, the shape of the dowel pin must be a perfect cylinder. When the dowel pin moves away from MMC, it can change shape yet be acceptable as long as it stays within the ∅12.2mm limit. For example, in Figure GDT 1B the pin is tapered with its maximum diameter at one end and its minimum diameter at the other. In Figure GDT 1C the dowel pin is at ∅11.8mm along its entire length so it can be bent slightly yet stay within the ∅12.2mm maximum diameter limit envelope. The pin can take on any shape as long as it meets the following requirements: 1.) the pin must meet the minimum diameter size requirement along its entire length and, 2.) regardless of the pin's shape, no part of the pin can extend beyond the envelope formed by the maximum and minimum diameters.

The only description for the internal hole of Figure GDT 1A is its limits of size Ø12.2mm to Ø11.8mm. If the hole is at its minimum diameter (or MMC), no part of the object can infringe into this cylindrical void. When it moves away from MMC it can change shape. For example, in Figure GDT 1B the hole is tapered, with one end of the hole at its maximum diameter and the other at its minimum. In Figure GDT 1C the hole has a Ø12.2mm diameter along its entire length. This allows the hole to deviate from a perfect shape yet still allows the minimum Ø11.8mm internal clearance along the entire length of the hole.

Figure GDT 2A, 2B, and 2C illustrate the "envelope" principle. The shaded areas illustrate two coaxial cylinders: one at the cylinder's maximum diameter limit and the other at the cylinder's minimum diameter limit. This theoretical tubular shape is the "envelope" the pin's shape and size must stay within.

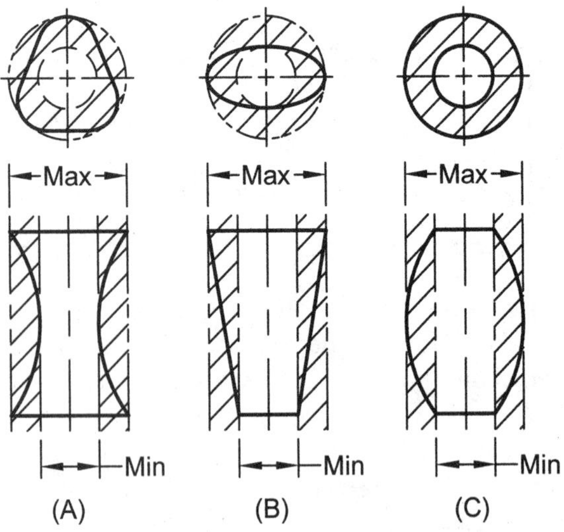

Figure GDT 2: A cylinder with very different shapes but all surfaces are within an imaginary tube formed by the maximum and minimum size dimensions.

In Figure GDT 3A and 3B, the shaded areas represent the extents of the limit dimensions for two rectangular prisms. The envelope for this rectangular prism can be envisioned as a small box sized by the lower values of the limit dimensions contained inside a larger box sized by the upper values of the limit dimensions.

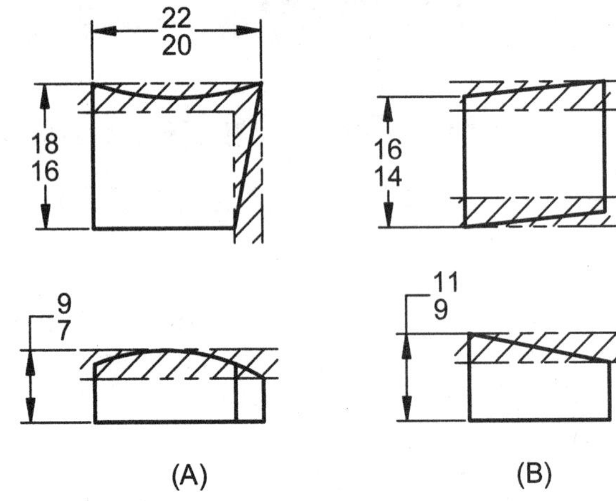

Figure GDT 3: Rectangular prisms with very different shapes, but all surfaces are within imaginary volumes formed by the maximum and minimum size dimensions.

Rule #1applies only to an individual feature and not to the relationship between two features.

Checking the size limits envelope:

A practical way to check if a part is within the required size envelope is to use a ring gauge or a plug gauge. Figure GDT 4A shows a short dowel pin inside a ring gauge. The hole in the ring gauge has a diameter equal to the MMC of the pin, or Ø8.0mm. As long as the dowel pin meets size specifications and also fits inside the ring gauge, then it is a good part. Figure GDT 4B shows a plug gauge inside a hole. The diameter of the plug gauge is equal to the hole's MMC, or Ø6.8mm. As long as the hole meets size specifications and also allows the plug gauge to pass through, then it is a good part.

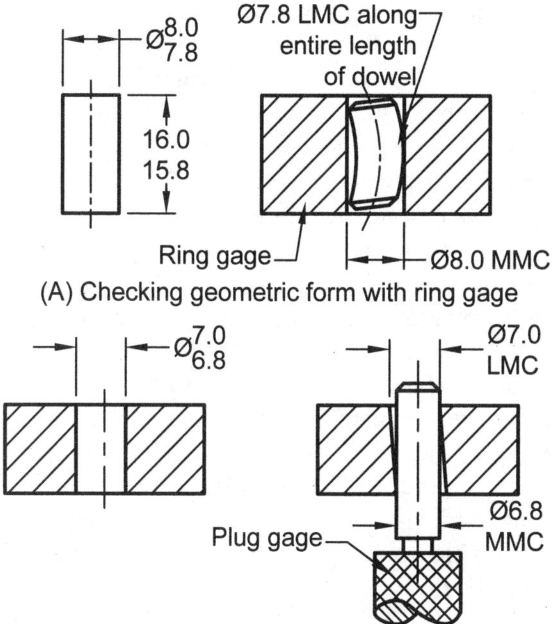

Figure GDT 4: An easy way to validate shape compliance is to use A) a ring gauge to check a cylinder, and B) a plug gauge to check a hole.

Standard Stock Items:

Items whose geometry is already controlled by established industrial or government standards, such as bars, sheet stock, tubing, or structural shapes, are not controlled by the limits of size as it would be redundant.

Datums:

A *datum* is an exact surface, line, point, axis, or cylinder from which measurements are taken. An example of a datum surface is a surface plate that is a polished slab of granite. When a part is placed on the surface plate's highly polished surface, it torches the surface at a minimum of three places, as three points define a plane. The surface of the polished surface plate is referred to as the *simulated datum*. The surface of the part in contact with the surface plate is referred to as the *datum feature*. For example, if the height of the part needs to be verified, a height gauge is required. A height gauge, as illustrated in Figure GDT 5, has a heavy base that slides on the polished granite surface and a dial faced readout (digital is also available) that rides up and down a vertical bar while calibrating the height from the plate's surface. Note that the height is measured from the top of the surface (simulated datum) and not the bottom of the part (datum feature). Other examples of simulated datums are the surface of a milling machine or any fence perpendicular to the machining table.

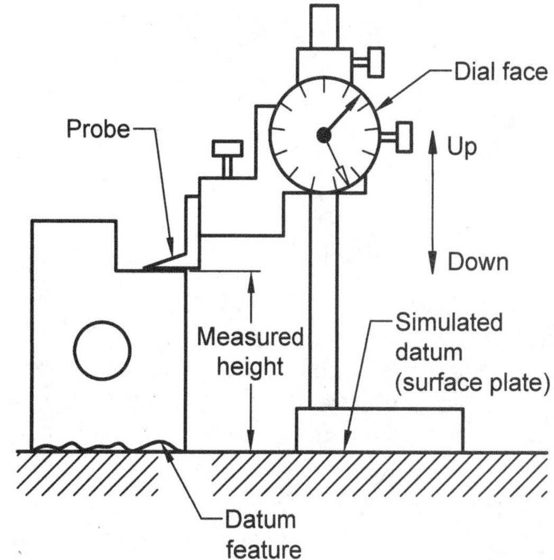

Figure GDT 5: The height gauge measures height from the surface of the simulated datum surface plate to the bottom of the probe.

In order to measure the geometric feature of a part, the motion of a part must be restricted. The X, Y, and Z axes in Figure GDT 6 illustrate the twelve degrees of freedom. A part can translate in two directions along each axis and rotate about each axis in a clockwise and counterclockwise direction. The orientation of the Y and Z axis may seem strange, but it is based on the axes of most metal machining equipment. For example, if you stand in front of a milling machine, the machining bed moves right (+X axis) and left (-X axis); the machine bed moves toward you (+Y axis) and away from you (-Y axis); and the spindle, holding the cutting tool, moves up (+Z axis) and down (-Z axis) into the work piece.

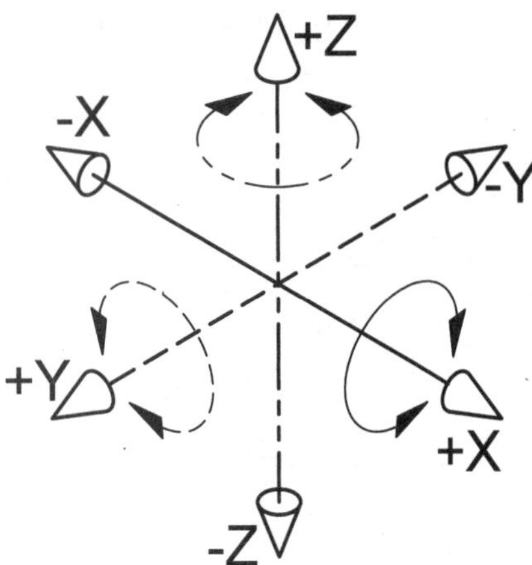

Figure GDT 6: Twelve degrees of freedom allows a part to transverse in both the positive and negative directions along each axis and rotate in a clockwise and counterclockwise direction around each axis.

A datum reference frame is the best way to visualize the restrictions of a part's motion. Three planes intersect at ninety degrees to one another as illustrated in Figure GD&T 7. The part in Figure GDT 7 is in contact with datum surfaces XY, XZ, YZ. But the part can still move along the positive X, Y, or Z axis. If the part is temporarily clamped in the illustrated position, measurements can be taken from the simulated datum planes: XY, XZ, and YZ.

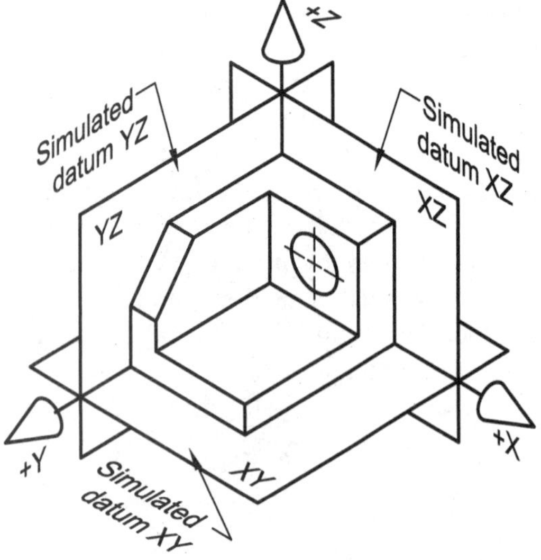

Figure GDT 7: A datum reference frame consists of three intersecting planes at 90° to each other.

Datum surfaces must be indicated on the drawing. Figure GDT 8A illustrates the symbol used to indicate a datum surface on a drawing. The height of the datum box is two times the lettering height, while the width is two times the lettering height for single letter callout and four times the lettering height for double letter datum callout. The datum triangle, which can be either filled or unfilled, has a height equal to the height of the lettering and an included angle of 60°. Figure GDT 8B illustrates the application of the datum symbol on plane surfaces where datums E and F are on the edge view of a plane's surface and datums G and H are on extension lines of the edge view of the surface. Figure GDT 9 illustrates how the datum feature surfaces XY, XZ, and YZ are noted on the orthographic views of the sample part illustrated in Figure GDT 7.

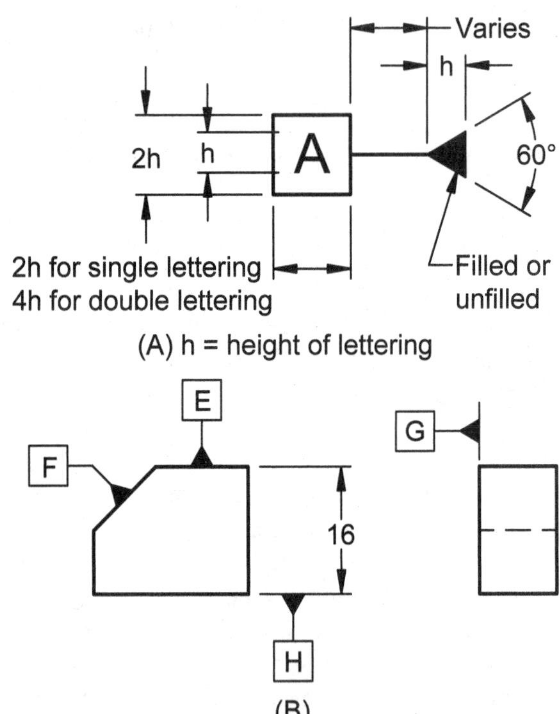

Figure GDT 8: A) Specifications for drawing the datum symbol; B) example applications of the datum symbol on planar surfaces.

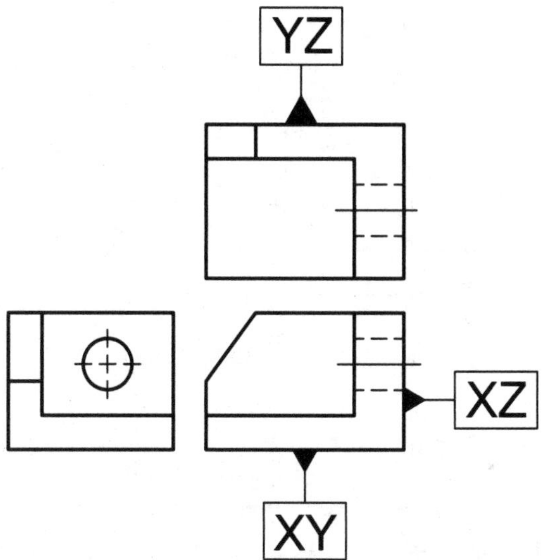

Figure GDT 9: Orthographic views of the part from Figure GDT 7 showing datum symbols for the datum features XY, XZ, and YZ.

Figure GDT 10A illustrates the application of datum symbols on solid cylinders. Datum J and its alternative both have the dimension line arrowheads inside the extension lines. With datum K one of the dimension line's arrowheads is placed on the outside of the extension lines while the datum triangle replaces the opposite dimension line arrowhead. Datum J in the right side view is noting the same surface as surface J in the front view but it is attached to the edge view of the cylinder.

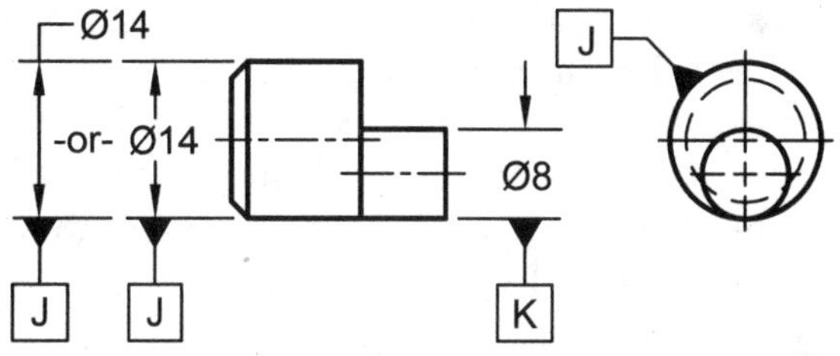

(A) Datum callouts on solid cylinders

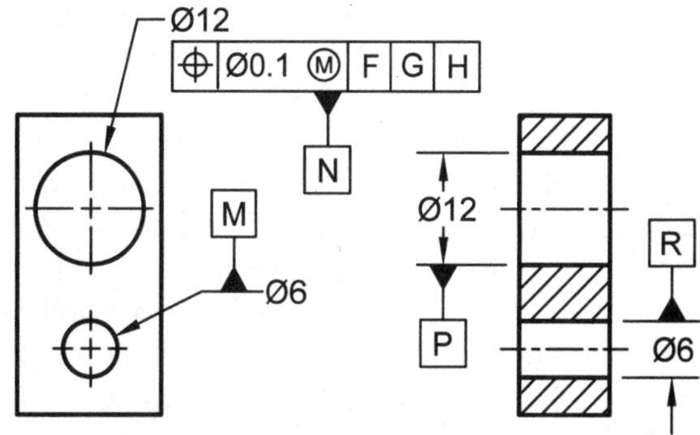

B) Datum callouts on holes

Figure GDT 10: Datum symbol callouts on A) solid cylinders, and B) holes and sectioned holes.

Figure GDT 10B illustrates the application of datum symbols to holes in their circular and sectional views. Datum M is extending from the horizontal "hook" line of the leader calling out the size of the ∅6mm hole. Datum N is extending off the bottom of the feature control frame which is a location modifier of the ∅12mm hole (feature control frames will be discussed later in this chapter). Datum P is the call out for the ∅12mm sectioned hole when both the dimension and arrowheads fit between the extension lines. Datum R is the callout for the ∅6mm sectioned hole where the arrows will not fit between the extension lines and the datum triangle replaces one of the inward facing arrowheads.

Note that the letters "I", "O" and "Q" are not used to indicate datums because they may be confused with numbers one or zero. If the number of single letters is inadequate, double letters are used, e.g., AA, BB, etc.

Order of Datums:

The order in which datum surfaces are selected is important. Figure GDT 11A illustrates the front and right orthographic views of a part with datum features S, T, and U labeled. The small holes are located using dimensions inscribed in rectangular boxes called basic dimensions. A *basic dimension* is a theoretical perfect dimension. In this example they are indicating the ideal location of the holes. The holes' sizes are labeled with the note 2X Ø6 ± 0.2mm. Below the holes' size specification is a long rectangular box called a feature control frame which will be discussed in detail later. In the feature control frame the three boxes containing S, T, and U indicate the order of datums. S is the primary datum, T the secondary datum and U the tertiary datum. The "3, 2, 1 point" rule helps when visualizing the order of datums. In Figure GDT 11B the primary datum feature S contacts the simulated datum S in at least three points. The secondary datum feature T contacts the simulated datum T in at least two points, and datum feature U contacts the tertiary simulated datum U in at least one point. Figure GDT 11B illustrates the 1-2-3 point contact rule by exaggerating the irregularities of the surfaces in contact with simulated datum surfaces S, T and U.

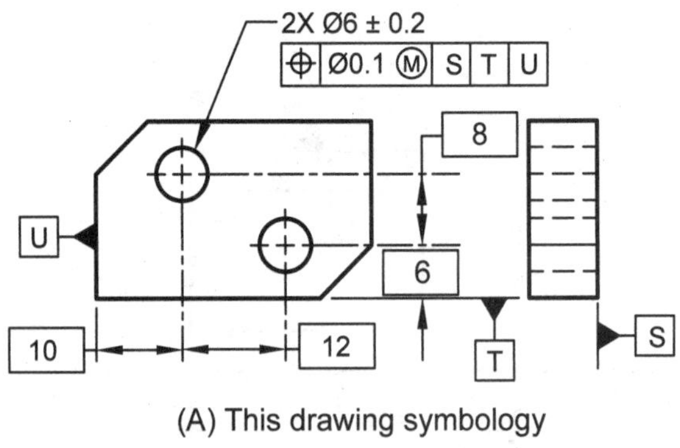

(A) This drawing symbology

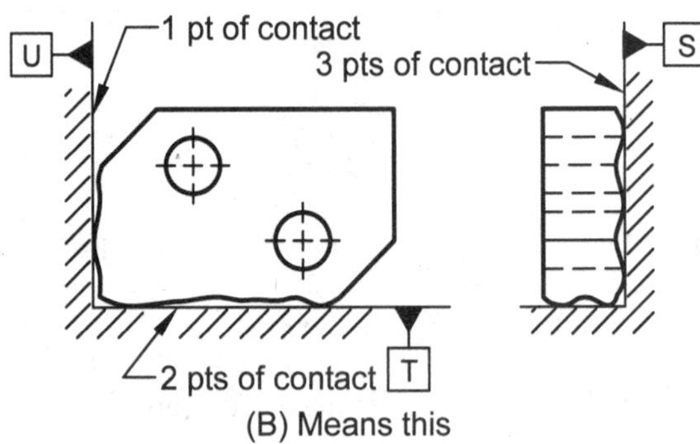

(B) Means this

Figure GDT 11: The order of the primary, secondary, and tertiary datum features in the feature control frame determines the points of contact with the simulated datums.

Selecting the correct order of datums is critical in order to manufacture a product to the desired specifications. It is also important for verification or inspection purposes. For example, the location of the two drilled holes in Figure GDT 12A needs to be verified for quality control purposes. In Figure GDT 12A the angle of surface X should be 90° to surface W, but the error of the angle is exaggerated to illustrate the importance of selecting the correct order of datums. The day shift quality control inspector correctly interprets the drawing as illustrated in Figure GDT 12B where V is primary (3pts), W is secondary (2 pts contacts) and X tertiary (1 pt contact). The height of the holes is measured from simulated datum W and their width from simulated datum X. The afternoon shift quality control inspector misinterprets the drawing as illustrated in Figures GDT 12C where V is primary (3 pts contact), X is misinterpreted as the secondary (2 pts contacts), and W is misinterpreted as the tertiary (1 pts contact). This misinterpretation would cause the hole's centers to be too high above simulated datum W and too close to simulated datum X. This would cause the part to be rejected because it does not meet specifications.

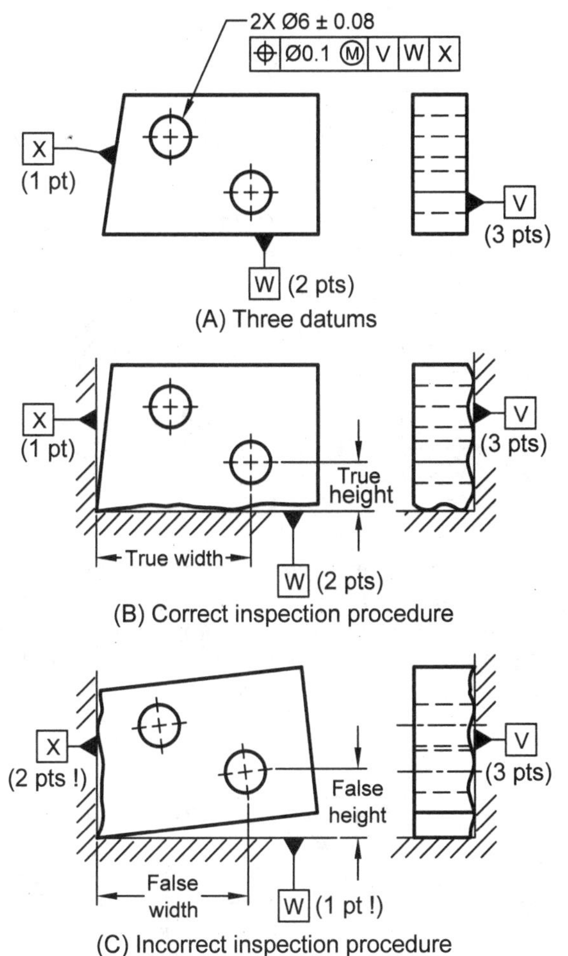

Figure GDT 12: A) illustrates the datum callouts on the drawing where V is primary, W secondary, and X tertiary datums; B) the correct interpretation of the order of datums when inspecting the location of the holes; C) misinterpretation of the order of datums when inspecting the location of the holes results in incorrect data.

Datum axis:

When a cylindrical feature such as the one in Figure GDT 13 is used as a datum, the true axis of the cylinder is used as the datum axis. In Figure GDT 13 the drawing has two datums. Datum G is the back surface and datum H is the exterior of the cylinder. In order to verify the location of the holes, a datum axis must be established. In Figure GDT 13B two theoretical planes, represented on the drawing by the primary center lines, are at 90° to each other and perpendicular to datum G. The datum axis is established by the line of intersection of the two theoretical planes. The datum axis acts as the origin from which measurements can be taken and the two theoretical intersecting planes are used to indicate the direction of the measurements.

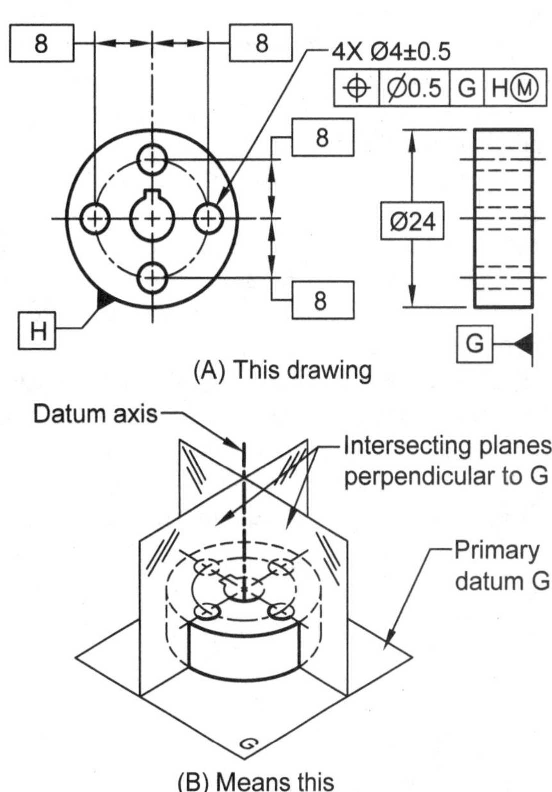

Figure GDT 13: The datum axis is formed by the two theoretical intersecting planes perpendicular to primary datum G.

Features at RFS:

RFS means ***regardless of feature size.*** It describes the condition when a tolerance is applied to a feature regardless of its size. A size feature called out as a datum and modified by RFS requires instruments that can be adjusted to come in contact with the datum features on a part. For an internal cylindrical feature such as datum C in Figure GDT 14A, the datum axis is established by the central axis of the dowel pin. The dowel pin is the largest diameter cylinder that fits in the hole (an expandable mandrel can also be used instead of a dowel pin). The surface of the cylindrical dowel pin is the simulated datum while the actual interior surface of the hole is the datum feature. For an external cylindrical feature such as datum D in Figure GDT 15 the datum axis is established by the central axis of the ring gage which has the smallest hole the cylindrical feature will fit into. The datum axis could also be established using a three jawed chuck (as used on an electric drill) that is tightened around the cylinder. The interior of the ring gage's hole is the simulated datum while the surface of cylinder D is the datum feature. The internal planar surfaces of the slot illustrated in Figure GDT 16 are used as datums of size at RFS. The central datum plane Z is established by the central plane of the largest gauge block that will fit into the slot while being in contact with both surfaces of the datum slot Z. The surfaces of the gauge blocks are the simulated datums and the surfaces of slot Z the datum features. External planar surfaces used as datums of size at RFS are illustrated in Figure GDT 17. The central plane B is established by two planes at minimum separation yet touching both surfaces of datum B. The surfaces of the movable parallel planes are the simulated datums and the external surfaces of datum B are the datum features.

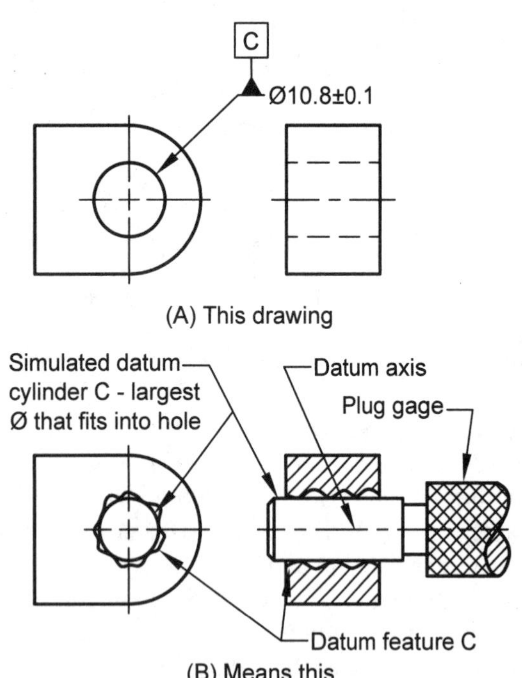

Figure GDT 14: A) a hole is designated as a datum feature at regardless of feature size; B) the largest cylinder fitting into hole C establishes the datum axis.

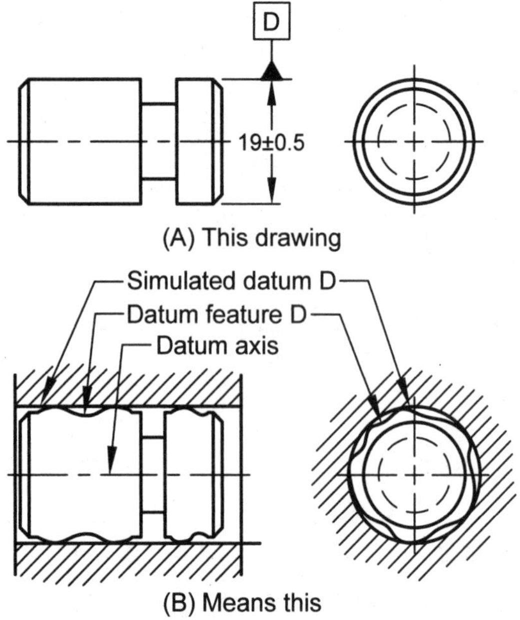

Figure GDT 15: If the exterior cylinder is designated as the datum feature D at regardless of feature size, B) the smallest circumscribed cylinder establishes the simulated datum and the datum axis.

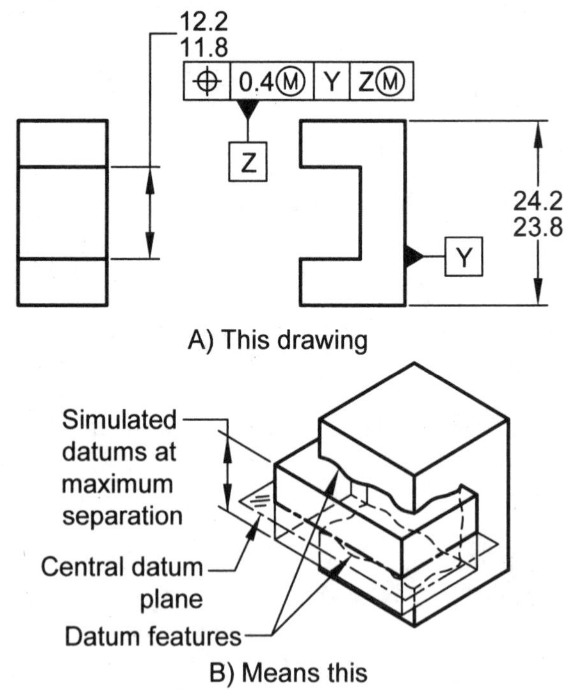

Figure GDT 16: The central plane is a plane midway between two simulated datums formed by the largest rectangular prism (gauge block) fitting between the two datum features.

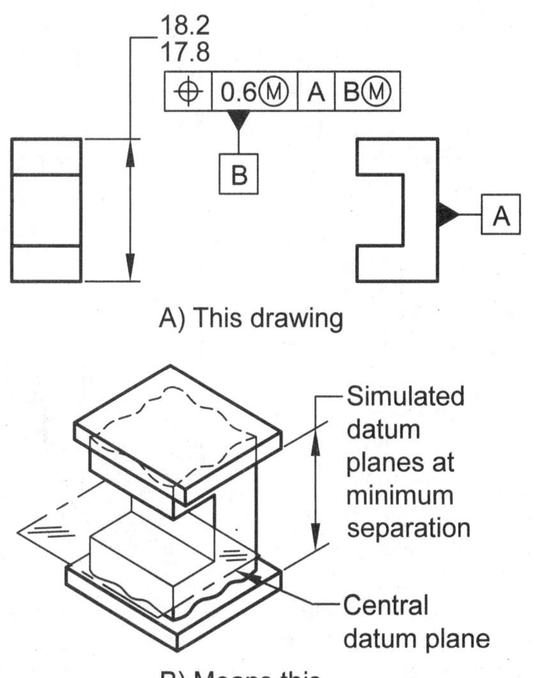

Figure GDT 17: The central plane is midway between the two simulated planes formed by the two rectangular prisms squeezing up against the datum features.

Compound and Coplanar Datums:

A runout modifier is applied to the part's largest cylindrical diameter in Figure GDT 18A. Runout is checked by placing a dial indicator on the outside diameter of the cylinder in question, then rotating the part 360°. The dial indicator's maximum reading is the amount of runout. In Figure GDT 18A the datum box in the feature control frame has a hyphenated datum E-F. This hyphenated datum is called a *compound datum* which uses two features to establish a single datum. This means that when runout is checked, the object must be held by both datum E and datum F. Figure GDT 18B illustrates two three jawed chucks that are coaxial.

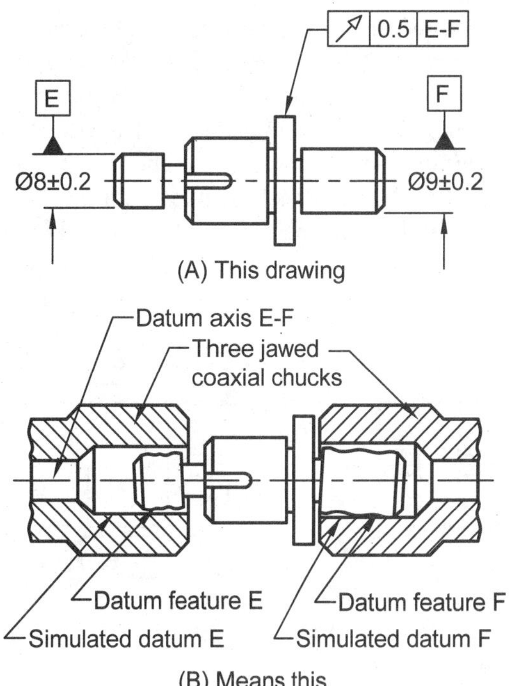

Figure GDT 18: The largest cylindrical portion needs to be checked for runout while being rotated and held between cylindrical datums E and F.

A *coplanar datum* uses two planar surfaces to establish the datum. The callout for a coplanar datum is two hyphenated letters in the feature control frame as illustrated in Figure GDT 19A. It is the same callout as for a compound datum. Figure GDT 19B illustrates datum feature X and datum feature Y resting on a simulated datum such as a surface plate which forms the coplanar datum.

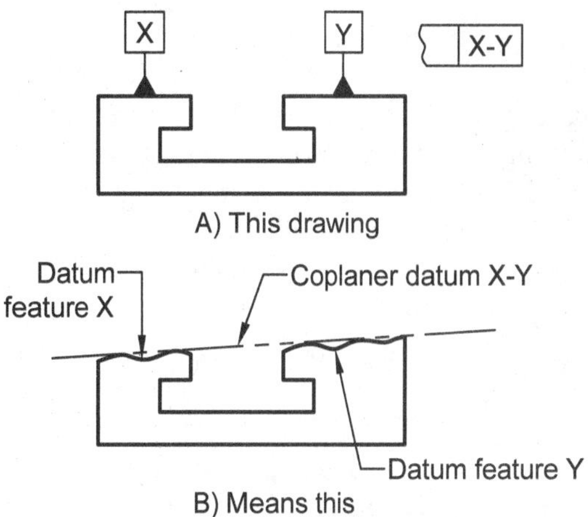

Figure GDT 19: A) the drawing specifies the coplanar datum X-Y; B) simulated coplanar datum X-Y rests on both the datum feature X and the datum feature Y.

Chain Lines and Partial Datum Surfaces:

A portion of a surface can be designated as a datum using a chain line. A *chain line* is a thick center line dimensioned using a basic dimension and having a datum symbol attached to it. Figure GDT 20A shows datum J limited to an area 22mm from the end of the part. Figure GDT 20B shows the partial datum surface J resting on an inspection block.

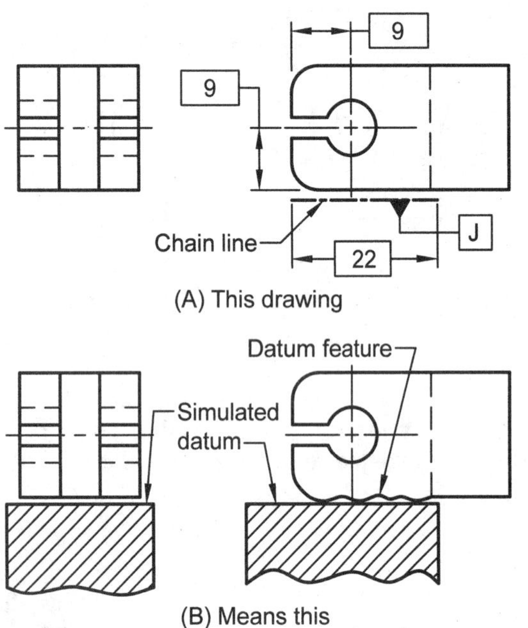

Figure GDT 20: A) Calls out the use of a chain line of a specified length to define the datum area; B) the part rests on only 22mm of the simulated datum.

Datum Target Points:

If it is necessary to define a datum surface using only select points on the surface instead of specifying the entire surface, datum target points are used. *Datum target points* are designated on the drawing with an "X" as illustrated in Figure GDT 21A. The X is at a 90° angle and twice as tall as the drawing's lettering.

Each datum target point is located using basic or toleranced dimensions and labeled with a leader and balloon as illustrated in Figure GDT 21B. The balloon is a split circle having a diameter 3.5 times the height of the lettering as illustrated in Figure GDT 21B.

The lower half of the balloon contains a letter and a number as illustrated in Figure GDT 21B. The letter represents the datum on which the target point is located. The number is the unique identifying number of the point.

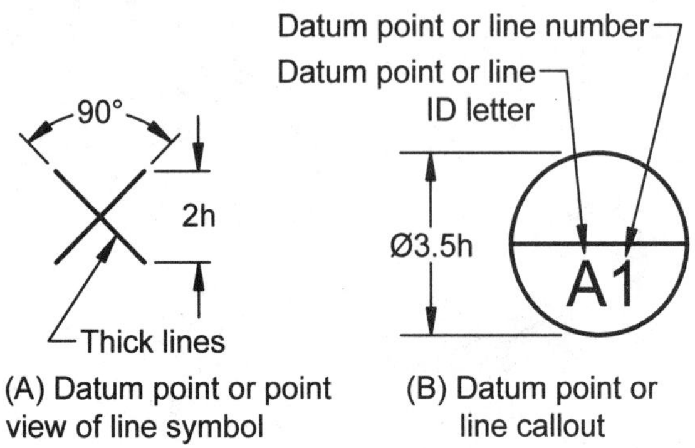

Figure GDT 21: A) Datum point or point view of a line and B) a point or line callout.
h = lettering height

In Figure GDT 22A the datum surface K is called out with a datum symbol on the right view. However, only the three datum target points K1, K2, and K3 designated by X's and balloon callouts are used to establish datum K. All three points are located using basic dimensions. Figure GDT 22B shows the part resting on a fixture having three spherical radius headed dowel pins (pointed head dowels would also work). Each dowel pin presents one of the three datum target points K1, K2 and K3.

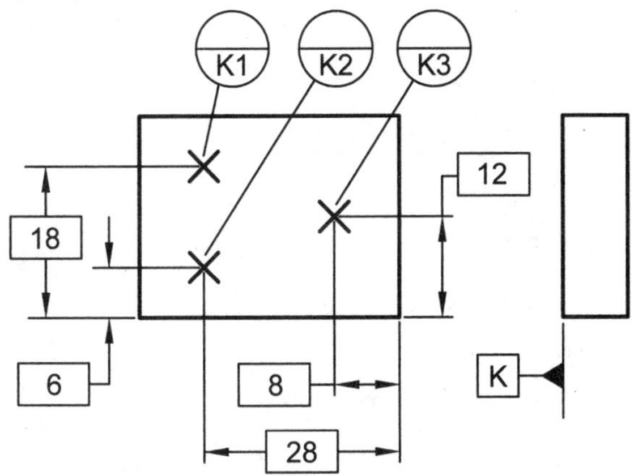

(A) Three datum points located and called out
X = Datum point

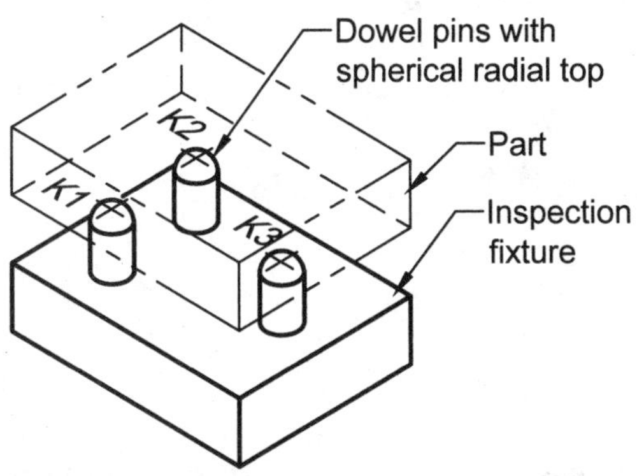

(B) An inspection fixture with three dowels with SR tops for three datum points

Figure GDT 22: A) a drawing with three datum points called out and located; B) an inspection fixture using three spherical headed dowels to establish the three datum points.

If there is no direct view of a datum target point, the point can be located using basic dimensions in two adjacent views as illustrated in Figure GDT 23.

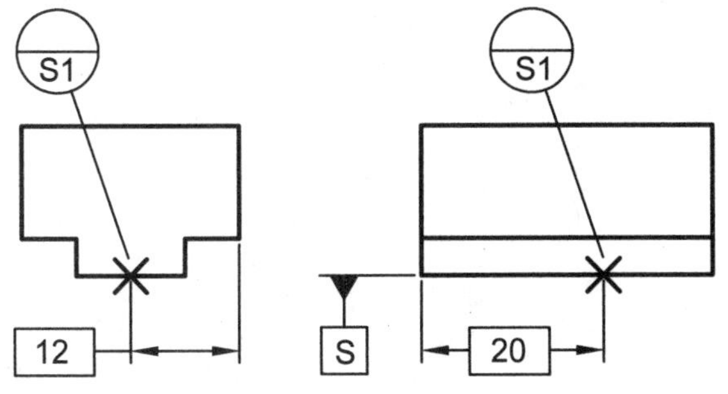

Figure GDT 23: When the view showing the location of a datum point is not drawn, the datum point is established by dimensioning the point in two adjacent views.

Datum Target Line:

A *datum target line* occurs when a flat surface rests against a cylindrical dowel pin. A datum target line is noted on a drawing with an "X" in the datum's edge view and a phantom line in the direct view as illustrated in Figure GDT 24A. In Figure GDT 24A datum M is established using two datum target lines M1 and M2. Datum N is established using the single datum target line N1. Figure GDT 24B shows a part in an inspection fixture having three dowel pins. Dowel pins M1 and M2 establish datum M and the one dowel pin N1 establishes datum N.

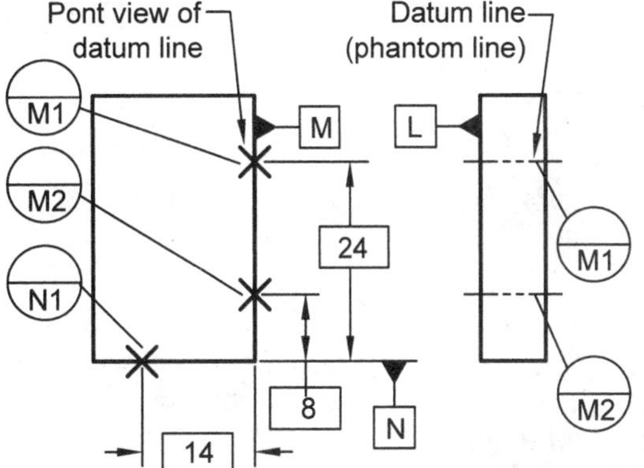

(A) Drawing with 3 datum lines called out

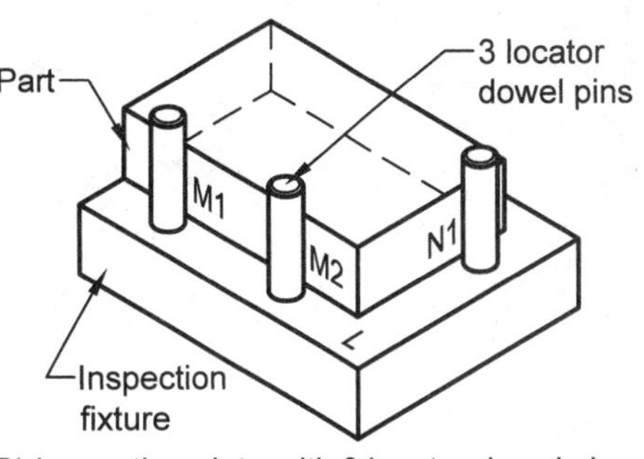

(B) Inspection plate with 3 locator dowel pins as simulated datums

Figure GDT 24: A) drawing specifies three datum lines: M1, M2, and establish the secondary datum and N1establishes the tertiary datum; B) datum lines are formed when a planar surface is pushed up against a dowel pin as with the datum lines M1, M2, and N1.

Datum Target Area:

When spherical radius or pointed dowel pins do not provide a stable datum for the purposes of inspection or machining, datum target areas of a defined shape are used. A *datum target area* is designated on a drawing by an area bordered by a phantom line and hatching inside as illustrated in Figure GDT 26A. A hatched datum target area is called out by a leader line and split balloon as illustrated in Figure GDT 25. If the datum target area has a circular shape, the top half of the split balloon specifies the diameter of the target area. If the diameter number will not fit in the upper half of the symbol's circle, it can be placed outside the circle as illustrated by the right symbol in Figure GDT 25. If the datum target area is not circular, the top half of the split balloon is left blank. In the bottom half of the split balloon, the letter represents the datum the target area is on, while the number is a unique number for the given datum target area.

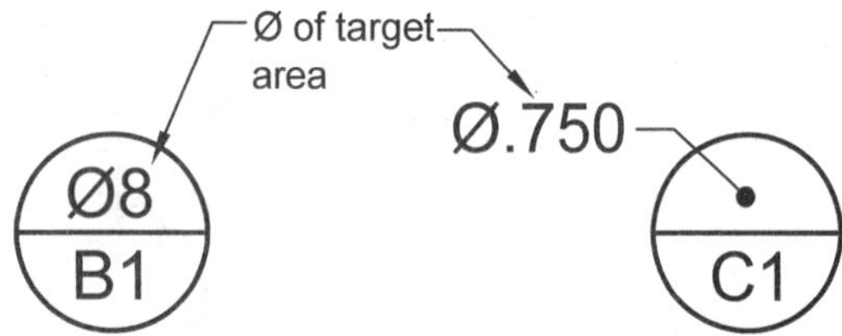

Figure GDT 25: Datum target area callouts use a leader and a split balloon with the diameter of the target area on the top half and the datum letter and the target areas unique number on the bottom half.

Figure GDT 26B illustrates three ∅6mm dowel pins of an inspection fixture which corresponds to the three datum target areas T1, T2, and T3 on the drawing. The three dowel pins establish datum T.

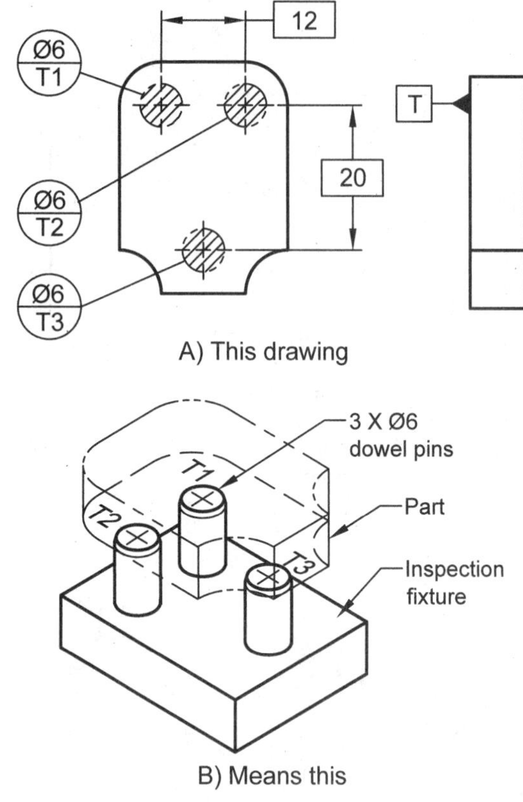

Figure 26: A) Three datum targets each at ∅6mm establishes datum T; B) inspection fixture using three ∅6mm dowel pins to establish the datum surface T.

If it is impractical to have all datum target areas on a single plane, they can be designated on multiple planes as illustrated in Figure GDT 27A and 27B.

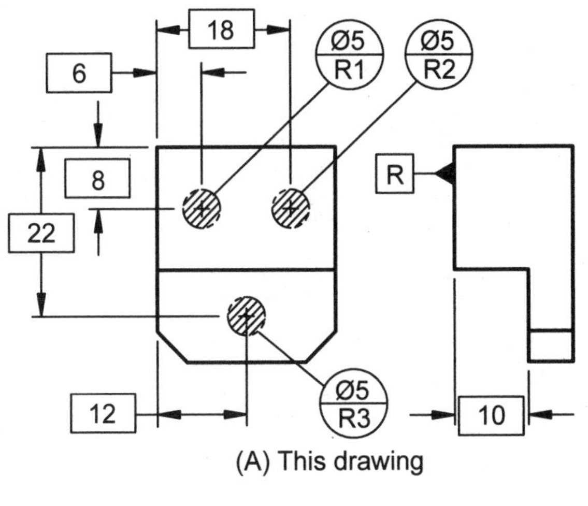

(A) This drawing

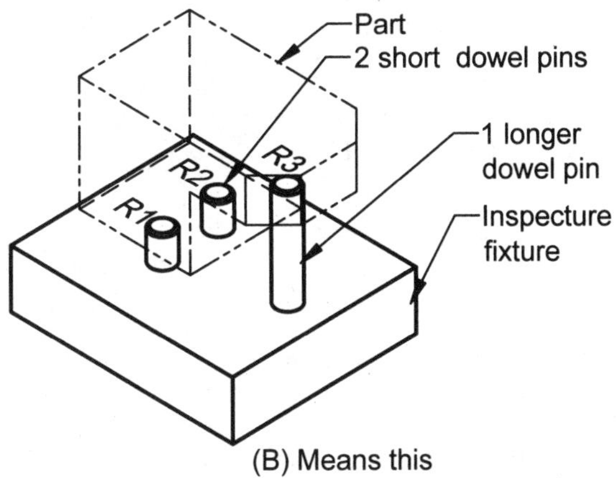

(B) Means this

Figure GDT 27: A) drawing with two datum points on one plane and one on a different plane; B) inspection fixture with two short dowel pins and one long dowel pin.

Figure GDT 28 designates the datum target area P1 which is located on the near side of the given view and called out with a solid lined leader. If a datum target area is located on the far side of a given view, it is noted with an "X" and a balloon having a hidden line leader as illustrated in Figure GDT 28. Datum target areas are not required to be circular. They can be any practical and definable shape.

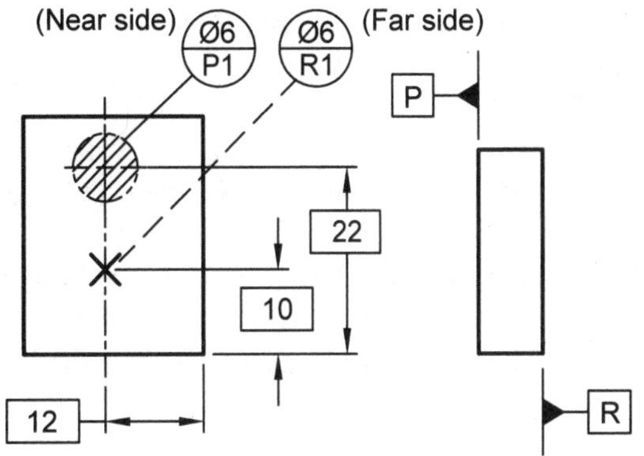

Figure GDT 28: Datum target area P1 is located on the near side while target area R1 is on the far side.

Datum target areas do not have to be circular in area. They can take on any shape required to define a datum plane. Figures GDT 29A, 29B, 29C and 29D illustrate examples of noncircular datum target areas.

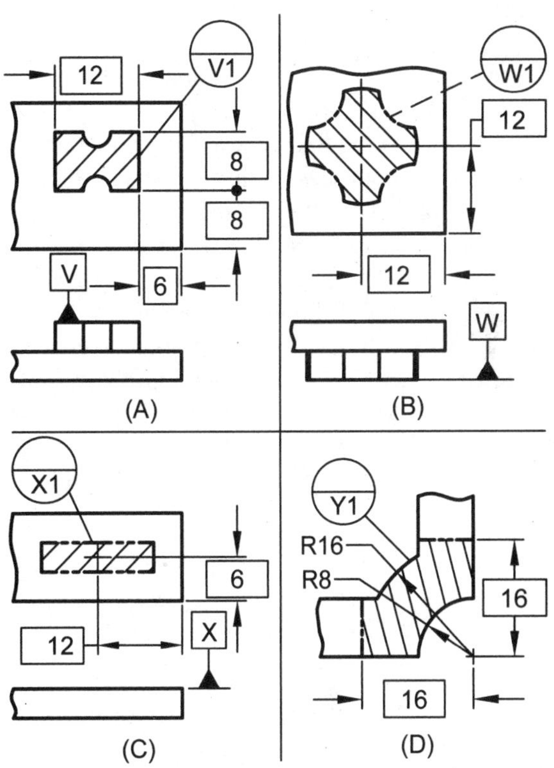

Figure GDT 29: Datum targets can take any shape. The upper half of the datum target symbol is left blank since the target area is not a circle.

Geometric Dimensioning and Tolerancing (GD&T):

GD&T is a method for defining a part's geometry which goes beyond the form description based simply on tolerance dimensions as directed under Rule #1. It is one factor that has allowed industries like the automotive industry to hold such tight tolerances in manufacturing and assembly. GD&T uses a system of symbols to express the type of geometric tolerance.

Most of the symbols in Figure GD&T 30 were discussed in Chapter 8 Dimensioning but are repeated in this chapter because they are used integrally with GD&T symbology. The dimensions in Figure GDT 30 provide the proportions of each symbol based on the lettering height (h). Two new symbols in Figure GDT 30 are the basic dimensioning and the statistical tolerance (dimensional) symbols. A basic dimension is a theoretical perfect size, location, or orientation dimension. The box around a basic dimension is its distinguishing characteristic. The statistical tolerance symbol is an elongated hexagon surrounding the capital letters "TS." Statistical tolerance means the tolerance value was determined using statistical process control (SPC) which is a process of sampling, measuring, charting measurements and statistical calculations to determine the tolerance. When the statistical tolerance symbol follows a tolerance value in a feature control frame it is a statistical geometric tolerance. When the statistical tolerance sample is placed next to a tolerance dimension it is a statistical size tolerance.

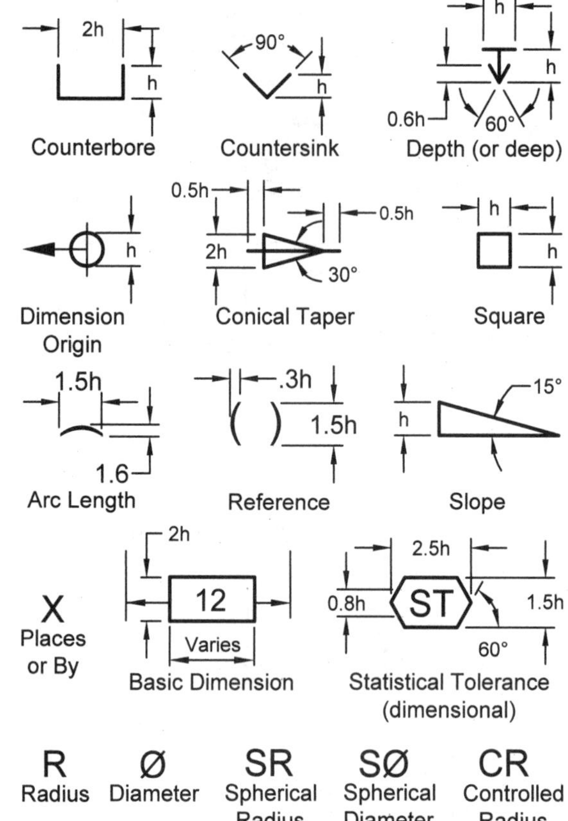

Figure GDT 30: Dimensioning note symbols
h = Lettering Height

Feature Control Frame:

The backbone of GD&T is the feature control frame. It is a series of connected rectangles as illustrated in Figure GDT 31. A complete feature control frame is illustrated in Figure GDT 31A. The first box holds the geometric characteristic symbol which is an adjective helping to define the part's geometric shape. Figure GDT 32 provides a list of geometric symbols each of which will be discussed in detail later in this chapter. The second box in Figure GDT 31A contains the actual numerical value of the geometric tolerance and may contain modifiers. In this example the ∅ symbol means the tolerance has a circular or cylindrical pattern and the circled M means maximum material condition (MMC). The next three boxes are for the datum designations. The first of the datum boxes is for the primary datum, the second datum box is for the secondary datum, and the third datum box is for the tertiary datum. Figure GDT 31B illustrates a short feature control frame stating that the surface must be flat within 0.08mm. Flatness is a type of independent form tolerance which means it does not need to reference a datum surface; therefore no datum boxes are required. Figure GDT 31C illustrates a feature control frame stating that a surface must be perpendicular within a dimetral tolerance of ∅0.05mm at MMC with respect to datum A.

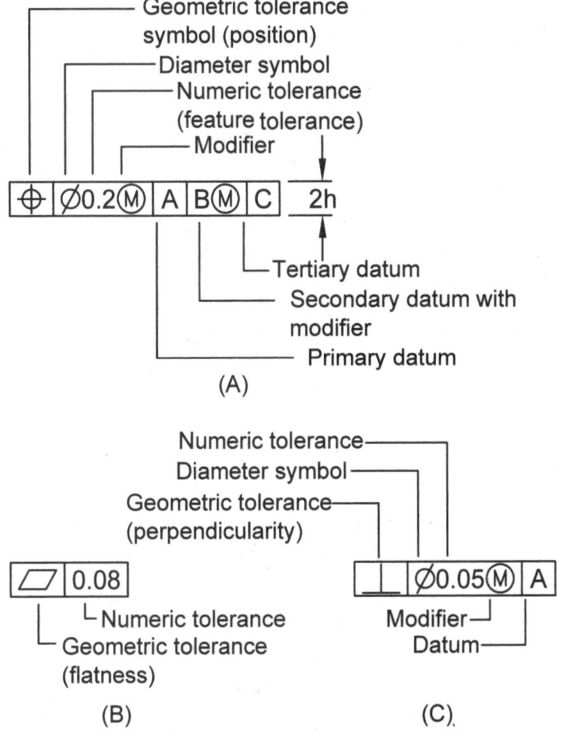

Figure GDT 31: Feature control frame.

Geometric Characteristic Symbols:

Geometric characteristic symbols GDT 32 lists and categorizes the geometric characteristic symbols. Figure GDT 33 illustrates the form and proportion of each of the geometric characteristic symbols based on lettering height.

The third column of the table in Figure GDT 32 categorizes the geometric symbols into five categories: form, profile, orientation, location, and runout. The fourth column further categorizes the geometric symbols by type of tolerance: individual features, individual or related features, or related features. Geometric tolerances in the individual features category do not refer to a datum or datums. For example, when flatness is applied to a planar surface, the plane must be sandwiched between two parallel planes that are separated by the geometric tolerance specified. The plane being checked for flatness and its accompanying tolerance zone planes do not need to be related to a datum as flatness does not require orientation from a datum. Profile of a line or profile of a plane can be either an individual or related feature. Depending on the situation, these tolerances could be referenced to a datum or simply have the tolerance applied locally. The remaining geometric symbols are related, meaning that the geometric tolerance is tied to a datum or datums. For example, to check perpendicularity a datum plane is required in order to measure a 90° angle from the datum to the plane being checked for perpendicularity.

Geometric Characteristic Symbols

Symbol	Description		Type of tolerance
⏤	Straightness	Form	Individual features
⏥	Flatness		
○	Circularity		
⌭	Cylindricity		
⌒	Profile line	Profile	Individual or related features
⌓	Profile surface		
∠	Angularity	Orientation	Related features
⊥	Perpendicularity		
//	Parallelism		
⌖	Position	Location	
◎	Concentricty		
⌯	Symmetry		
*↗	Runout Circular	Runout	
*⌰	Runout Total		

* Either filled or unfilled

Figure GDT 32: GD&T basic geometric characteristic symbols and classifications.

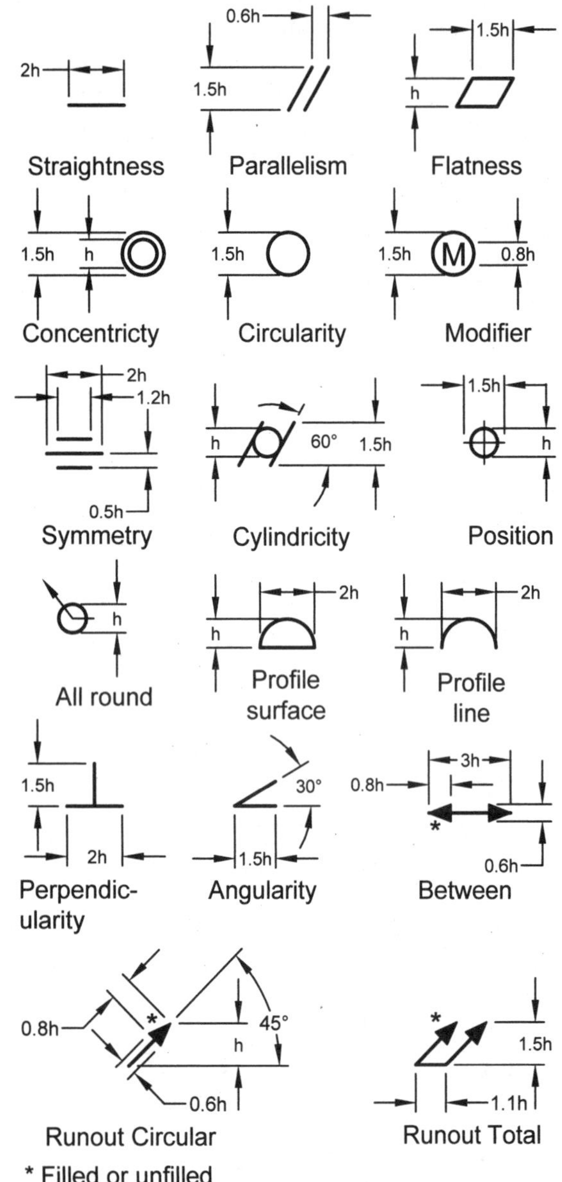

Figure GDT 33: Form and proportions based on lettering height (h) for GD&T geometric characteristic symbols.

Modifiers in a feature control frame:

Figure GDT 34 lists the modifiers used in a feature control frame or, in the case of the "Between" symbol, below a feature control frame.

Modifiers	
Ⓜ	Maximum Material Condition
Ⓛ	Least Material Condition
Ⓟ	Projected Tolerance Zone
Ⓕ	Free State Variation
Ⓣ	Tangent Plane
⟨ST⟩	Statistical Tolerance (geometric)
◄► *	Between

* Either filled or unfilled

Figure GDT 34: Table of modifiers used in the feature control frame or, in the case of the "Between" symbol, below the feature control frame.

Maximum material condition (MMC) should be taken literally. It occurs when the part has the most amount of material within its size limits. MMC, designated by a capital "M" inscribed within a circle, means the tolerance value is applied only when the geometric feature of size is as large as it can be or in the case of a hole as small as it can be. The "circled M" is placed inside the feature control form after the geometric tolerance value and/or after a datum reference.

Least material condition (LMC) should be taken literally. It occurs when the part has the least material within its size limits. LMC, designated by a capital "L" inscribed within a circle, means the tolerance value is applied only when the geometric feature of size is as small as it can be or in the case of a hole as large as it can be. The "circled L" is placed inside the feature control frame after the geometric tolerance value and/or after a datum reference.

A projected tolerance zones designated by a capital "P" inscribed within a circle, means that the existing tolerance zone is extended a specified distance beyond the part's surface. Projected tolerance zones are used mainly with fixed fasteners and press fits. The "circled P" is placed inside the feature control frame after the geometric tolerance value.

Free state variation is the amount of distortion which allows the dimensional limits to vary upon being released from manufacturing or inspection fixture. Free state variation is common with thin walled parts referred to as "non-rigid" meaning they have some degree of flexibility. The "circled F" is placed inside the feature control frame after the geometric tolerance value.

A ***tangent plane*** designated by a capital "T" inscribed within a circle is a theoretical perfect plane that is placed on the high surfaces of the plane being tested for compliance. The tangent plane must be within the geometric tolerance zone. This allows individual points to fall outside the geometric tolerance zone. The "circled T" is placed in the feature control frame directly after the geometric tolerance value. Target planes are used with angularity, parallelism, and perpendicularity.

Between arrows are used with profile of a line and profile of a plane to designate exactly from where and to where the profile tolerance is to be applied.

Angularity:

Angularity is an orientation tolerance that controls the angular relationship between two planes or a plane and an axis. The relationship cannot be at zero or 90°. Angularity is a relational control; therefore at least one datum must be referenced. The value of the angle is specified with a basic dimension.

In Figure GDT 35A the stepped surface of the serrated edge clamp has an angularity callout with a tolerance zone of 0.2mm, and a basic angle of 45° with respect to datum A. Figure GDT 35B interprets the angularity callout showing a tolerance zone of two parallel planes 0.2mm apart and at 45°. The surface being called out and controlled must fall between the two theoretical planes.

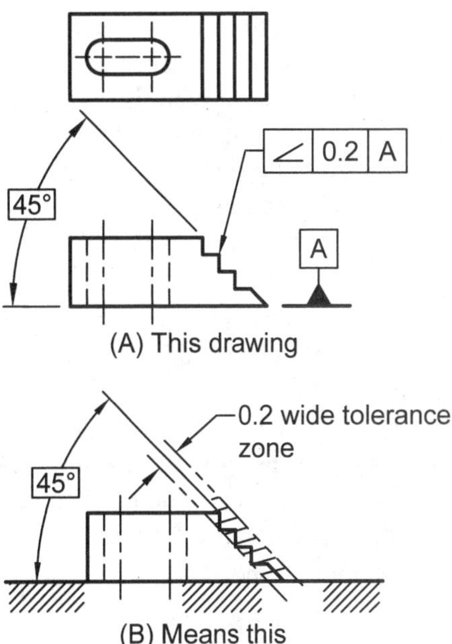

Figure GDT 35: A) the stepped surface has an angularity callout of 0.2mm; B) the tolerance zone is two parallel planes 0.2mm apart and on a basic 45° angle.

Figure GDT 36 illustrates one method for inspecting the 45° angularity callout from Figure GDT 35. The clamp is placed on a sine bar which is propped up by gauge blocks until it is at a perfect 45° angle. A flat bar is placed on the stepped surface and a dial indicator is slid across to obtain a FIM (full indicator movement) reading.

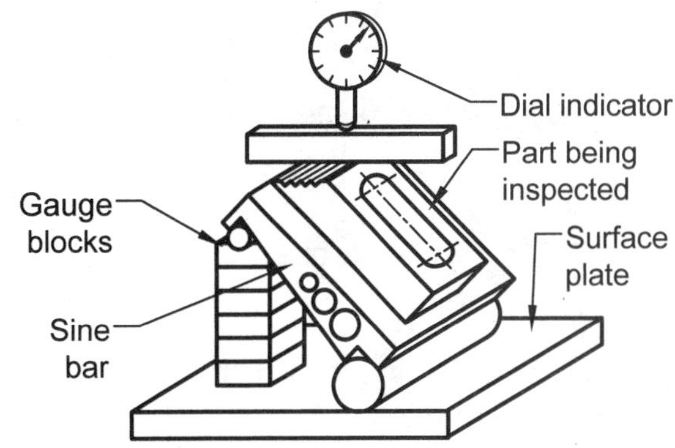

Figure GDT 36: The serrated edge clamp is checked for angularity using a sine block propped up by gauge blocks at a 45° angle and measured with a dial indicator.

Angularity is also used to control the angle between a plane and an axis such as the center axis of a hole. The tolerance zone is cylindrical with a diameter equal to the angularity tolerance value and a height equal to the depth of the hole. The axis being controlled must stay within this cylindrical tolerance zone. Figure GDT 37A illustrates a feature control frame modifying a ∅8.0mm to ∅8.2mm hole having a basic angle of 60° and an angularity tolerance of ∅0.2mm with respect to both datum C and D. Figure GDT 37B illustrates the central axis of the hole which can take on any orientation as long as it stays within the ∅0.2mm cylindrical tolerance zone. The hole in Figure GDT 37B is shown slightly off 60° but still within the ∅0.2mm tolerance zone.

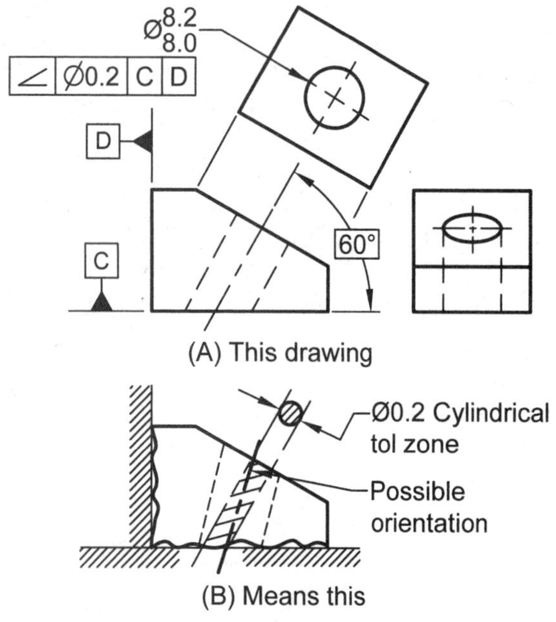

Figure GDT 37: Angularity of a cylindrical hole.

Parallelism:

Parallelism is an orientation tolerance that controls a flat surface, a median plane, or center axis. Parallelism is a relational control, therefore at least one datum must be referenced. If the datum is a flat surface and the feature being controlled is a flat plane, the parallelism tolerance zone is two parallel planes parallel to the datum surface and separated by the value of the parallelism tolerance. Figure GDT 38A illustrates a hinged base (a part from a grinding vice assembly) with a feature control frame calling out a parallelism tolerance of 0.4mm with respect to datum X. Figure GDT 38B illustrates the tolerance zone made up of two parallel planes 0.4mm apart. Every portion of the called out surface must stay between these two tolerance planes. Figure GDT 38C illustrates one method for checking parallelism of the flat surface. Datum feature X of the hinged base is placed on a surface plate. Then a dial indicator slides on the surface plate to check the entire controlled surface looking for points outside the tolerance zone.

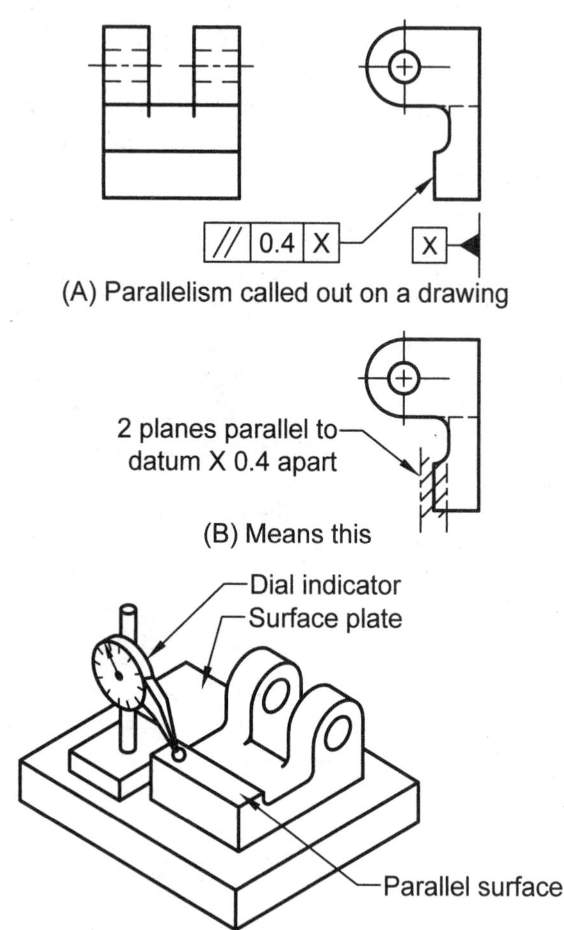

Figure GDT 38: A and B) the called out surface on the hinge base must stay within the 0.4 tolerance zone; C) parallelism is being checked using a surface plate and dial indicator.

If a datum is a center axis and the feature being controlled is a center axis, the parallelism tolerance zone is a cylinder parallel to the datum axis and a diameter equal to the parallelism tolerance. In

Figure GDT 39A the feature control frame calls for the center axis of the ∅6mm hole to be parallel to the center axis of the ∅12mm hole which is acting as datum T. The tolerance zone for this parallelism callout is a cylinder having a diameter of ∅0.4mm and length of 7mm as illustrated in Figure GDT 39B. The center axis of the ∅6mm hole can take on any orientation as long as it stays inside the ∅0.4 x 7mm long cylindrical tolerance zone.

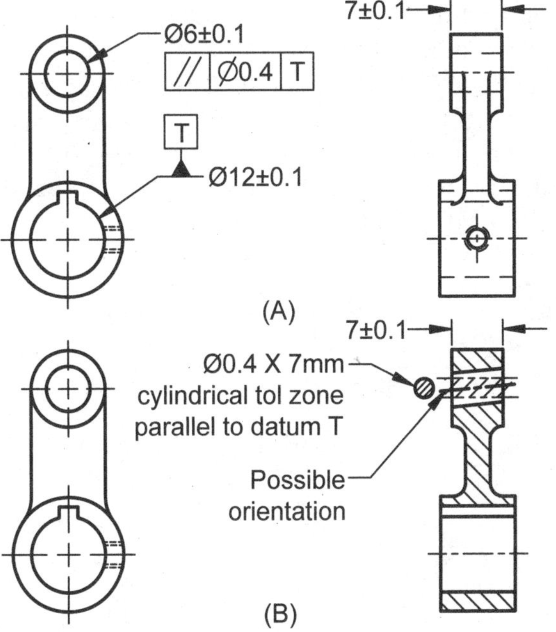

Figure GDT 39: Parallelism cylinder to cylinder: A) the center axis of the ∅6mm hole is called out as parallel to the center axis of the ∅12mm hole; B) the center axis of the ∅6mm hole must stay within the ∅0.4mm cylindrical tolerance zone.

If the datum is a planar surface and the feature being controlled is a center axis, the geometric tolerance zone is defined by two parallel planes. Figure GDT 40A illustrates a feature control frame calling out a hole to be parallel to the planar datum surface G. The tolerance zone is two parallel planes 0.2mm apart and parallel to datum G as illustrated in Figure GDT 40B. In Figure GDT 40B the hole's parallelism is checked by inserting the tightest possible gauge pin or expandable mandrel into the hole. Then 1) the dial indicator is zeroed at one entrance of the hole and 2) the dial indicator is moved to the opposite entrance of the hole and a full indicator movement reading is taken at top dead center. Readings should be taken as close as possible to the hole's entry points.

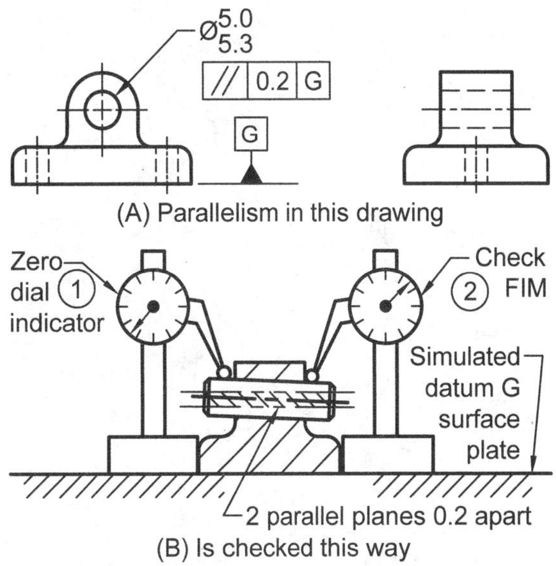

Figure GDT 40: A) A feature control frame calls out a hole as being parallel to planar surface datum G; B) parallelism is checked using a gauge pin (or mandrel) and a dial indicator.

Perpendicularity:

Perpendicularity is an orientation tolerance that controls the relationship between two planes, a plane and an axis, or two axes. Perpendicularity is a relational control; therefore a datum(s) must be designated.

If the perpendicular relationship is between two planes, the tolerance zone is two parallel planes separated by the value of the perpendicular tolerance and perpendicular to a datum reference plane. Figure GDT 41A illustrates a feature control frame calling out the surface on a right base (from an assembly of a chisel sharpening tool) as being perpendicular to datum H. Figure GDT 41B illustrates a tolerance zone made up of two parallel planes 0.4mm apart and perpendicular to datum H. Every portion of the called out surface must be between these two tolerance planes. Figure GDT 41C illustrates a setup for checking the perpendicularity of the called out surface. The part has been clamped to an angle plate with the help of a spacer block. This setup allows the dial indicator to check the called out surface in the horizontal position.

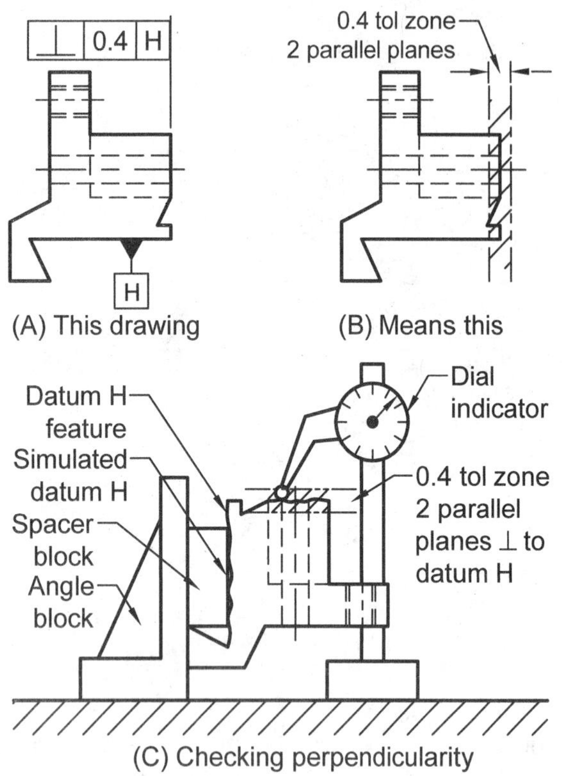

Figure GDT 41: A) a surface is called out as being perpendicular to datum H; B) illustrates the two parallel plane tolerance zone perpendicular to datum H; C) a setup checking the surface's perpendicularity.

If the perpendicular relationship is between a plane and an axis, the tolerance zone is a cylinder perpendicular to the referenced datum plane and its diameter is equal to the perpendicular tolerance. Figure GDT 42A illustrates a feature control frame calling out a cylindrical pin to be perpendicular to the planar surface of datum J. The tolerance zone, in which the axis of the pin must lie, is equal to the length of the 30mm pin and a diameter of 0.3mm. Figures GDT 42C, 42D, and 42E show the pin in a functional gauge. The diameter of the functional gauge's hole must be a diameter that will accept any pin meeting both size and geometric tolerance specifications. A ∅18.3mm hole in the functional gauge will meet this condition which is called the hole's *virtual condition (VC)*. The virtual condition diameter is calculated by adding the pin's MMC diameter plus the geometric tolerance or:

VC = Virtual Condition
VC = MMC + geometric tolerance
VC = ∅18 + ∅0.3mm = ∅18.3mm.

Figure GDT 42C illustrates the pin with no perpendicularity error at MMC (∅18mm). Figure GDT 42D illustrates the pin at MMC (∅18mm) but the pin has a slight perpendicular error yet still within specification. As listed in the table in Figure GDT 43B, the maximum amount the pin can be out of perpendicular at MMC is the cylindrical tolerance zone of ∅0.3mm. Figure GDT 43E illustrates the pin at LMC (∅17.7mm). As listed in the table in Figure GDT 43B, the diameter of the cylindrical tolerance zone increases as the part moves from MMC to LMC. The tolerance as given in the table for LMC is ∅0.6. The decrease in the pin's diameter increases the perpendicularity geometric tolerance. This increase in the geometric tolerance as the pin moves away from MMC toward LMC is called "*bonus tolerance*".

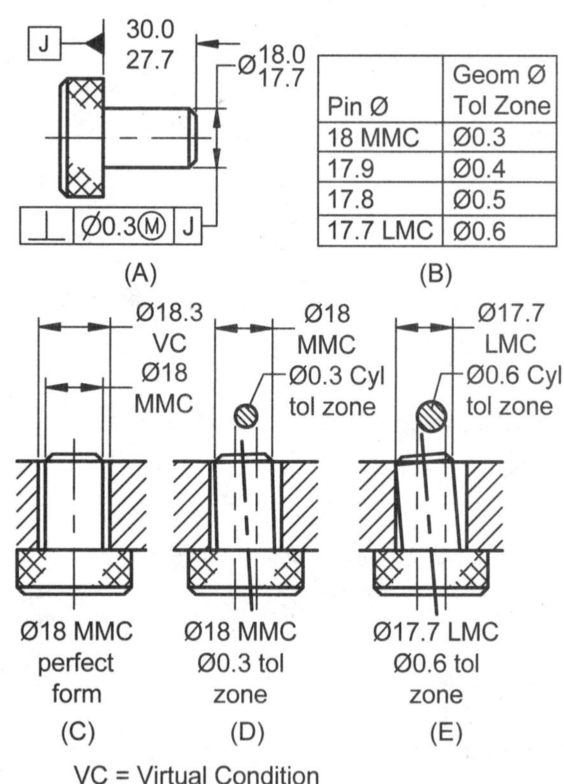

Pin Ø	Geom Ø Tol Zone
18 MMC	Ø0.3
17.9	Ø0.4
17.8	Ø0.5
17.7 LMC	Ø0.6

Figure GDT 42: The perpendicularity tolerance is applied at MMC, therefore the amount of actual perpendicularity tolerance depends on the diameter of the pin. At MMC (∅18mm) the tolerance is ∅0.3mm while at LMC (∅17.7mm) the tolerance is ∅0.6mm.

If the perpendicular relationship is between two axes, the tolerance zone is two parallel planes perpendicular to the datum axis and separated by the perpendicular tolerance value. Figure GDT 43A illustrates a hole called out as being perpendicular to the datum axis G. Figure GDT 43B shows two parallel planes defining a tolerance zone with 0.2mm between the planes.

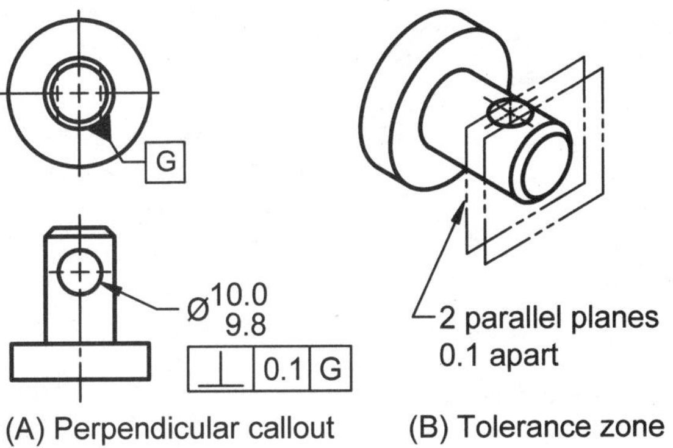

Figure GDT 43: Perpendicularity between a cylinder and a hole.

Straightness:

Straightness is the tolerance of form which specifies tolerance limits for line elements on a surface, a central axis, or a median plane. The elements on a surface or a center axis must be in a straight line. Since straightness is a tolerance of form no references are made to datums. The straightness tolerance must be less than the size tolerance.

Straightness of Surface Elements:

The leader of a straightness feature control frame is applied in the view where the line elements appear as a straight line and will be checked parallel to this view. For example, in Figure GDT 44A the feature control frame can: point to the edge view of the surface as shown in the rectangular shaped front view; be placed below the dimension figures; or hang from an extension line in the front view. Line elements will be checked along the length of the cylinder using a tolerance of 0.03mm. Figure GDT 44B illustrates the 0.03mm tolerance zone along the cylinder's length in the front view while the "I"s in the right view illustrate the end view of the 0.03mm tolerance zones and the five locations straightness will be checked.

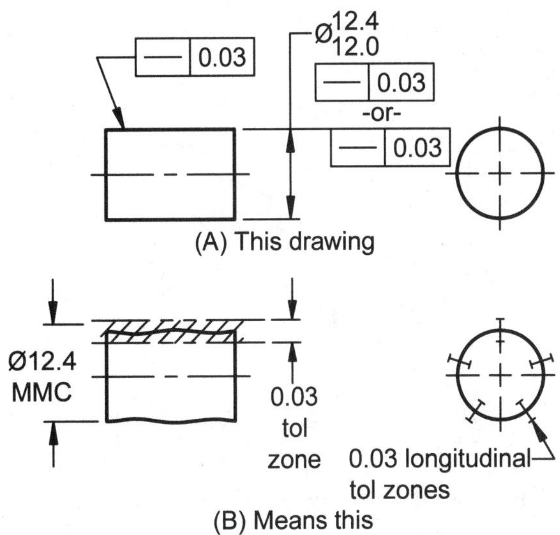

Figure GDT 44: A) line elements on the surface will be checked for straightness; B) the front view shows the 0.03mm tolerance zone while the right view shows the end view of five longitudinal checks.

Checking Surface Elements on a Cylinder:

Figure GDT 45 illustrates one method for checking line surface elements on a cylinder. The cylinder is cradled in a V-block which is resting on a level bar. The bar is leveled using adjustable jacks beneath the bar and V-block. The dial indicator is run along the cylinder at top dead center. The cylinder is rotated and the dial indicator is run along top dead center again. This process is repeated as required for manufacturing and quality control purposes.

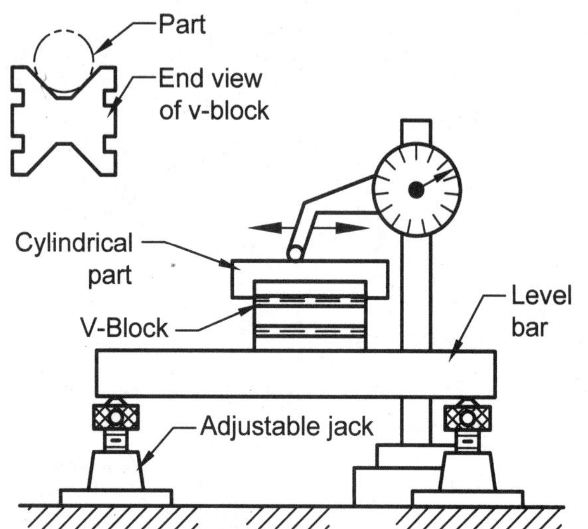

Figure GDT 45: Adjustable jacks level the bar and v-block so the dial indicator can check for straightness at top dead center.

Straightness is used to check a variety of surface shapes. For example, in Figure GDT 46A the top surface of the "treasure chest" is called out to be straight within 0.04mm. Figure GDT 46B illustrates a dial indicator taking a number of line element checks along the curved surface.

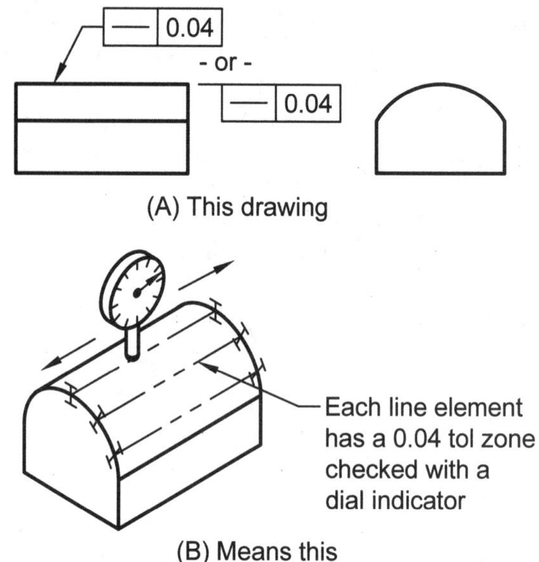

Figure GDT 46: A) callout for straightness within 0.04mm; B) straightness of the "treasure chest" top is checked by taking separate FIM dial indicator readings.

Central Axis:

Straightness of a center axis, as with a cylinder or hole, does not affect surface elements. The feature control frame in Figure GDT 47A hangs below the cylinder's dimension and has a diameter symbol preceding the tolerance value. This means that the tolerance zone is a cylinder with a diameter equal to the tolerance value and a length equal the length of the control cylinder. Figure GDT 47B illustrates a cylindrical tolerance zone ∅0.05mm by 24mm long. The cylinder's central axis can take on any orientation as long as it stays within this cylindrical tolerance zone. Figure GDT 48 illustrates one method for checking the straightness of a center axis. The cylinder is held vertical in a spindle. Surface error is measured using opposing dial indicators and making readings at several intervals along the cylinder. The axis error is calculated to be one-half the total surface error. In addition the cylinder must meet all size specifications.

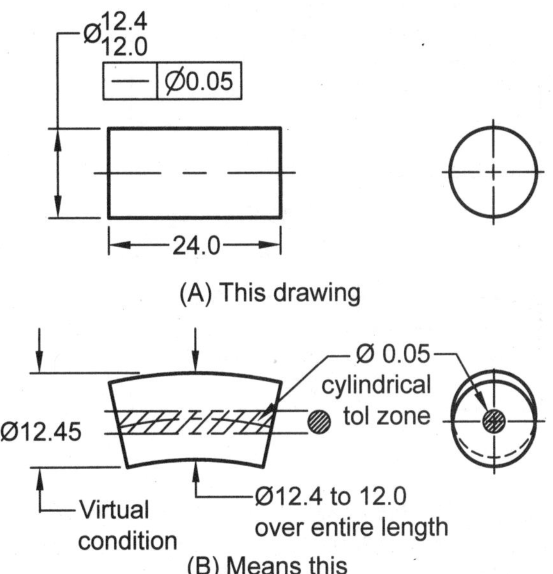

Figure GDT 47: A) callout for straightness of the cylinder's central axis; B) a cylindrical tolerance zone runs the length of the part with a diameter equal to the straightness tolerance.

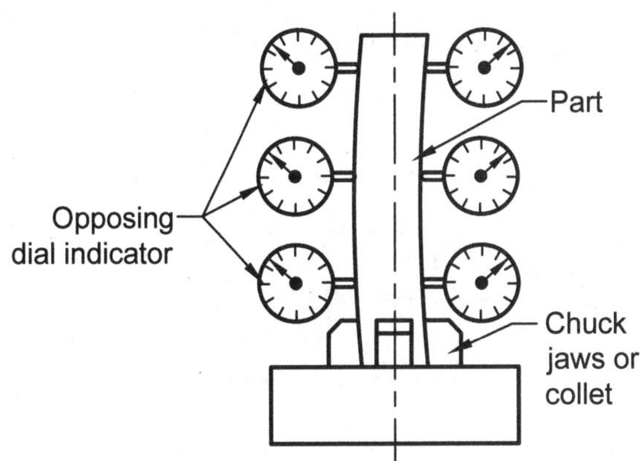

Figure GDT 48: Straightness of a central axis can be checked by taking several surface error readings using opposing dial indicators.

Figure GDT 49 illustrates a pin with a straightness callout of ∅0.4 at MMC. A simple way to check the straightness of an axis is to use a functional gauge. Each part being checked must first be checked for size compliance. The gauge is simply a hole into which the cylinder must fit. The diameter of the hole is its virtual condition (VC) which is equal to the MMC of the cylinder plus the geometric tolerance: VC = ∅6.0mm (MMC) + ∅0.4 (geometric tolerance) = ∅6.4. The length of the functional gauge is equal to the length of the pin. The table in Figure GDT 49 lists the varying geometric tolerance or bonus tolerance values as the part moves from MMC to LMC.

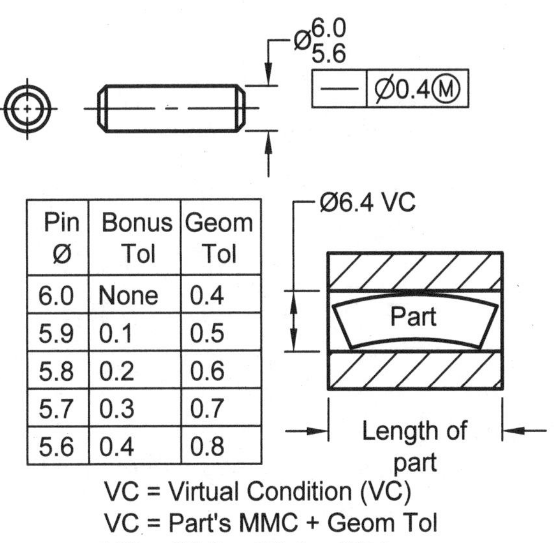

Pin Ø	Bonus Tol	Geom Tol
6.0	None	0.4
5.9	0.1	0.5
5.8	0.2	0.6
5.7	0.3	0.7
5.6	0.4	0.8

Figure GDT 49: To check a cylinder's straightness axial error, a functional gauge can be used. The gauge's length is equal to the pin's length and the diameter is equal to the hole's VC.

Geometric Tolerance Modifiers:

When the modifier MMC or LMC is placed in a feature control frame modifying a straightness tolerance as illustrated in Figure GDT 50A, the geometric tolerance varies. The table in Figure 50B and 50C tabulates the amount of straightness tolerance for different diameters of the cylinder when modified by either B) MMC or C) LMC.

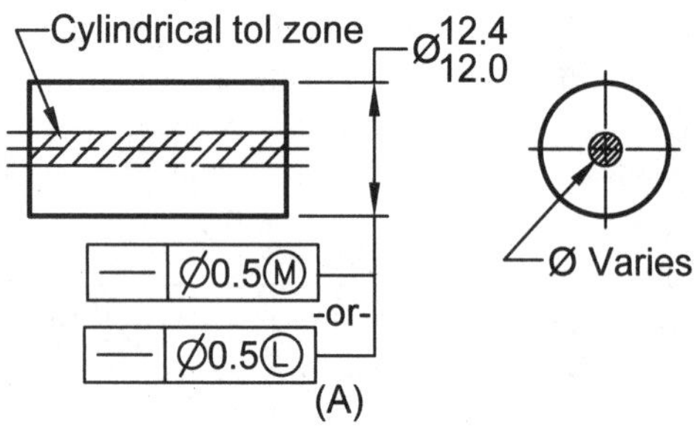

— Ø0.05 Ⓜ

Feature Ø	Ø cyl tol zone
12.4 MMC	0.5
12.3	0.6
12.2	0.7
12.1	0.8
12.0 LMC	0.9

(B)

— Ø0.05 Ⓛ

Feature Ø	Ø cyl tol zone
12.0 LMC	0.5
12.1	0.6
12.2	0.7
12.3	0.8
12.4 MMC	0.9

(C)

Figure GDT 50: A) Straightness is called out on a cylinder's central axis with the tolerance being modified using B) MMC, or C) LMC.

Flatness:

Flatness is a form tolerance which requires all elements on a surface to be within a tolerance zone that is made up of two parallel surfaces separated by the value of the flatness tolerance. Since flatness controls surface elements only, no datums are referenced nor are the material conditions MMC or LMC used. A flatness feature control frame should point to the view showing the surface as a line. Figure 51A illustrates a feature control frame calling out the top surface as being flat within 0.3mm. The controlled surface must stay within the tolerance zone made up of two planes 0.3mm apart as illustrated in Figure 51B. Since the part's height is given with a limit dimension 14.0mm to 14.6mm, the flatness tolerance zone can float anywhere between the 14.0mm to 14.6mm size tolerance zone. Since datum references are not used, the tolerance planes are not necessarily parallel to the part's bottom surface.

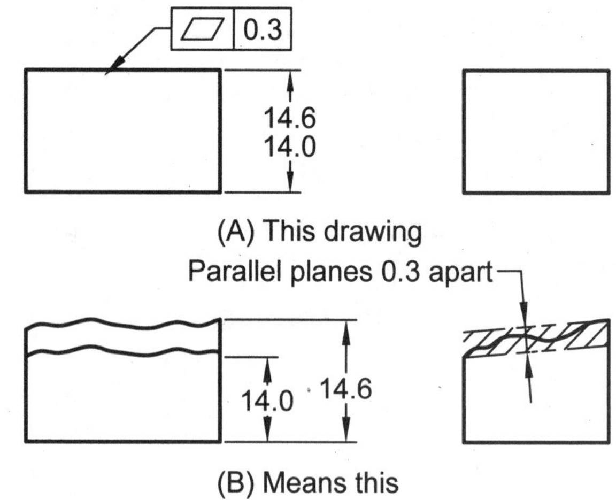

Figure GDT 51: A) the top surface is called out to be flat within 0.3mm; B) two parallel planes form the tolerance zone. The entire surface feature must lie within the tolerance zone formed by the two parallel planes 0.3mm apart.

Figure GDT 52 illustrates one method for checking flatness using a dial indicator coming up through the bottom of an inspection plate while the part being checked slides on top of the inspection plate over the dial indicator.

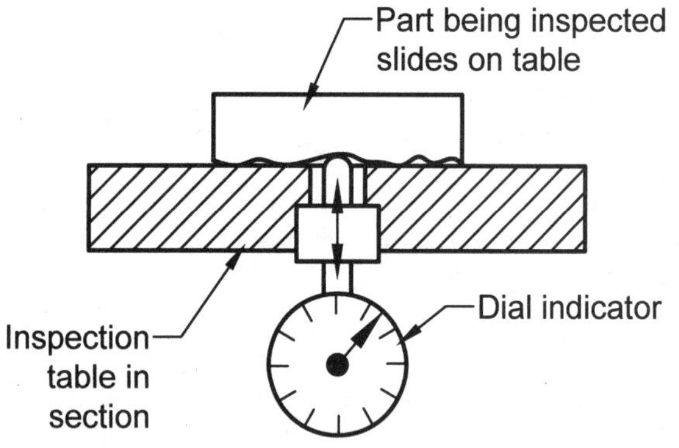

Figure GDT 52: Inspection plate has a hole through which the dial indicator protrudes. The entire surface of the part is run over the dial indicator.

Figure GDT 53A illustrates a surface being called out for flatness while Figure 53B shows the part on adjustable jacks that are resting on a surface plate (at least three jacks are required but only two are shown in the illustration). Once the surface being checked is leveled, a dial indicator is run over its surface.

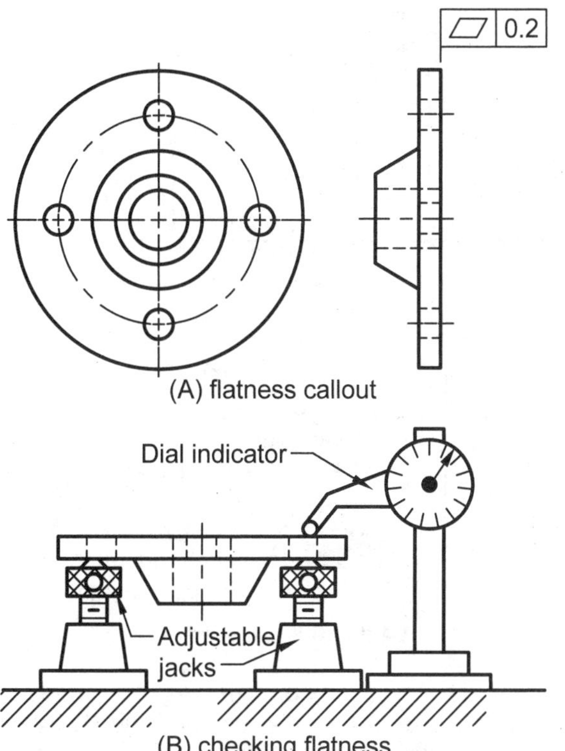

Figure GDT 53: Adjustable jacks level the surface, then flatness is checked with the dial indicator.

Flatness of a Specific Area:

If only a limited area of a surface needs to be checked for flatness, the area is outlined using phantom lines, filled with cross hatching, and located using basic dimensions. Figure GDT 54 is an example of a limited area being called out for flatness.

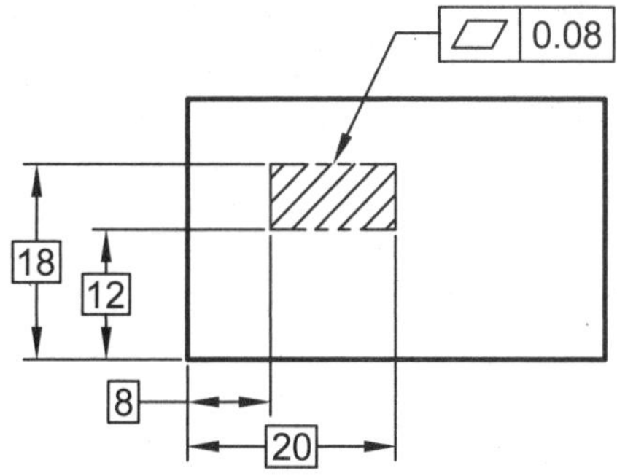

Figure GDT 54: Only the shaded area is checked for flatness.

Circularity:

Circularity is a tolerance of form that checks the periphery of any circular cross section taken perpendicular to the axis of a cylinder, cone, or sphere to ensure that all surface elements are within a two concentric circle tolerance zone as illustrated in Figure GDT 55B. The two concentric circles are separated by the circular tolerance value which is measured on radius. After first checking for size compliance, each circularity check is made and evaluated independently of each other. Circularity measures surface elements so no references are made to datums. Since it is not associated with a feature of size, MMC and LMC modifiers are not used. The circulation tolerance must be less than the size tolerance except for parts in free state variation. Figure GDT 55A illustrates a cylinder with a circularity call out. The leader of a circularity feature control frame can be attached to the profile of the surface or to its circular view as illustrated in Figure 55A. Figure GDT 55B shows a cross section of the cylinder at A-A with a tolerance zone of two concentric circles separated by the circularity tolerance of 0.2mm measured on radius. All surface elements of the cross section must stay within the concentric circular tolerance zone.

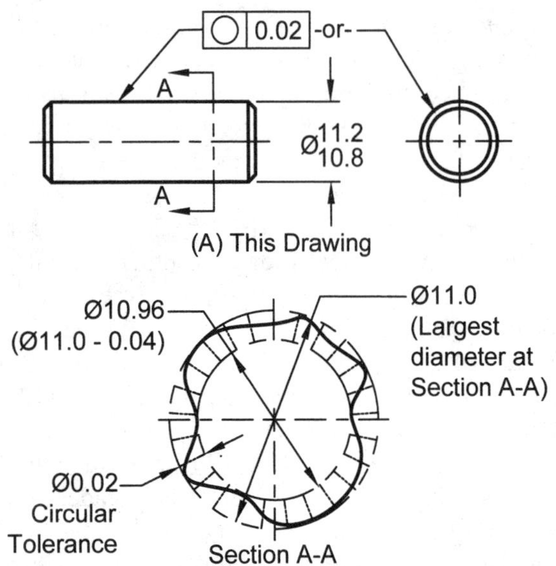

Figure GDT 55: A) a circularity tolerance of 0.02mm is called out; B) each circularity check is evaluated independently to ensure surface elements are within the two concentric circles separated by 0.02mm on radius.

Figure GDT 56 illustrates two methods for checking circularity: A) between centers, and B) using a three jawed chuck or collet. Individual checks are made along the cylinder at intervals determined by manufacturing and quality control requirements. The dial indicator can be replaced by a probe which sends the surface data to a plotter which records the cross sections on polar graph paper for analysis. Figure GDT 57 illustrates a cross section plotted on polar graph paper.

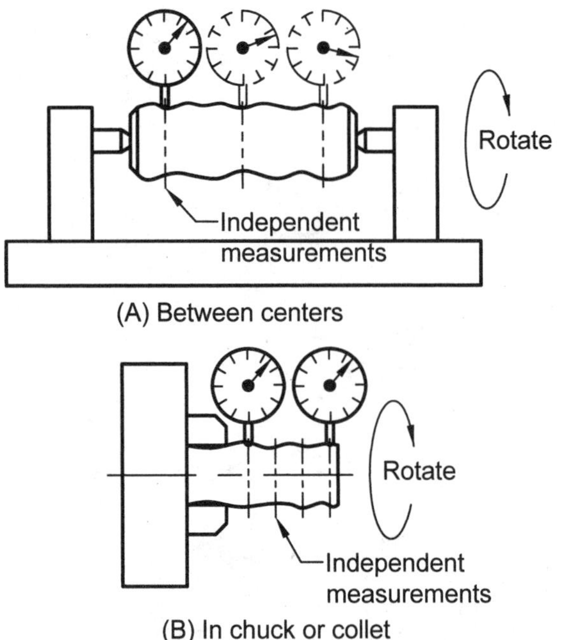

Figure GDT 56: Two methods for measuring circularity are A) between centers and B) with the use of a chuck or collet.

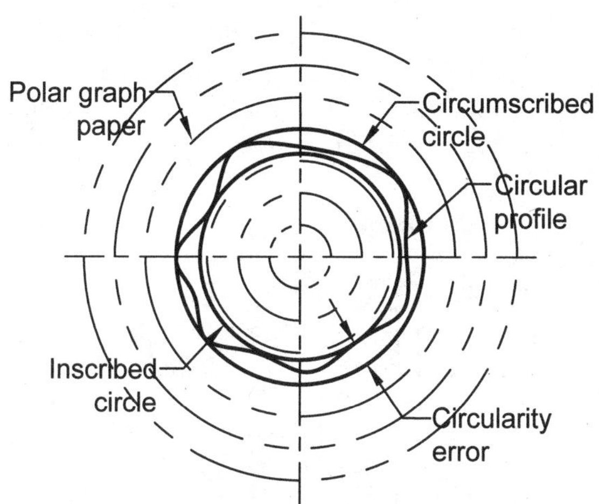

Figure GDT 57: A probe sends circular profile data to a plotting device that plots the profile on polar graph paper for analysis.

Examples of other shapes that can be controlled by circularity are forms such as the cone on the nail set or the curves on the garden tool handle as illustrated in Figure GDT 58.

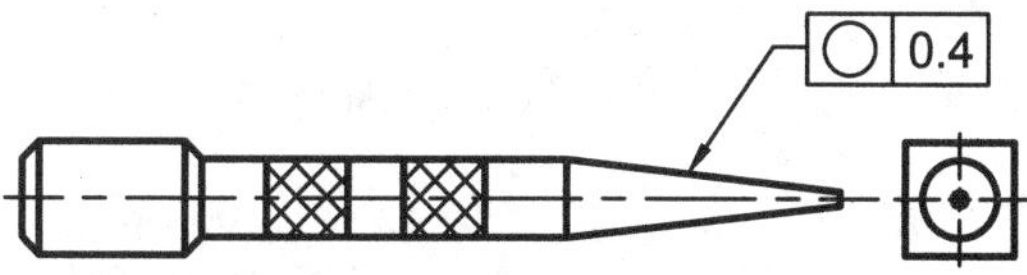

(A) circularity callout on the cone of a nail set

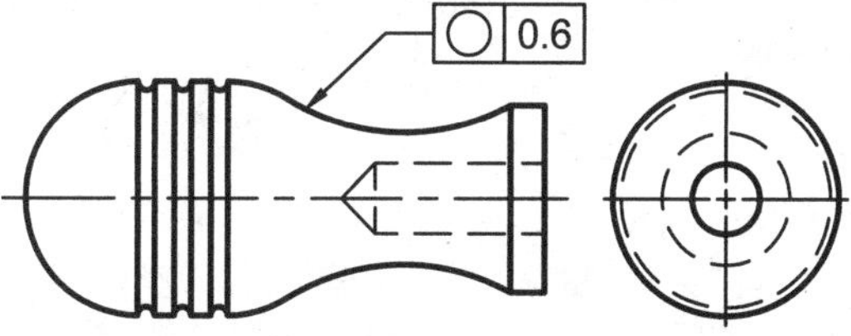

(B) circularity callout on the curves of a garden tool handle

Figure GD T 58: Circularity can control shapes other than a cylinders.

Cylindricity:

Cylindricity it is a tolerance of form that checks cylindrical features only. All circular and longitudinal elements on the surface of the cylinder must lie within two concentric cylinders separated by the value of the cylindricity tolerance which is measured on radius. Cylindricity controls surface elements so it does not use datums and since it is not associated with a feature of size no MMC or LMC modifiers are used. The cylindrical tolerance must be less than the size tolerance. Figure GDT 59A illustrates a cylinder with a feature control frame calling out a cylindricity tolerance of ∅0.06. Figure GDT 59B illustrates the cylinder's surface irregularities within the concentric cylindrical tolerance zone of 0.06mm measured on radius.

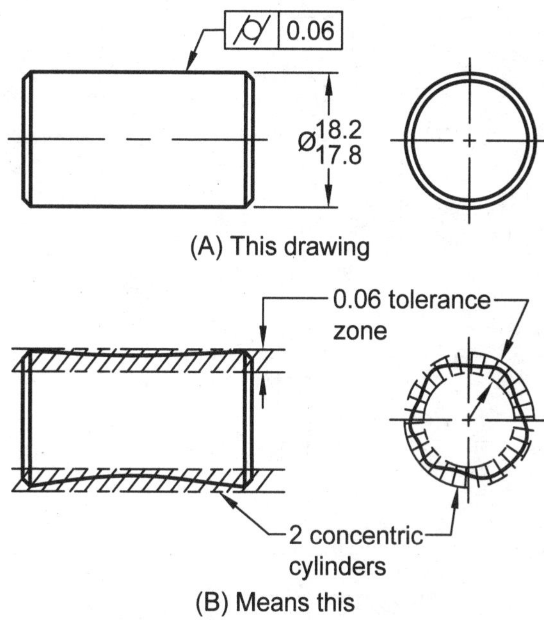

Figure GDT 59: Cylindricity requires that the entire face of the cylinder be contained between two concentric cylinders separated by the cylindrical tolerance measured on radius.

Figure GDT 60 and Figure GDT 61 illustrate two methods of checking cylindricity. In Figure GDT 60 the cylinder is rotated between centers as the dial indicator moves parallel to the rotating axis while taking a continuous reading. In Figure GDT 61 the cylinder is rotated in a spindle while the dial indicator moves vertically up and down while taking a continuous reading. The inspection set up used in Figure GDT 60 and Figure GDT 61 can be used for both circularity and cylindricity. However, circularity takes individual circular surface element readings while cylindricity takes a continuous reading in order to get both circular and longitudinal readings.

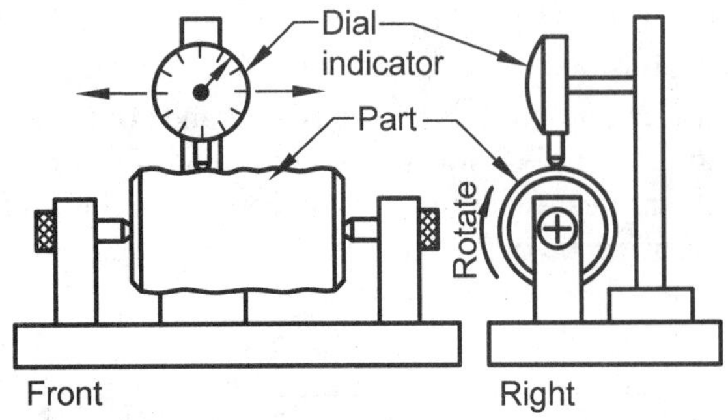

Figure GDT 60: Cylindricity can be checked by rotating the cylinder between centers while the dial indicator takes a continuous reading as it transverses parallel to the cylinder's central axis.

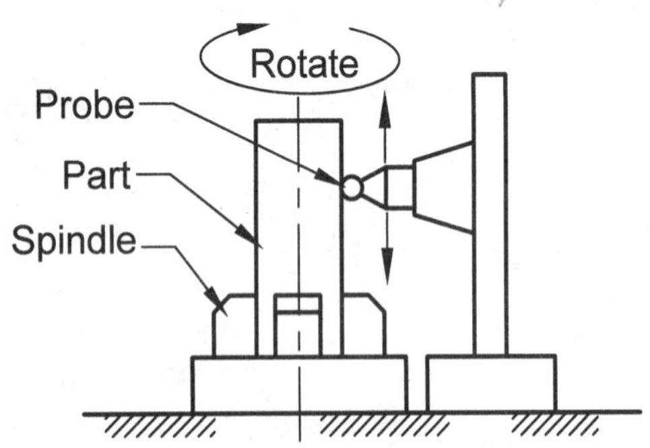

Figure GDT 61: Cylindricity can be checked using a rotating spindle while a probe moves up and down the cylinder collecting continuous data.

Profile's Basic Concepts:

A profile is the 2-D outline of a part. In order to properly see a profile a full section may be required. Profile tolerance controls the shape of a surface by comparing individual elements on the surface to a perfect reference profile. The perfect profile is the theoretically desired or perfect shape defined using basic dimensions. Profile is used to control surface elements, therefore no diameter symbols or MMC/LMC modifiers appear in the feature control frame.

Depending on the information given on a drawing, a profile can control form, orientation, and/or location. If no datums are specified, as illustrated in Figure GDT 62A, only form is controlled. Profile of a surface normally requires datums. So if a datum(s) is added to the feature control frame, as illustrated in Figure GDT 62B, both form and orientation are controlled. The limit dimension in Figure 62B allows the tolerance zone to float within its limits. If a basic dimension is used to locate the perfect form curve, as illustrated in the Figure GDT 62C, form, orientation, and location are controlled. The profile tolerance zone is locked at the basic dimension location.

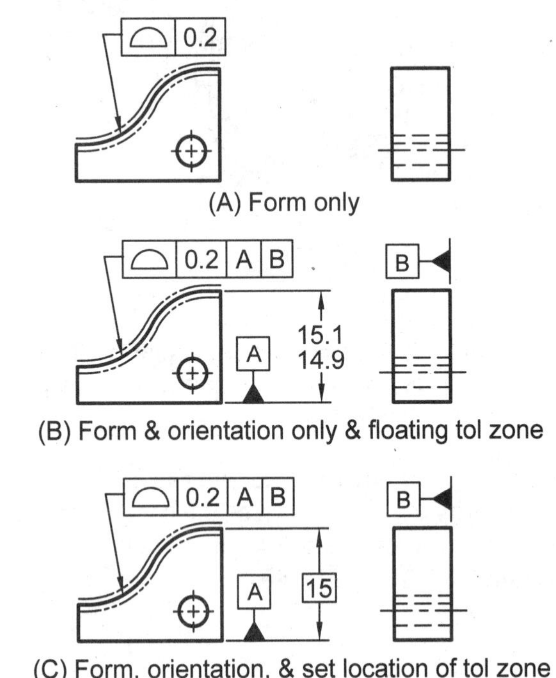

Figure 62: A) without a datum only form is controlled; B) with datums, form and orientation are controlled; and C) with datums and a basic dimension, form, orientation and location are controlled

Profile of a Line:

Profile of the line is used to control single 2-D line elements on a surface when control over the entire surface is not warranted. Each line element runs parallel with the profile, and each line element is checked independently of each other. Profile of a line is commonly used on parts that have a changing cross sectional profile. For each line element 1, 2, and 3 in Figure GDT 63 a separate dial indicator reading is taken while keeping the dial indicator normal to the surface. The numbers of readings would be the minimum number to ensure compliance with the manufacturing and quality control requirements.

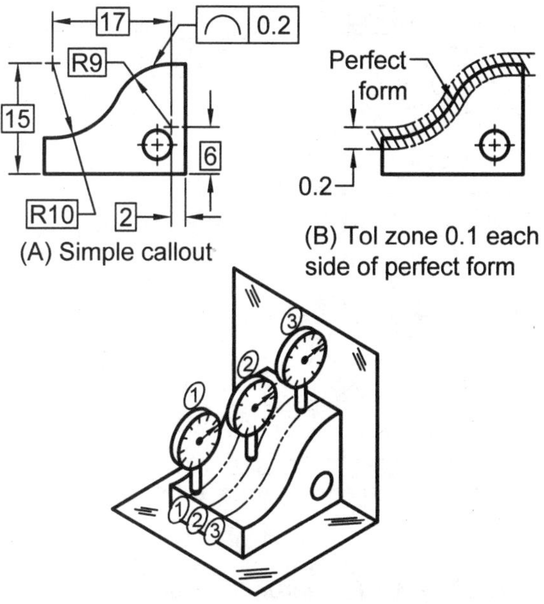

(C) Each line element checked separatly

Figure GDT 63: A) and B) the tolerance zone is defined by two planes parallel to perfect form and 0.2 apart with 0.1 on each side of perfect form; C) each element checked must lie between the tolerance zone.

In order for a dial indicator to check a given path it must be able to follow the path of perfect form. This is done by using a follower template as illustrated in Figure GDT 64B. The dial indicator is connected to a follower tip which runs along on the follower template. Theoretically the dial indicator moves in a perfect form path. Figure GDT 64A illustrates a feature control frame for profile of a line with respect to datums A, B, and C. In Figure 64B, a hold down knob keeps the part steady during inspection. The same setup can be used for profile of surface.

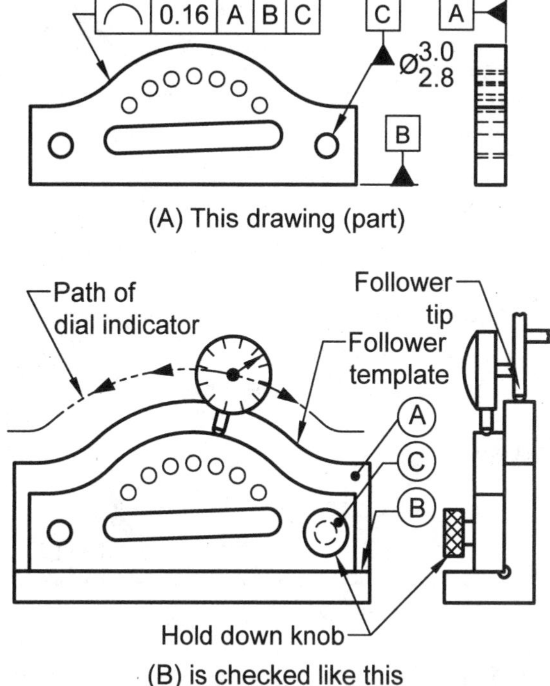

Figure GDT 64: A) The top curved surface of the guitar bridge and strings template is called out for a profile of a line control; B) a follower template keeps the dial indicator on the perfect form path.

Profile of the line and profile of a surface are considered to be bilateral unless specified. Bilateral means the tolerance is divided equally on both sides of the perfect form surface. Figure GDT 65A illustrates a typical profile of a line callout resulting in a bilateral tolerance. Unilateral tolerance places the tolerance to one side of the perfect form surface indicated on the drawing. Figure GDT 65B illustrates the method used to call out a unilateral tolerance on the outside of the part. Figure 65C illustrates a unilateral tolerance on the inside of the part.

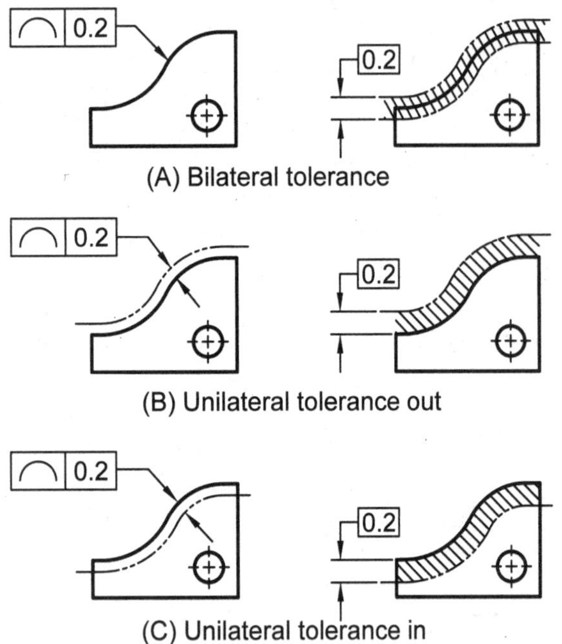

Figure GDT 65: Profile of a line and profile of a surface have three different methods of locating the tolerance zone: A) bilateral, B) unilateral out, and C) unilateral in.

Floating Tolerance Zone:

If the location of a profile curve is dimensioned with limit dimensions, it establishes a zone in which the profile curve can float. Figure GDT 66A has a limit vertical dimension of between 20mm to 18mm which is the only dimension affecting the height of the profile. That sets up a 2mm tolerance zone for the profile curve to float in as illustrated in Figure GDT 66B. However, the curve also has the additional restriction of a profile of a line callout with a tolerance of 1.0mm. So line elements on the profile are restricted to a 1.0mm wide profile of a surface tolerance zone which can then float inside the size tolerance zone of 2mm. The floating tolerance zone also works with profile of a surface.

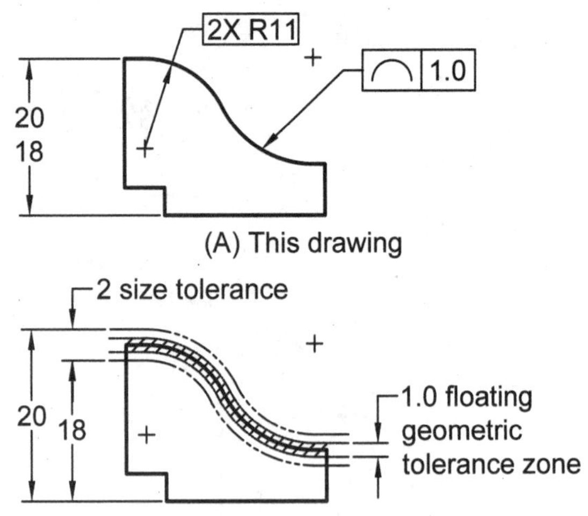

Figure GDT 66: The 1.0mm profile of a line tolerance zone floats within the 2mm size tolerance zone.

Profile of a Surface:

Unlike profile of a line, which controls individual line elements, profile of a surface controls every element on the surface. Profile of a surface can be used on most surfaces but is commonly used on surfaces formed by extruding a curved profile line along an axis and sweeping a curve profile line around an axis. Profile of a surface usually requires datum references. Figure GDT 67A illustrates the edge view of a curved surface sized and located with basic dimensions and displaying a profile of a surface feature control frame. Figure 67B illustrates how to visualize the tolerance zone, which is made up of two parallel planes separated by the tolerance value of 0.4mm with 0.2mm on either side of the perfect form curve (bilateral).

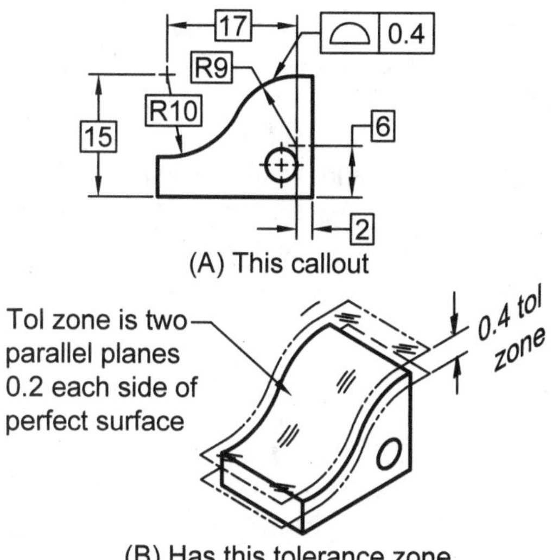

Figure GDT 67: A) Profile of a surface calls out a 0.4 tolerance; B) two parallel planes form the tolerance zone.

Profile of a Surface on a Spherical Radius or Planar Surface:

Figure GDT 68A illustrates a knob with a spherical radius on the left end and an angled flat surface on the right end. The spherical end has a profile of a surface callout with the tolerance of 0.1mm with respect to datums A and B. The flat end has a profile of a surface callout with a tolerance of 0.2mm with respect to datums A and B. Figure GDT 68B illustrates how the surfaces are verified. The spherical radius tolerance zone is made up of two spherical radius shaped planes parallel to the perfect surface and 0.1mm apart with a 0.05mm bilateral tolerance on each side of the perfect surface. All elements on the spherical surface must stay between the two curved plane tolerance zones. The angled flat end has a profile of a surface tolerance with two flat planes parallel to the angled surface and separated by the 0.2mm tolerance value with a 0.1mm bilateral tolerance on each side of the perfect surface. All the elements on the angled surface must stay within this tolerance zone.

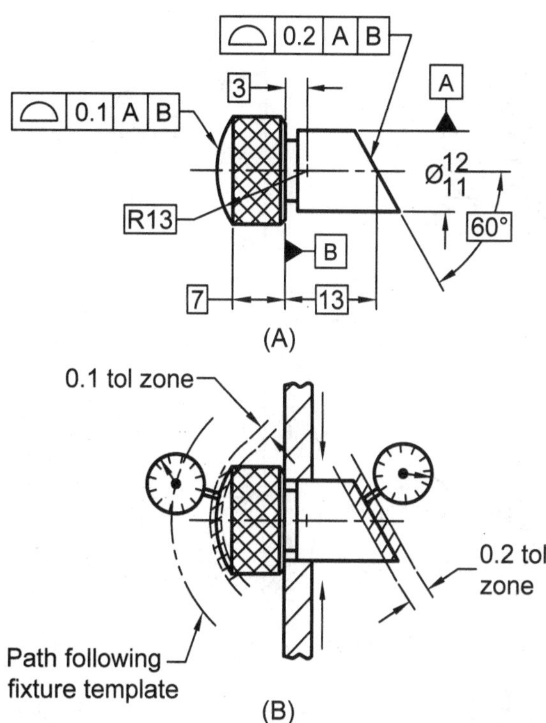

Figure GDT 68: Profile of a surface can be used to control spherical radius surfaces and planar surfaces.

Profile of a Surface on a Conical Surface:

The router bit collet in Figure GDT 69A shows the cone shaped portion of the collet with a profile of a surface callout. The tolerance zone for the cone-shaped surface is two concentric cones separated by the tolerance value of 0.08mm. Figure GDT 69B shows the cone-shaped tolerance zone in cross section.

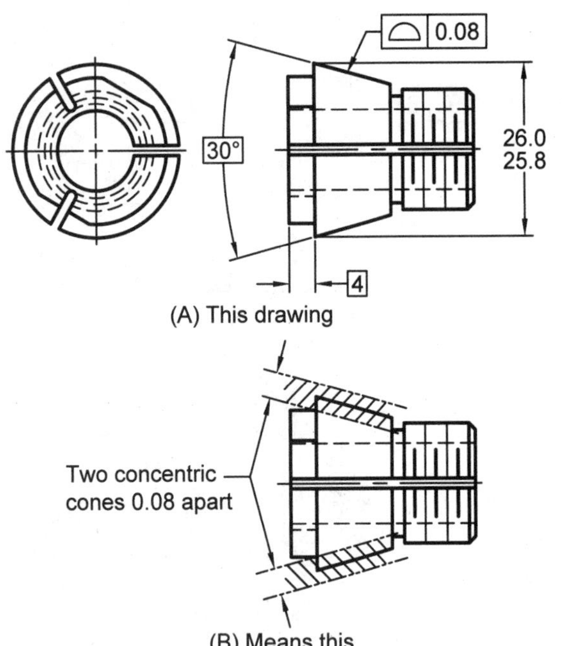

Figure GDT 69: Profile of a surface on a conical shaped router bit collet has a tolerance zone of two concentric cones.

Profile All Around:

Profile of a surface can be applied all the way around a surface by simply adding an all around circle at the elbow of the feature control frame's leader. Figure GDT 69 illustrates the use of the all around circular symbol to modify the entire interior surface of the bridge slot of the guitar string/bridge template.

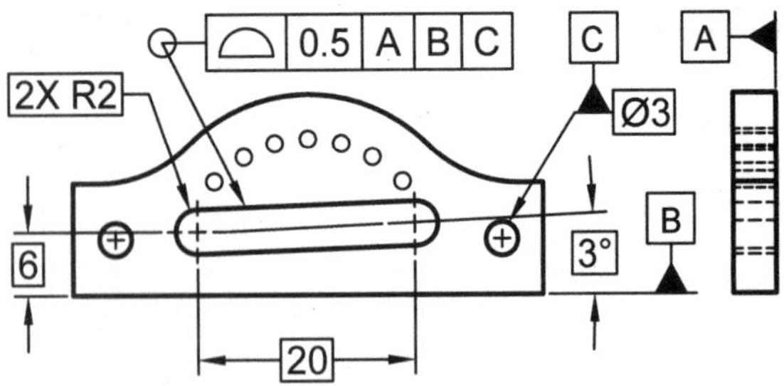

Figure GDT 70: The all around circle is added to the elbow of the profile of a surface feature control frame's leader so the entire interior of the angled slot is controlled.

Limits of Application:

When a profile of a line or profile of a surface feature control frame is directed toward a surface, the tolerance applies until there is an abrupt change in direction. When it is unclear where to apply the tolerance, the limits of the tolerance must be specified on the drawing using the "between" modifier. Figure GDT 71 illustrates a king from a chess set with a profile of a line tolerance applied between points X and Y. The between modifier is positioned below the feature control frame and consists of a double headed arrow with the limiting letters on either side as illustrated in Figure GDT 71.

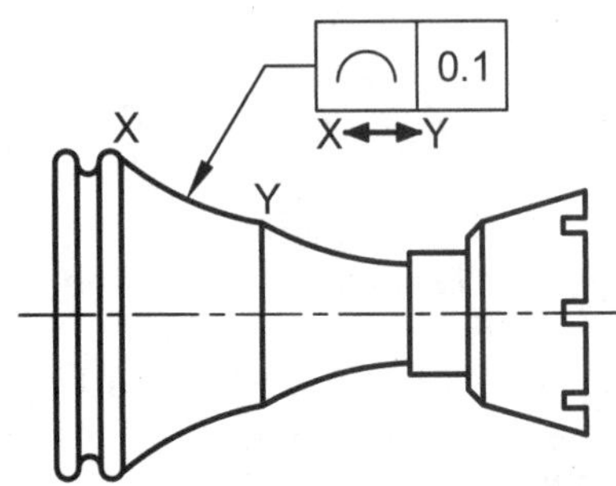

Figure GDT 71: The between symbol is placed below the feature control frame while points are labeled on the drawing to specify exactly where the inspection of the profile is to start and stop.

Positional Tolerance:

Positional tolerance uses a feature control frame to specify the tolerance zone required to accurately locate a feature or group of features such as center point(s), center axis, or the center of a plane. Positional tolerance modifies only features of size so it is understood that the positional tolerance applies regardless of feature size (RFS) unless the modifier MMC or LMC is in the feature control frame. A single feature control frame can modify a single feature such as the single hole in Figure GDT 72 or a group of features such as the cluster of related holes illustrated in Figure GDT 73. A positional tolerance zone is referenced to true position, which is the theoretical exact location of a feature. A feature's true position is located by using basic dimensions and reference datums.

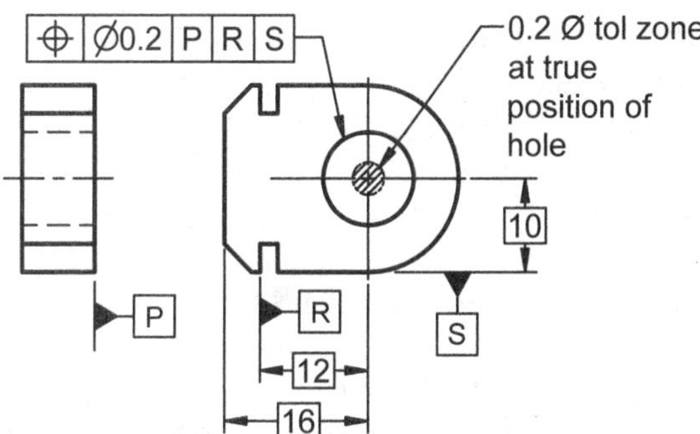

Figure GDT 72: The true position of the hole is given with basic dimensions and referenced to datums. The center of the hole can be anywhere inside the ∅0.2 tolerance zone.

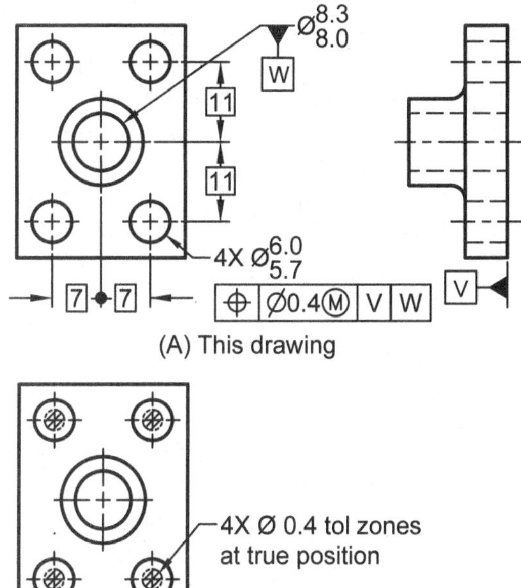

Figure GDT 73: A) four small holes are located as a group with the single positional tolerance feature control frame; B) a ∅0.4mm tolerance zone at MMC is at the true position center of each hole.

The shape of a positional tolerance zone is not simply a circle, but it is also a cylinder. The three holes in Figure GDT 74 illustrate the positional tolerance zones of the cylinders with heights equal to the thickness of the part and a diameter equal to the value of the positional tolerance. Three conditions are illustrated: A) the hole's center axis is exactly aligned with the true position axis; B) the hole's center axis is at the extreme edge of the positional tolerance zone (extreme position variation); and C) the hole's center axis is at the maximum attitude yet still within this cylindrical tolerance zone (maximum attitude variation).

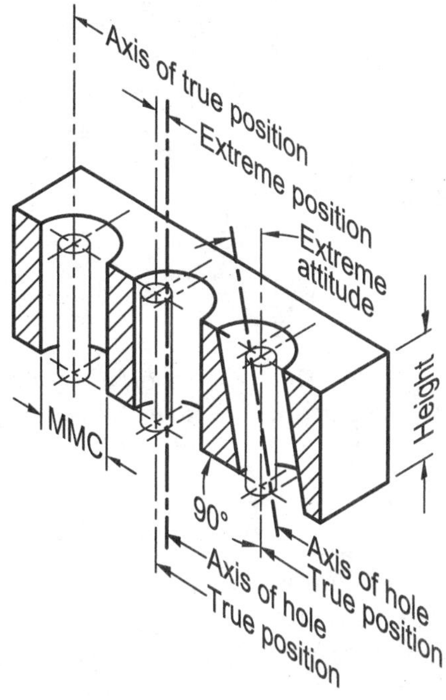

Figure GDT 74: The central axis of the hole is: A) at true position, B) at extreme position, and C) at extreme attitude.

Positional Tolerance at MMC:

Since positional tolerance modifies only features of size, the use of the MMC or LMC symbol is applicable. Figure GDT 75 illustrates a part with a hole located at the center of a "T" slot. The positional tolerance of ∅0.4mm is applied at MMC. As the whole moves away from MMC, or increases in size, the actual positional tolerance value increases proportionally as illustrated in the table in Figure GDT 75. The increase in the positional tolerance as the hole moves away from MMC is called *bonus tolerance.*

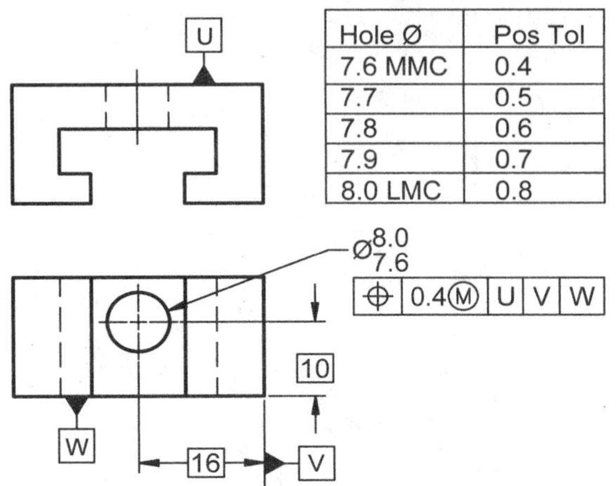

Hole Ø	Pos Tol
7.6 MMC	0.4
7.7	0.5
7.8	0.6
7.9	0.7
8.0 LMC	0.8

Figure GDT 75: The table shows the positional tolerance increasing as the hole moves away from MMC (increases in size).

Composite Positional Tolerance:

The composite positional tolerance feature control frame illustrated in Figure GDT 76 uses a single positional tolerance symbol for two rows of positional tolerance. Composite positional tolerance is used for a pattern of features such as the holes in Figure GDT 77. Composite positional tolerance is applied when the location of the pattern is less restrictive than the hole-to-hole requirements. The top line of the composite tolerance control frame is the *pattern locating tolerance* and the bottom line is the *feature relating tolerance.*

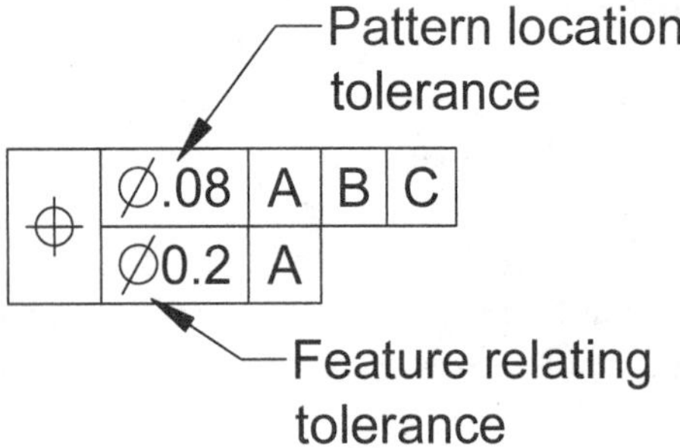

Figure GDT 76: A composite positional tolerance feature control frame is used when the location of a pattern of features is less restrictive than the hole-to-hole requirements.

Figure GDT 77A illustrates a pattern of four holes modified by a composite positional feature control frame. The top line of the composite positional feature control frame specifies a pattern locating tolerance of Ø0.4mm applied at MMC with respect to datums A, B, and C. Figure GDT 77 illustrates the pattern locating tolerance zone using lightly hatched circles. The pattern locating tolerance is located at true position using basic dimensions referenced to the specified datums. The bottom line of the composite positional feature control frame specifies the feature relating tolerance as 0.2mm applied at MMC with respect to datum A. Figure GDT 77 illustrates the feature relating tolerance using dark hatched circles. Datum A in the second line is not used as a reference for dimensioning purposes but simply requires the orientation of the feature relating tolerance zone to be perpendicular to datum A. Remember that positional tolerance is cylindrical in shape. Figure GDT 77B illustrates the pattern of four holes in an alternate yet acceptable location. The pattern of the center axes of the holes can float around within the feature relating tolerance zone which in turn can float inside the pattern locating tolerance zone. The center axes pattern must simultaneously meet both the requirements of the feature relating tolerance zone and the pattern locating tolerance zone.

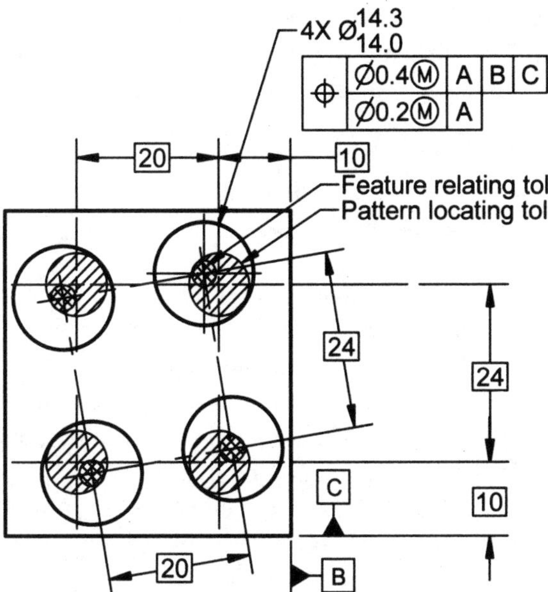

(A) Visualizing a composite positional tolerance (Datum A is the back surface)

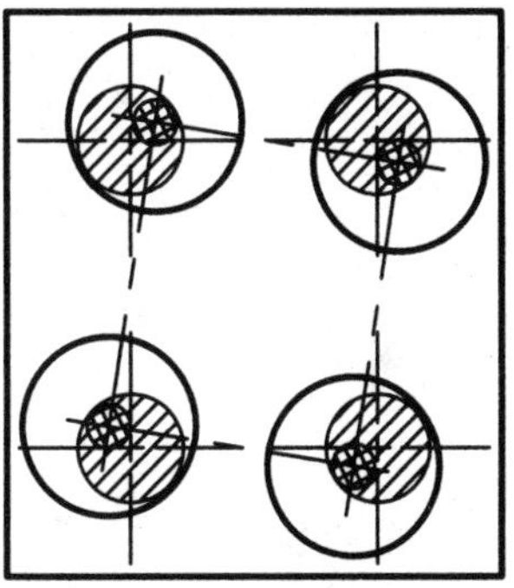

(B) Alternate yet acceptable location of the set of holes

Figure GDT 77: The top row of a composite tolerance pattern is the pattern location tolerance and the bottom row is the feature relating tolerance. The center axes of the holes must meet the pattern locating and the feature relating tolerances simultaneously.

In Figure GDT 78, part of the feature relating tolerance zone falls outside the pattern locating tolerance zone. This is acceptable as long as the central axis of the pattern of circles falls within the feature relating and pattern locating tolerance zones simultaneously. The center axis of the hole in Figure GDT 78 falls exactly on the perimeter of the pattern locating tolerance zone. This means that the central axis must be vertical, otherwise the axis would not stay within this cylindrical pattern tolerance zone.

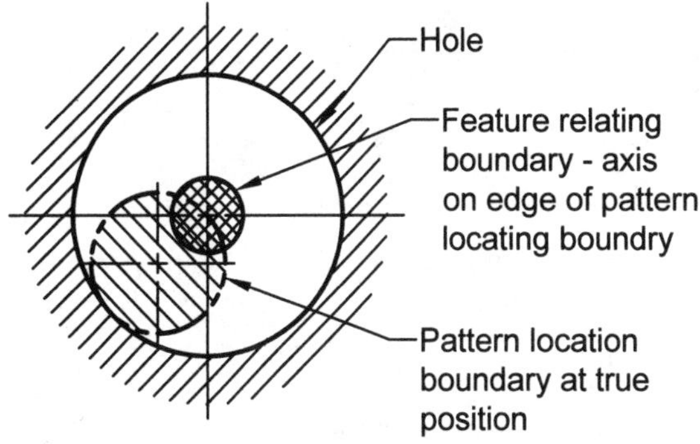

Figure GDT 78: A portion of the feature relating tolerance zone can spill over beyond the pattern tolerance zone as long as the axis of the hole falls within both tolerance zones simultaneously.

Datums in the Second Line of a Composite Positional Tolerance:

Datums in the second line of a composite tolerance are not used for location requirements of the feature relating tolerance. Figure GDT 79 illustrates three conditions of how datums are used in the second line of a composite tolerance. Figure GDT 79A illustrates a composite tolerance with no datum specified in the feature relating tolerance row. With no datum specified in the second row, the central axes of the hole pattern have no orientation requirements. Figure GDT 79B illustrates the use of a primary datum in the second line of the composite tolerance. With a primary datum specified in the second row, the orientation of the pattern's central axes must be perpendicular to the specified primary datum. Figure GDT 79C illustrates the use of both the primary and secondary datum. With two datums specified in the second row, a central axes of the hole pattern must be perpendicular to the primary datum and parallel to the secondary datum.

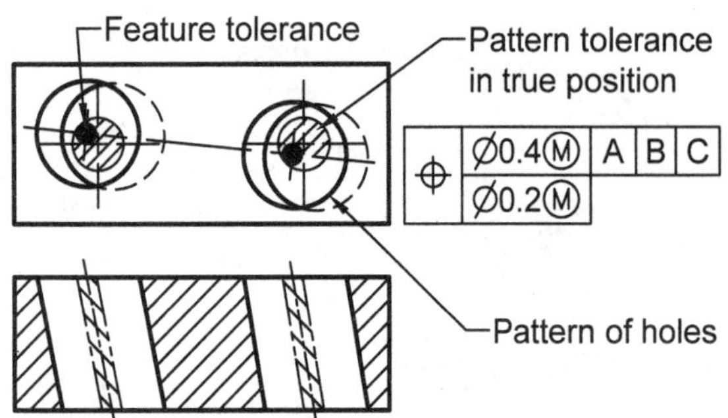

(A) No datum reference - no orientation requirements

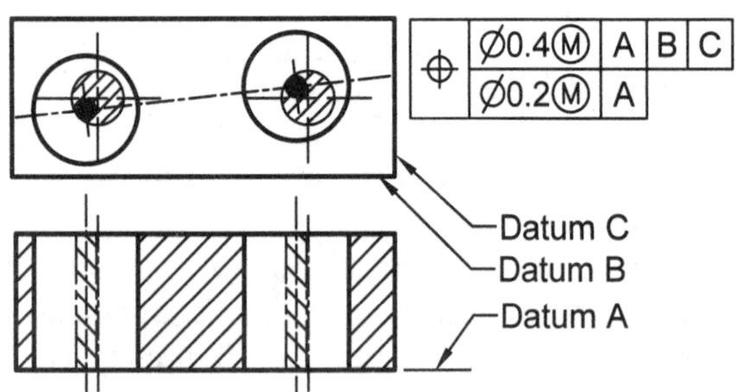

(B) Primary datum A referenced: axes ⊥ to datum A

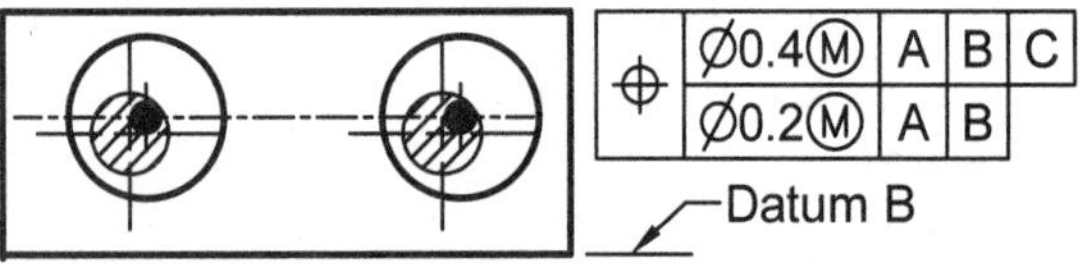

(C) Primary A and secondary B datums referenced: axes of holes ⊥ to A and // to B

Figure GDT 79: The use of datum references affects the orientation of the pattern's axes but does not imply location requirements.

Virtual Condition:

Design requirements determine the size and geometric tolerance requirements of parts to be assembled. For example, in Figure GDT 80 the 2 Bevel Gear must freely slide onto the 1 Base & Stud. The size of the cylinder on 1 Base & Stud can vary between Ø8.0mm and Ø7.6mm and it has a perpendicular geometric modifier of Ø0.4mm at MMC. What is the minimum diameter hole in 2 Bevel Gear necessary to allow it to always slide onto the 1 Base & Stud? The answer is the GD&T concept called *virtual condition (VC)*. VC can be visualized as an envelope of perfect form whose size is determined by a combination of a feature's MMC plus or minus the applied value of the geometric tolerance. The virtual condition of the 2 Bevel Gear is calculated by taking the Ø8.0mm MMC of the shaft of 1 Base & Stud and adding its perpendicular geometric tolerance Ø0.4mm.

VC = MMC + geometric tolerance
VC = Ø8.0mm + Ø0.4mm
VC = Ø8.4mm

The VC of Ø8.4mm is the minimum diameter for the hole in the 2 Bevel Gear. This example calls for the hole to have a tolerance of Ø0.2mm resulting in a hole with a size of Ø8.4mm to Ø8.6mm. No part of 2 Bevel Gear can intrude into the Ø8.4mm virtual condition cylindrical envelope. As the hole in 2 Bevel Gear moves away from the MMC of Ø8.4mm toward LMC of Ø8.6mm it will still slide onto 1 Base & Stud. If the shaft of 1 Base & Stud moves away from MMC of Ø8.0mm toward LMC of Ø7.6mm, 2 Bevel Gear will still slide onto the shaft.

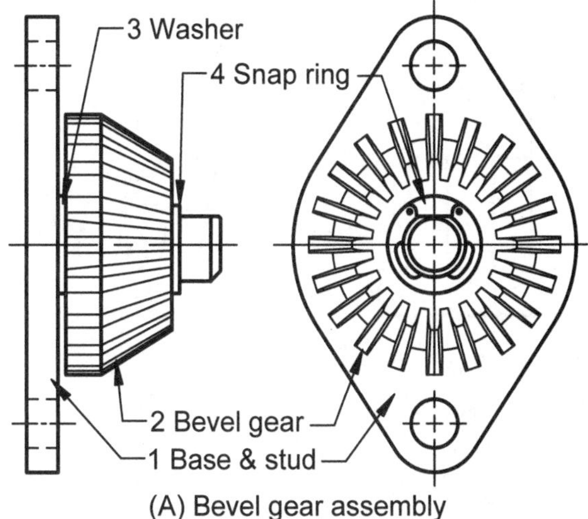

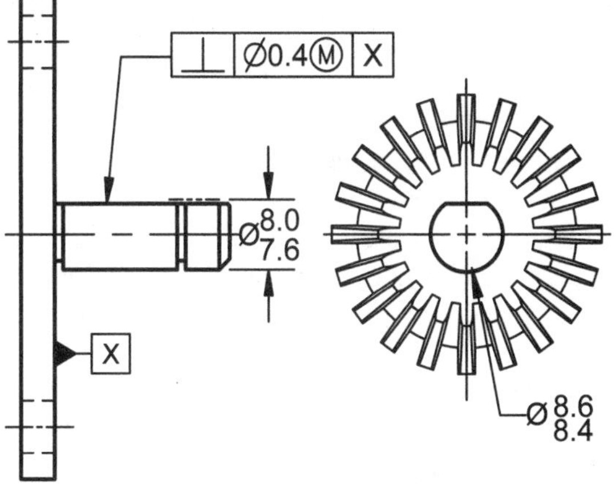

(B) 1 Base & stud (C) 3 Bevel gear

VC = Virtual condition
VC = MMC of 1 Base & Shaft + geometric tol
VC = Ø8.0 + Ø0.4
VC = Ø8.4 Minimum hole size for 3 Bevel Gear

Figure GDT 80: Bevel gear assembly: A) four part assembly; B) ∅8.0 is the shaft's MMC; C) ∅8.4 is the hole's VC

Concentricity:

Concentricity is very difficult to verify and should be used only when controlling the axis is critical. Other options such as positional tolerance or runout should be considered. Concentricity controls the center axes of features formed around a center axis. The center axis of the modified feature must align with the center axis of a datum axis lying within a cylindrical tolerance zone whose diameter is equal to the concentricity tolerance as called out in the feature control frame. Figure GDT 81A illustrates a smaller cylinder being called out as concentric to the cylindrical datum A. Figure GDT 81B illustrates the smaller cylinder with a slight attitude but still within the ∅0.4 specified cylindrical tolerance zone. Figure GDT 80C illustrates the smaller cylinder as not concentric to datum axis A, as evidenced in the right view, but still with the ∅0.4 specified cylindrical tolerance zone.

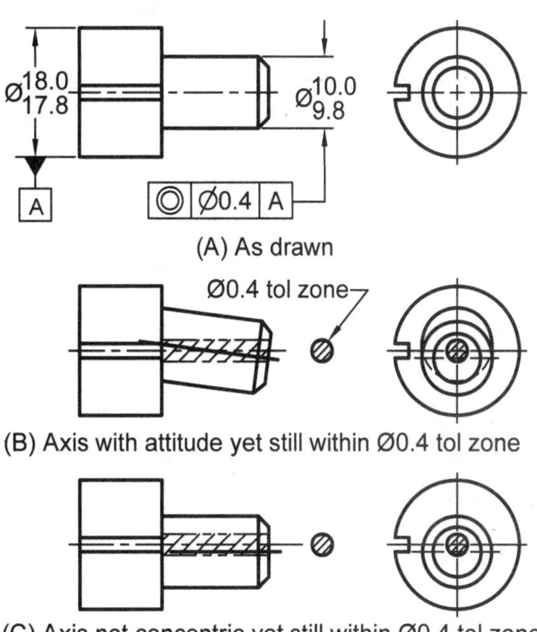

Figure GDT 81: A) step cylinder with a concentricity geometric tolerance callout; B) and C) the central axis of the smaller shaft must stay within the cylindrical concentric tolerance zone of ∅0.4mm.

Coaxial Control Using Positional Tolerance:

Coaxial features can be controlled using positional tolerance when the assembly of parts is the major concern. Figure GDT 82A illustrates a step cylinder with a positional tolerance callout on the smaller cylinder while datum A is assigned to the larger cylinder. The tolerance zone is a ∅0.4mm cylinder, the length of the smaller cylinder. Figure GDT 82B illustrates the sectional view of a functional gauge used to test the part for compliance. The diameter of the larger cylinder in the functional gauge is equal to the MMC of the larger cylinder or ∅18.0mm. The diameter of the smaller cylinder is equal to the VC = MMC of the smaller cylinder plus the geometric tolerance or ∅10.0mm + ∅0.4mm = ∅10.4mm. In order for the step cylinder to fit into the functional gauge the smaller cylindrical feature must be coaxial within specification.

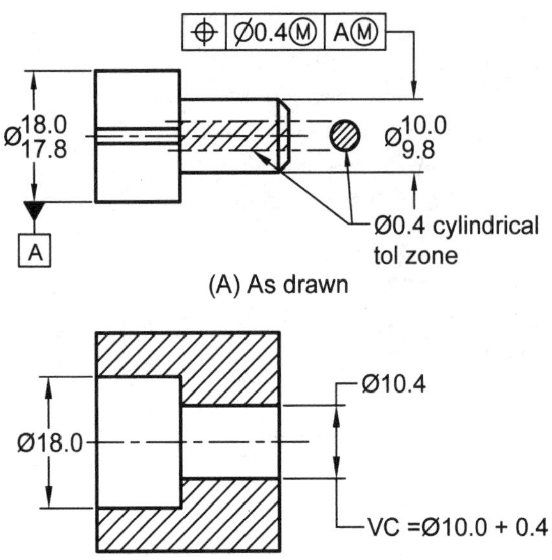

Figure GDT 82: Coaxial features can be controlled using positional tolerance. A) the tolerance zone is a Ø0.4 cylinder; B) a functional gauge simulates the mating part.

Symmetry:

Symmetry controls identical features of size that are located on each side of a central plane and are equally distant from the central plane. There are two methods for controlling symmetry: 1) positional tolerance, and 2) symmetry tolerance. The use of the symmetry geometric characteristic modifier is illustrated in the feature control frame of Figure GDT 83A along with the symmetry numeric tolerance value, and a datum reference. The feature control frame modifies the upright rails 16.0mm wide. The location of the rails in Figure GDT 83B illustrates a 0.2mm wide tolerance zone formed by two parallel planes that are parallel to the central plane of datum feature D and runs the entire length of the rail feature. The central plane of the rail feature must lie within this three dimensional tolerance zone.

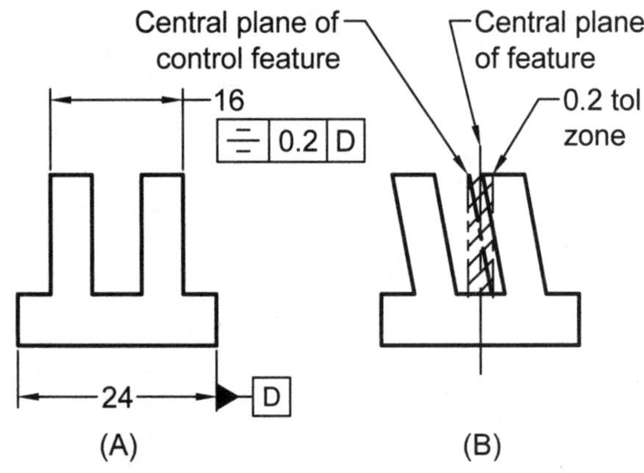

Figure GDT 82: A) the feature control frame calls for the upright rails to be symmetric about the central plane of the datum feature D; B) the central plane of the rail has a slight attitude yet is still within the 0.2 tolerance zone.

Projected Tolerance Zone:

A projected tolerance zone can be designated using two different methods. In Figure GDT 84A the feature control frame contains a circled "P" and the value for the height of the projected tolerance zone is placed after the positional tolerance. In Figure GDT 84B a circled "P" is placed in the feature control frame after the positional tolerance and a phantom line and dimension are placed on the drawing by the hole. The advantage of the second method is that there can be no misinterpretation as to which side the projection tolerance zone applies.

A projected tolerance zone is used in conjunction with threaded fasteners and press fit parts. The projected tolerance zone lies completely outside the actual part as illustrated in Figure GDT 84C. If the centerline of the bolt in Figure GDT 84C is not exactly perpendicular to datum T, as illustrated, then the bolt could interfere with a mating part. The mating part has a clearance hole which could be increased to accommodate the bolt but this has limits. The projected tolerance zone controls the attitude of the bolt. The diameter of the projected tolerance zone and the height of the projected tolerance zone determine the bolt's attitude. The smaller the diameter and/or the taller the projected tolerance zone, the more upright the bolt must be.

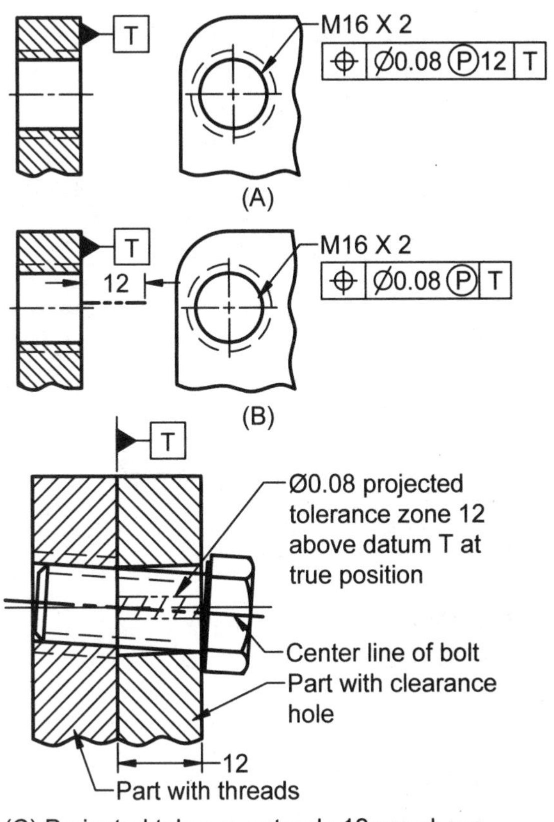

Figure GDT 84: Projected tolerance zone: A and B) methods for calling out a projected tolerance zone; C) assembly showing an application of the projected tolerance zone.

Runout:

Runout controls the amount of surface variation on a part rotating about a datum axis. Two types of surface variation can be controlled: 1) a surface rotating around a datum axis such as the surface of a cylinder or cone, or 2) a rotating surface perpendicular to the datum axis of revolution such as the end surface of a cylinder or cone. Runout controls surface elements so the diameter symbol is not used and RFS is always assumed, meaning MMC and LMC are not used with either the tolerance value or the datum references. Since runout references datums, at least one datum must be specified in the feature control frame.

Circular Runout:

Circular runout controls individual circular elements on a surface. When applied to elements about a central axis, runout controls both the circular and the coaxial features (coaxial means that two features share the same axis). When applied to surface elements at a right angle to an axis, runout controls wobble. Figure GDT 85A illustrates three different methods for calling out runout: 1) with a leader pointing to a surface, 2) placing it below the size diameter, or 3) hanging it from an extension line. Note that in the feature control frame the runout symbol uses a single arrow. In Figure GDT 85B, cylindrical datum feature A is held in a chuck and is rotated about datum axis A. Several individual circularity dial indicator readings are taken at locations along the cylindrical surface. Each full indicator movement (FIM) reading is evaluated separately. The largest reading of the separate dial indicator readings is the amount of circular runout error.

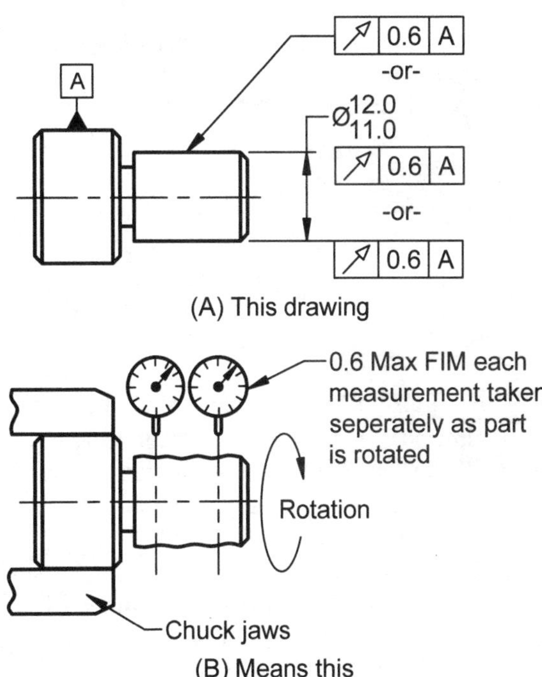

Figure GDT 85: A) three methods for calling out runout on a cylindrical surface, B) each measurement is taken separately with each 360° rotation.

Circular runout control is not limited to cylinders, cones, and spheres. Figure GDT 86A illustrates a chess pawn having three circular runout callouts. One is directed to a cone-shaped surface, a second to a curved surface, and a third to a spherical surface. As illustrated in Figure GDT 86B, each circular dial indicator reading is evaluated independently and each must be measured perpendicular to the part's surface.

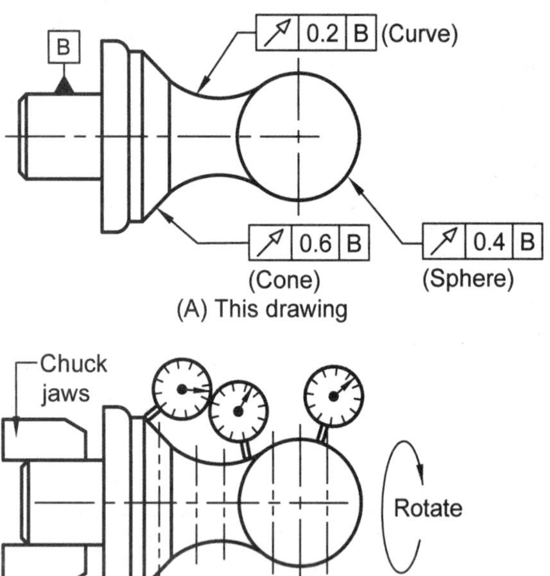

Figure GDT 86: A) profile of a line calling out a curve, cone, and sphere; B) circular runout takes each measurement separately with one 360° rotation while the dial indicator is oriented normal to the surface.

In Figure GDT 87A, the surface perpendicular to datum axis J receives a circular runout callout. Figure GDT 87B illustrates circular runout being measured by rotating the part about datum axis J while taking several individual circular dial indicator readings perpendicular to the flat surface. Runout does not control the entire shape of the flat surface but only each individual FIM reading. Figure GDT 87B illustrates a surface that is not perpendicular to the datum axis and will result in an error referred to as *wobble*.

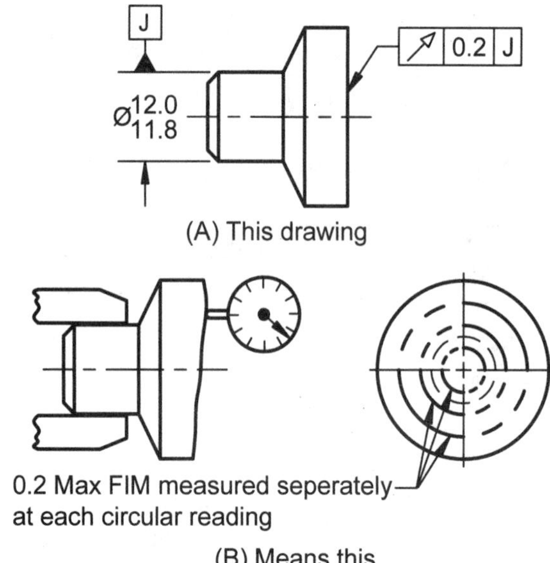

Figure GDT 87: A) profile of a line callout on a flat surface perpendicular to the datum axis J; B) several circular measurements are taken and evaluated for runout individually looking for wobble.

Limited Application:

A runout control applies to the entire area up to any abrupt change in the surface's direction. If only a limited area needs to be controlled, a dimensioned chain line is used to limit the control. Figure GDT 88 illustrates a cylinder with a runout feature control frame whose leader is pointing to a chain line (phantom line). The chain line is dimensioned to show the limits of the feature control frame's application.

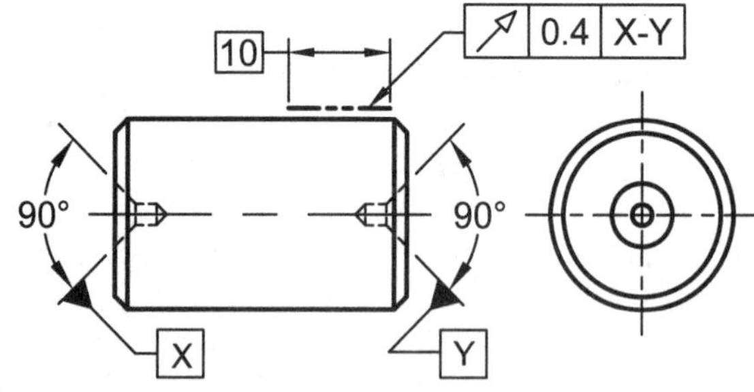

Figure GDT 88: The chain line (phantom line) and dimension designates the limit of the circular runout control.

Total Runout:

Total runout controls the entire surface of a revolving feature not simply circular elements as with circular runout. Total runout can only be applied to cylindrical surfaces rotating about a datum axis or a flat surface perpendicular to the rotating datum axis. For a cylinder rotating about the datum axis, total runout inherently measures circularity, cylindricity, straightness, taper, and surface profile. For a surface perpendicular to the datum axis, total runout inherently measures wobble, perpendicularity, and flatness.

Figure GDT 89A illustrates a cylinder with four feature control frames calling for total runout with respect to compound datum X-Y. Figure GDT 89B illustrates the setup used to measure the total runout of the cylinder. The cylinder is constantly rotating while the dial indicator transverses along the cylinder's top dead center. The difference between the lowest and the highest dial indicator reading is the total runout error.

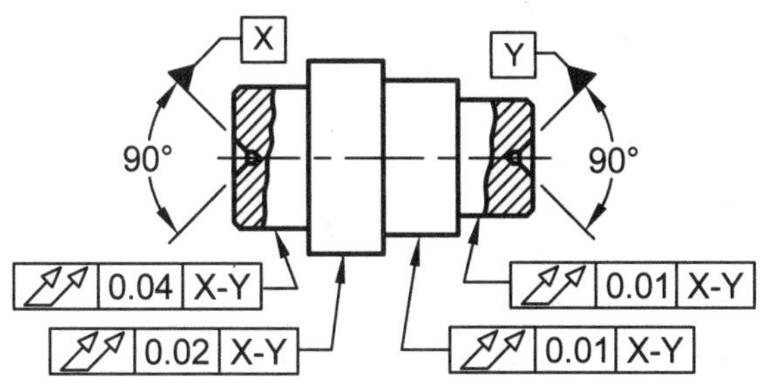

(A) This drawing

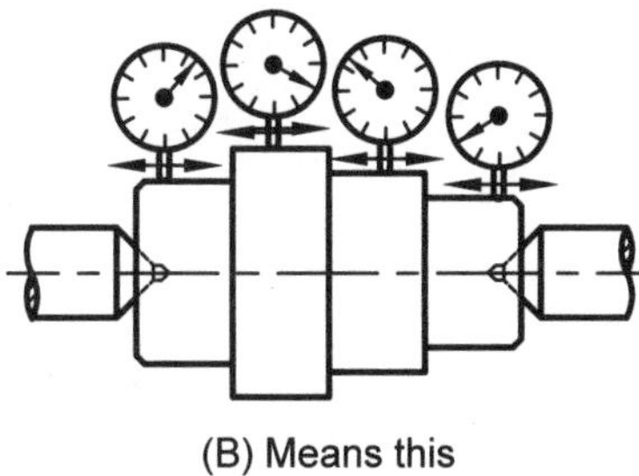

(B) Means this

Figure GDT 89: Total runout on a cylindrical surface is measured by rotating the part while the dial indicator transverses along top dead center.

In Figure GDT 90A, each cylinder has a total runout callout with respect to datums G and H. In Figure GDT 90B, the part is placed in the chuck jaws and lightly tightened, and then the part is tapped until the jaws are pressing against datum feature H. The part is then rotated while the dial indicator is transverse along the top dead center of each cylinder. The difference between the highest and lowest reading is the total runout error.

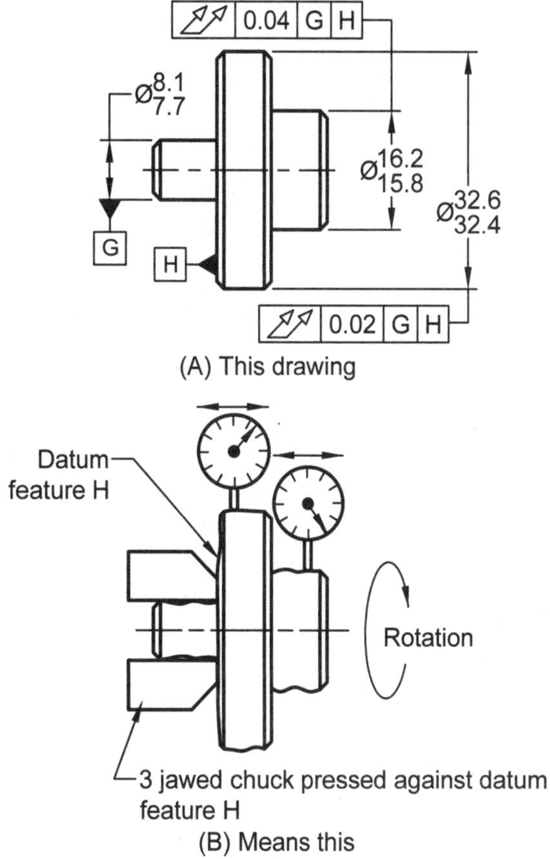

Figure GDT 90: This total runout setup uses two datums. The chuck jaws grip cylindrical datum G and then the part is pressed up against datum surface H.

Measuring total runout on a surface perpendicular to the rotating axis is similar to the way runout was measured in Figure GDT 87, except the dial indicator takes continuous reading as it transverses across the revolving surface perpendicular to the rotating axis. This measures the entire surface, not just individual line elements on the surface.

NOTES:

Chapter 12

Working Drawings

A working drawing, also known as a production drawing, provides the complete size and shape descriptions of a part or group of parts.

Working drawings are used to manufacture parts according to specifications or assemble several parts in the correct sequence.

Detail Working Drawings:

A detail working drawing delineates the exact shape and size of an individual part. The part is fully described using dimensions, notes, finish marks and GD&T symbology. Figure WKD 1 illustrates a detail working drawing of a sliding jaw from a metal vice.

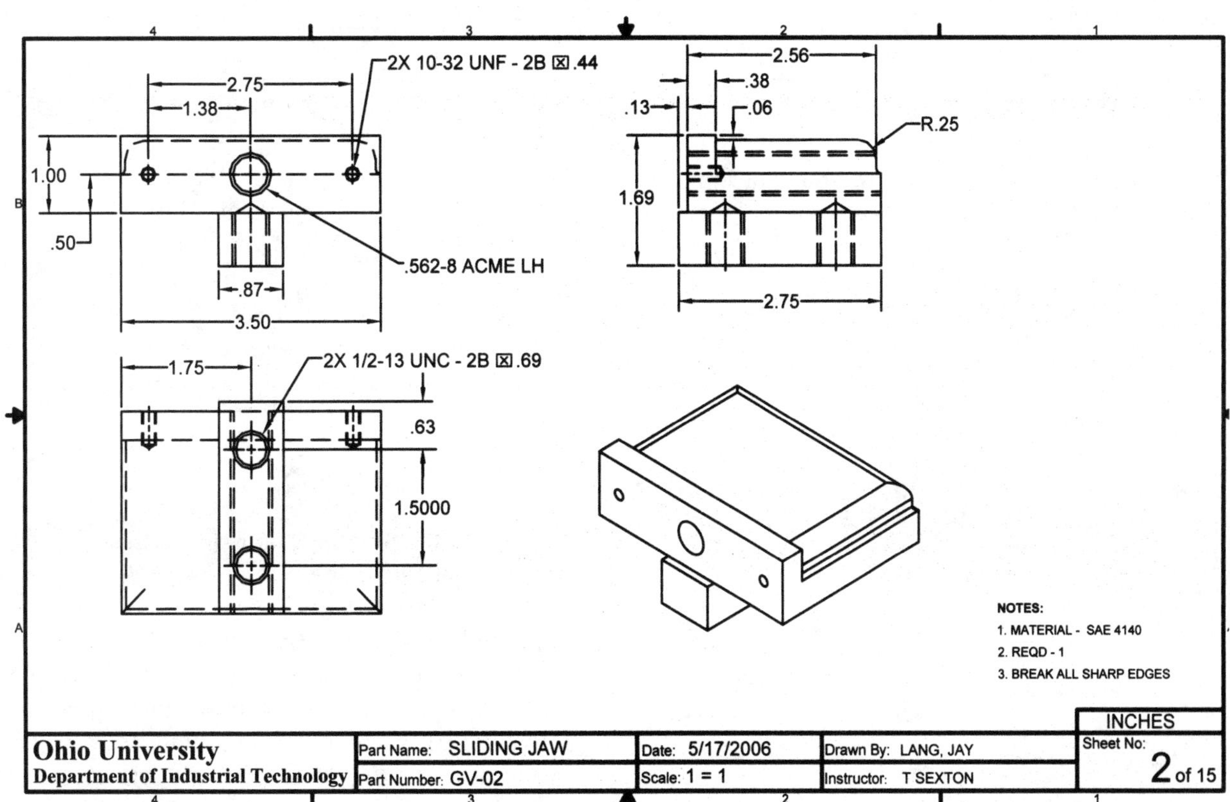

Figure WKD 1: Detailed working drawing of a sliding jaw from a metal grinding vice having the front, top, right, and isometric views. Dimensions and notes have been applied.

Assembly Drawings:

An assembly drawing shows a product completely put together. An assembly drawing is intended to show how parts relate to one another. There are two basic categories of assembly drawings: orthographic and pictorial.

Orthographic assemblies view the assembly from an orthographic viewpoint. Figure WKD 2 illustrates an orthographic assembly of a tub cutter used to cut copper pipe. The orthographic assembly can also be viewed as a section as illustrated by the water controlling gate valve in Figure WKD 3.

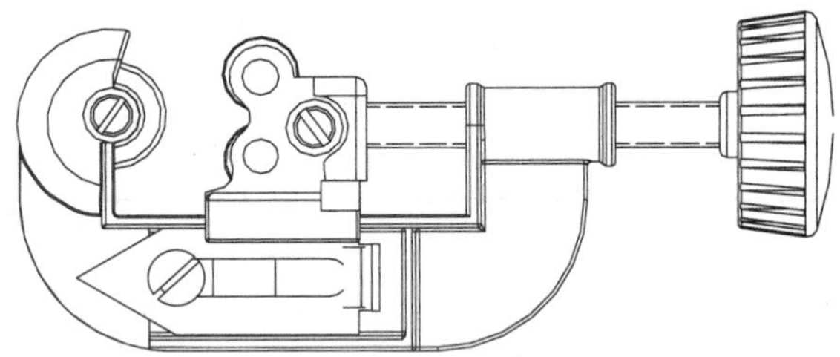

Figure WKD 2: Orthographic assembly drawing of a tube cutter used to cut copper pipe.

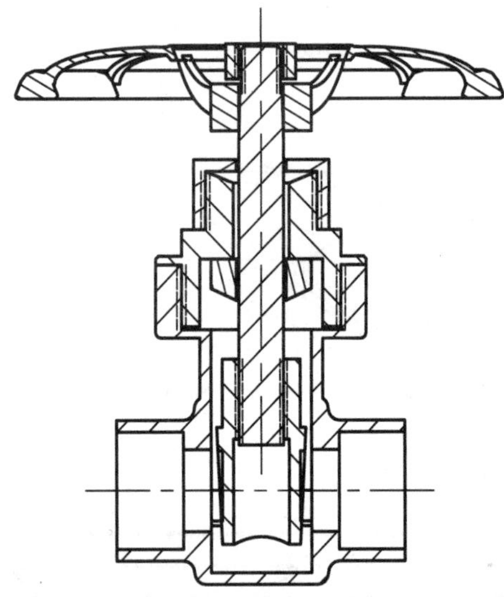

Figure WKD 3: Orthographic sectioned assembly of a water controlling gate valve.

Pictorial assemblies are normally drawn as isometric views. Figure WKD 4 illustrates an isometric pictorial assembly of a small metal bench vice. The pictorial assembly can also be viewed as a section as illustrated by the gate valve in Figure WKD 5. Isometric assemblies are invaluable when trying to explain a product to an audience unfamiliar with reading traditional orthographic drawings.

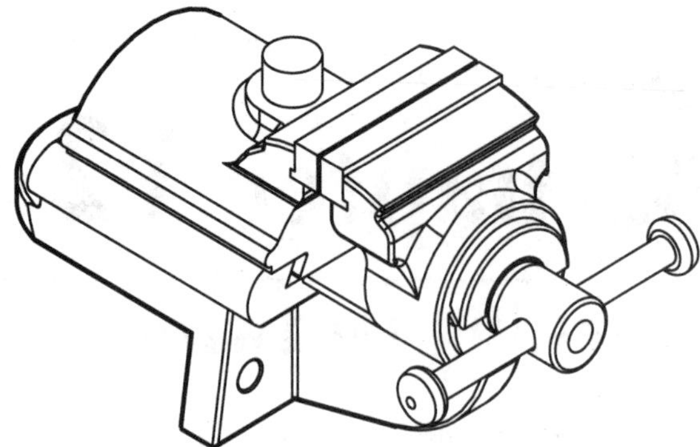

Figure WKD 4: Isometric assembly of a small metal bench vice

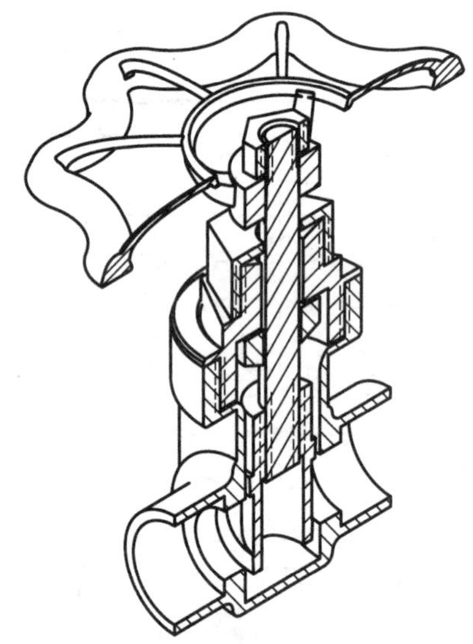

Figure WKD 5: Sectioned isometric assembly drawing of a gate valve.

Exploded Isometric Assemblies:

And exploded isometric assembly drawing has the advantage of showing the relationship between individual parts and at the same time showing each part on its own. Figure WKD 6 illustrates the same metal bench vice from Figure WKD 4 but as an exploded isometric assembly drawing. It helps to envision the exploded assembly as if an air hose is connected to the assembly's main component. When vacuum pressure is turned on, all the parts get sucked back into their functioning positions. When the air pressure is reversed all the parts are pushed or "exploded" along isometric flow lines back into their exploded assembly positions. Example applications of the exploded assembly are:

1. new power tools that come with an exploded assembly drawing so you can locate a bad part and its part number in order to purchase a replacement;

2. when you purchase a part in an automobile parts store, the clerk looks up the part in a catalog filled with exploded isometric assemblies of engines or chasses.

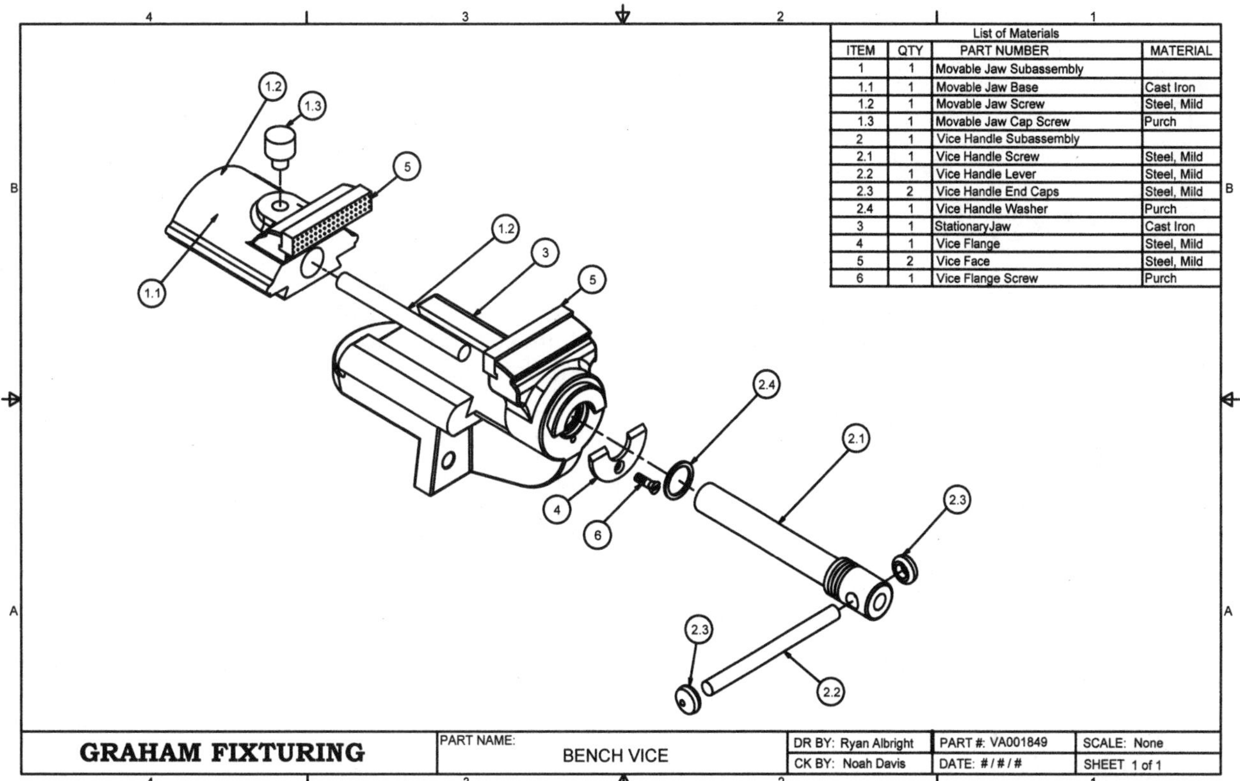

List of Materials			
ITEM	QTY	PART NUMBER	MATERIAL
1	1	Movable Jaw Subassembly	
1.1	1	Movable Jaw Base	Cast Iron
1.2	1	Movable Jaw Screw	Steel, Mild
1.3	1	Movable Jaw Cap Screw	Purch
2	1	Vice Handle Subassembly	
2.1	1	Vice Handle Screw	Steel, Mild
2.2	1	Vice Handle Lever	Steel, Mild
2.3	2	Vice Handle End Caps	Steel, Mild
2.4	1	Vice Handle Washer	Purch
3	1	StationaryJaw	Cast Iron
4	1	Vice Flange	Steel, Mild
5	2	Vice Face	Steel, Mild
6	1	Vice Flange Screw	Purch

Figure WKD 6: Exploded isometric assembly drawing with part numbering balloons and a list of materials.

Flow Lines:

The center lines between the exploded parts in an exploded assembly, called flow lines, help visually organize the drawing. It is important to always keep these flow lines as isometric lines meaning they can only be 30° left, 30° right, or vertical. Figure WKD 7 illustrates the use of isometric flow lines. Note that even when the flow lines change direction they stay isometric lines.

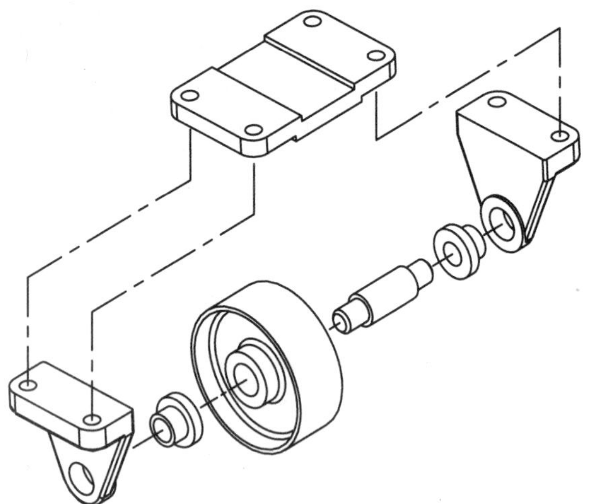

Figure WKD 7: Exploded isometric assembly with flow lines following 30° right and vertical isometric lines.

Balloons:

In order to identify each part in an assembly, the number of the part is placed inside a circle called a "balloon". A leader radiates from the balloon and ends on the part. In Figure WKD 6 balloons identify each part. In Figure WKD 6 two subassemblies are used: the first is the 1 Movable Jaw and the second the 2 Vice Handle. Parts 1.1, 1.2, and 1.3 complete the Movable Jaw's subassembly while 2.2, 2.2, 2.3, and 2.4 completes the Vice Handle subassembly. Balloon circles vary in size from approximately 12mm (.5 in) to 24mm (1in) depending on the length of the numbering system. The leaders are normally straight, but can be curved as illustrated in Figure WKD 8. Leaders can terminate with: 1) the end of the line, 20 a dot, or 3) an arrow as illustrated in Figure WKD 8. Some CAD software automatically terminates the balloon's leader with an arrowhead if it pointing to the edge of the part and a dot if it lands on the surface of the part. A balloon can also be split in half as illustrated in Figure WKD 8. The top half contains the part's number and the bottom half the page number where the part is detailed.

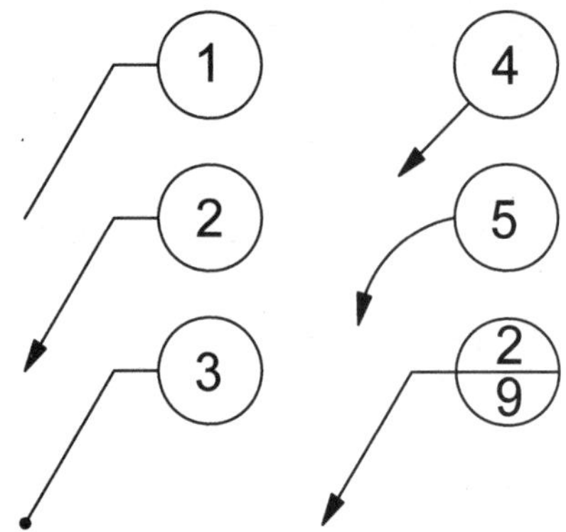

Figure WKD 8: Balloons, leader types, and a slit balloon with the top half representing the part's number and the bottom the page number on which the detail is found

General Sheet Layout:

A drawing sheet normally has the general layout as illustrated in Figure WKD 9. The components are:

1. title block
2. page number
3. list of materials
4. revision block
5. general notes area
6. drawing area
7. zone markings

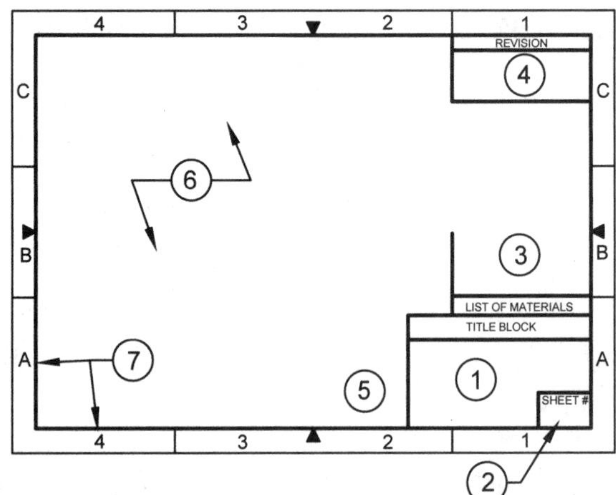

Figure WKD 9: General layout of a drawing sheet with 1) title block, 2) page number, 3) list of materials, 4) revision block, 5) general notes area, 6) drawing area, and 7) zone markings.

Title Blocks:

The title block in Figure WKD 10 contains important organizational information such as the:

- o company name (BLAST FOUNDRY)
- o part name (FORKED YOKE)
- o DWG NO: drawing number (DY00678)
- o SHEET NO: sheet number (1 of 5)
- o SHEET, sheet size (A2)
- o DATE: date (#/#/#)
- o SCALE: scale (1:1)
- o DR: drawn by (RA)
- o CHK: checked by (TS)

The title block can contain additional information but the above list of items are the most important components. The title block is located in the lower right corner of the sheet as illustrated in Figure WKD 9. Note that the sheet number is located in the lower right corner to allow you to find a specific sheet in a set of drawings without needing to flip over entire sheets. In the sheet number box of Figure WKD 9 the 1 is the sheet number and the 5 is the number of sheets in the entire set of drawings. Without specifying the total number of sheets you would never know if a page has been left off.

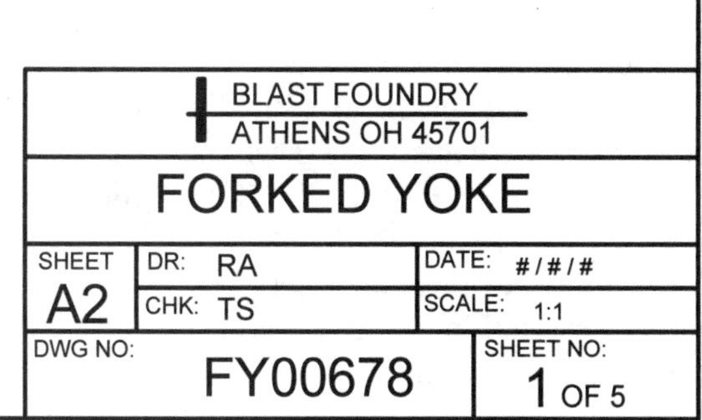

Figure WKD 10: Sample title block containing: company name, part name, drawing number, sheet size, date, scale, drawn by (DR), and checked by (CHK).

Architectural drawings can use a title block similar to Figure WKD 10 but can also run the title block up the right side of the page as illustrated in Figure WKD 11. The sheet number is always placed in the lower right corner and always reads from the bottom of the page. In Figure WKD 11 even the drawing title reads from the bottom of the page which makes it easier to find a specific page.

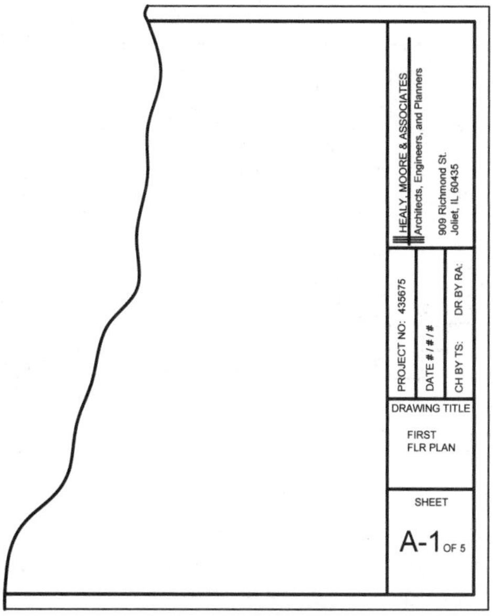

Figure WKD 11: Architectural drawing sheets sometimes use a title block that runs vertically up the right side of the drawing sheet.

List of Materials:

The list of materials, also known as the bill of materials or the parts list, is a table containing every part in the product. The list of materials normally contain: part number, part name, quantity required, material, and possibly notes such as the specification for purchased fasteners.

The largest or most important critical part receives the lowest number and the parts that will be purchased the highest numbers. The number of each part listed in the list of materials corresponds to the balloon identifying the part in an assembly drawing as illustrated in Figure WKD 6. Figure WKD 12 illustrates a list of materials for the bench vice in Figure WKD 6. The list of materials is often placed on the sheet containing an assembly or exploded assembly as illustrated in Figure WKD 6. If the list of materials is positioned in the top right corner, the part numbers are assigned in ascending numerical order reading from top to bottom. If located in the lower right corner above the title block, parts are listed in ascending order reading from bottom to top. If the product is modeled using 3-D parametric modeling software, each part has properties which would include among other things its number, name, quantity, and material. Based on each part's properties, a list of materials can be automatically generated in spreadsheet form and positioned on the drawing sheet.

List of Materials			
ITEM	QTY	PART NUMBER	MATERIAL
1	1	Movable Jaw Subassembly	As Noted
1.1	1	Movable Jaw Base	Cast Iron
1.2	1	Movable Jaw Screw	Steel, Mild
1.3	1	Movable Jaw Cap Screw	Purch
2	1	Vice Handle Subassembly	As Noted
2.1	1	Vice Handle Screw	Steel, Mild
2.2	1	Vice Handle Lever	Steel, Mild
2.3	2	Vice Handle End Caps	Steel, Mild
2.4	1	Vice Handle Washer	Purch
3	1	Stationary Jaw	Cast Iron
4	1	Vice Flange	Steel, Mild
5	2	Vice Face Pads	Steel, Mild
6	1	Vice Flange Screw	Purch

Figure WKD 12: A list of materials for the parts of the bench vice in Figure WKD 6. Each part has its unique number and name along with the quantity required and the material of the part.

If a product is made up of several subassemblies, a master part list can specify the drawing number of each assembly, subassembly, and detail drawing. The parts can be listed in order of size or importance or in the order the parts will be assembled. A system of indentations is used to help visually organize: the major assemblies, subassemblies, detailed drawings of each part, and purchased parts. Figure WRK 13 illustrates such a master parts list.

MASTER PARTS & DRAWING LIST				
NO	DRAW NO	DESCRIPTION	MATERIAL	QNT REQD
0	9100	BENCH VICE ASSEMBLY	AS NOTED	1
1	9110	MOVABLE JAW SUBASSEMBLY	AS NOTED	1
1.1	9120	MOVABLE JAW BASE	CI	1
1.2	9130	MOVABLE JAW SCREW	MILD STL	1
1.3	PURCH	.25-20 UN X .25 LG HEX HEAD CAP SCREW	MILD STL	1
2	9200	VICE HANDLE SUBASSEMBLY	AS NOTED	1
2.1	9210	VICE HANDLE SCREW	MILD STL	1
2.2	9220	VICE HANDLE LEVER	MILD STL	1
2.3	9230	VICE HANDLE END CAPS	MILD STL	1
2.4	PURCH	WASHER .500 X 1.250 X .083	MILD STL	1
3	9210	STATIONARY JAW	MILD STL	1
4	9310	VICE FLANGE	CI	2
5	9410	VICE JAW PADS	MILD STL	1
6	PURCH	5 (.125)-40 X .25 LG FLAT HEAD MACH SCREW	MILD STL	1

Figure WKD 13: A master parts list with an indentation system to help visually organize major assemblies, their subassemblies, individual parts, purchased parts and corresponding drawing numbers.

Revision Block:

The revision block is normally located in the upper right corner of the sheet as illustrated in Figure WKD 9. Figure WKD 14 illustrates a revision block containing the columns:

- REV for revision letter
- description of the revision
- date of revision
- initials of the person the revision is approved by

The revision letter is placed on the drawing at the location of the change and is normally circumscribed by a symbol such as the triangle to make it easier to locate. Figure WKD 15 illustrates the location of two revision letters positioned near to the actual change. The revision block in Figure WKD 14 tracks revisions A and B by noting the previous dimension and the current dimension.

REV	DESCRIPTION	DATE	BY
A	88 WAS 90	# / # / #	TS
B	54 WAS 52	# / # / #	RA
C			

Figure WKD 14: A revision block keeps track of changes made to a drawing by assigning a REV or revision letter and logging a description of the revision, date of the revision, and by whom the revision was approved.

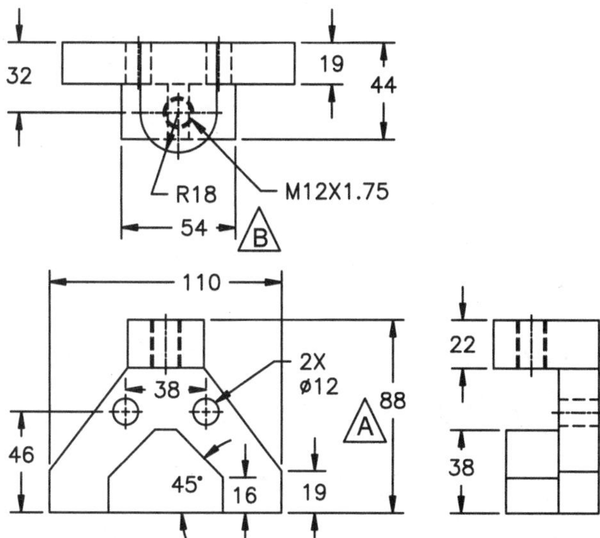

Figure WKD 15: Revisions A and B listed in Figure WKD 14 are noted on the drawing by letters inscribed in triangles.

Architectural drawings use the revision block described above but they also use drawings called "as-built" when renovating existing buildings. Over the life of a building many changes can be made. When an architect is hired to make a major alteration to a building, an updated or as-built drawing is needed. An as-built is developed by taking many field measurements and photographs to assist in the development of drawings.

General Notes Area:

The area to the left of the title block, as illustrated in Figure WKD 9, is used for general notes, for example: linear, angular, chamfer or surface finish tolerances, and the material of the project. General notes can also be a title block and are normally located on the left side of the title block. General notes on architectural drawings are located on one of the first pages in the set of drawings and normally include: a list of abbreviations, a legend of material indications, and the system for locating wall sections.

Zone Markings:

Around the edges of a drawing sheet are letters on the vertical edges and numbers on the horizontal edges. In Figure WKD 16 the shaded area is zone B3. The zone is located by finding the appropriate letter on the vertical edge and the appropriate number on the horizontal edge and visually finding their intersecting point. Zones make it easier to locate desired details on a sheet. For example, if you are working on a detail with someone on the phone, zones help you direct the person to a specific area of the sheet.

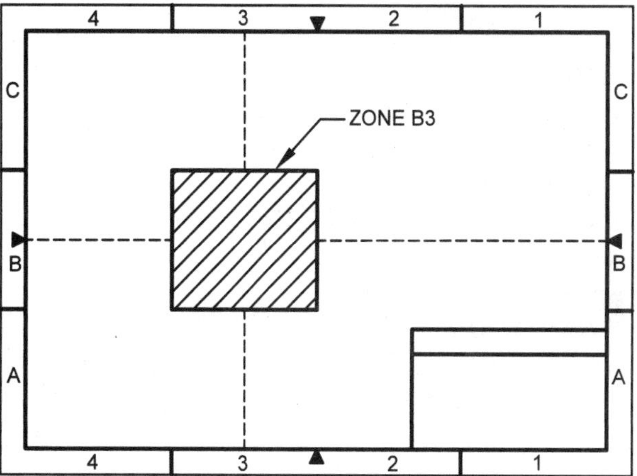

Figure WKD 16: Zones help you locate a detail on the drafting sheet by intersecting a letter on the vertical and number on the horizontal.

Chapter 12 Working Drawings

NOTES:

Chapter 13

Threads

Figure THR 1: Example of threaded fasteners.

A thread is a ridge of uniform profile in the form of a helix that wraps itself around the exterior or interior of a cylinder or cone. The threaded helix can be envisioned as a circular stair. As you climb up the stairs you are moving in both a circular and a vertical direction. Threads can be used for three basic purposes: 1) to fasten objects together such as with a nut and bolt, 2) to make fine adjustments such as focusing binoculars or for leveling a surveyor's transit, and 3) transmitting motion such as when a clutch squeezes the lead screw of a metal lathe forcing the apron, holding the cutting tool, to move along the axis of the lead screw.

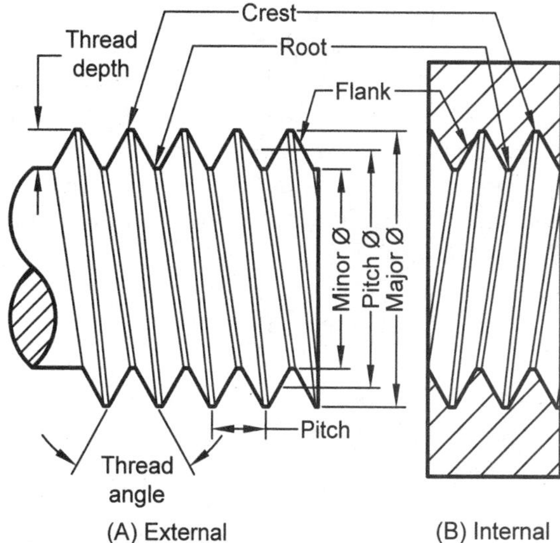

Figure THR 2: Basic thread terminology for A) exterior or B) interior threads.

Thread Terminology:

The understanding of thread terminology is essential to fully understand threads. Figure THR 2 illustrates the definitions.

Major diameter is the diameter of an imaginary cylinder that bounds the crests of the thread for both internal and external threads.

Minor diameter is the diameter of an imaginary cylinder that bounds the roots of the thread for both internal and external threads.

Pitch diameter is the diameter of an imaginary cylinder that passes through the location where the thickness of the thread is equal to the space between two adjacent threads.

Crests are the peak edges of the helical thread formed by two intersecting flanks. They are the portion of the thread farthest from the central axis of the cylinder or cone as illustrated by the external and internal thread in Figure THR 2.

Roots are the bottom valley of the helical thread formed by two adjacent flanks. They are the portion of the thread closest to the central axis of the cylinder or cone as illustrated by the external and internal thread in Figure THR 2.

Thread angle is the included angle measured between two adjacent flanks of the thread that is formed by a cutting, grinding, or a rolling process.

Flank is the surface connecting the crests and the roots.

Pitch is the distance measured from the same point on two adjacent threads such as from crest to crest. The pitch for inch based threads is indirectly noted by specifying the threads per inch. For inch threads the pitch is equal to the reciprocal of the number of threads per inch. The pitch for metric threads is specified in the thread's callout note.

Depth of thread is the distance between the major and minor cylinders measured normal to the threaded cylinder's axis.

Lead is the distance a screw (or nut) will advance axially with one 360° turn.

Thread Profiles:

A thread profile is the outline shape of the thread when it is sliced by a cutting plane passing normal to the threaded cylinder's axis. Each thread profile is designed for a specific function.

Unified profile threads, as illustrated in Figure THR 3, have a thread angle of 60° which is the most common inch based thread profile. This thread form was adopted by the United States, Britain, and Canada (ABC countries) in 1948. The adoption insured interchangeability of parts between these countries in the case of a national emergency. Note that the crests can be either flat or rounded. Unify threads are used for both fasteners and adjustment mechanisms.

Sharp V profile threads, as illustrated in Figure THR 4A, have a thread angle of 60° with crests and roots theoretically sharp. But in practice a very small flat is formed. They are used when additional friction is desired such as with set crews, threads used for adjustment, or on piping transmitting gas or fluid where an air tight seal is necessary.

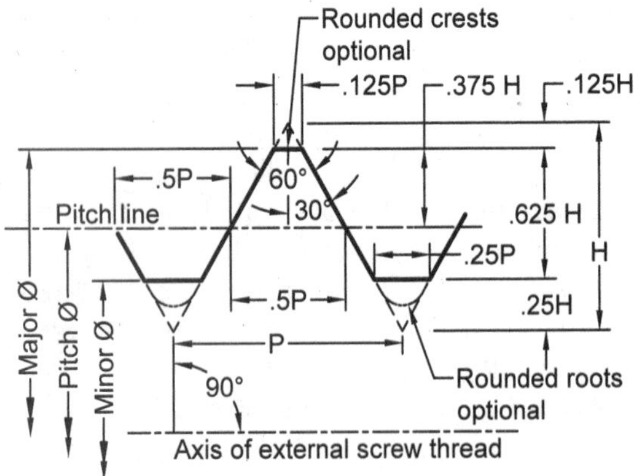

Figure THR 3: Unified thread profile.

Metric M profile threads are illustrated in Figure THR 4B. They have a thread angle of 60°. The crests and roots are shown flat but rounded roots and crest are possible. Both the International Standards Organization (ISO), which governs standards in most countries outside the U.S., and the American National Standards Institute (ANSI), which governs standards in the U.S., are in basic agreement on the M profile thread.

Square profile threads, as illustrated in Figure THR 4C, are designed to transmit power along the axis of the threaded cylinder. For all practical purposes the square thread has been replaced by the acme thread because the square thread is difficult to produce.

Acme profile threads, as illustrated in Figure THR 4D, are designed to transmit power along the axis of the threaded cylinder. Acme threads are used on devices such as wood and metal vices, the scissor jack used when changing an automobile tire, or on the lead screw of a metal lathe which moves a tool support apron along the axis of the lead screw.

Buttress Error profile threads, as illustrated in Figure THR 4E, are designed to resist high stresses along the axis of the threaded cylinder in one direction only.

Knuckle profile threads, as illustrated in Figure THR 4F, are typically formed by a rolling process but can also be cast. Knuckle threads can be found on light bulbs and light sockets.

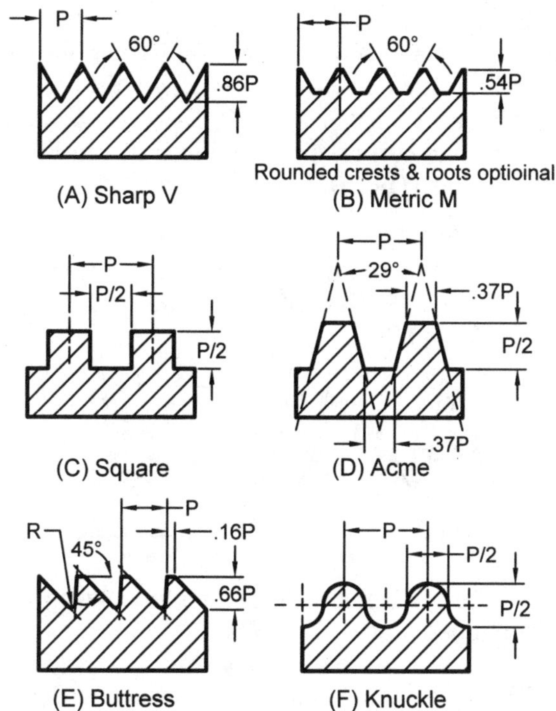

Figure THR 4: Common thread profiles: A) sharp V, B) metric M, C) square, D) acme, D) buttress, F) knuckle.

Unified Thread Series:

For each major diameter in the unified thread system there is a series of three threads: course (UNC), fine (UNF), and extra fine (UNEF). The unified series also includes constant pitch threads: UN4, UN6, UN8, UN12, UN16, UN20, UN28, and UN32. Constant pitch threads keep the threads per inch the same for a variety of different major diameters. When threads are formed by the rolling process, an "R" is added to the thread form abbreviation, e.g., UNRC, UNRF, UNREF, UNR4, UNR6…,. etc. Unified specialty threads (UNS) involve special combinations of major diameters, pitch, and lengths of thread engagement. Finally, unified miniature (UNM) threads involve thread with diameters from .0118 to .0557 inches.

Metric Thread Series:

The metric M thread profile has a series made up of a course and a fine pitch. With simply a course and a fine thread series, the number of thread options is much simpler than the inch base unified thread series. The form of the metric M profile is similar to the inch based unified thread. The metric MJ thread profile series is used for special diameter and pitch combinations designed for the aerospace industry.

Class of Fit:

How precisely a threaded fastener fits into a mating part is determined by the minimum clearance between the two. This minimum clearance is called allowance. For inch based unified threads the allowance is designated by a class of fit number which uses the numbers 1, 2, or 3 in the thread note.

Class 1 is a loose fit which allows easy assembly and disassembly. In this class, fasteners with slightly damaged threads can be used without a problem.

Class 2 is an average fit which is the most common class of fit. Almost everything you purchase in a hardware store is class 2. It is used in mass production.

Class 3 is a tight fit which allows for precision assembly, fine adjustment, and the ability to withstand vibration.

The actual numeric value of the class of fit allowance can be looked up by major diameter and threads per inch in the *Machinery's Handbook*. The *Machinery's Handbook* is a 2500 page resource "bible" covering any topic a machinist encounters.

Right vs. Left Hand Threads:

A right hand thread advances (tightens) when a fastener is turned in a clockwise direction. Because aright hand threads are far more common than left hand threads, thread are assumed to be right hand unless specifically labeled "LH" for left hand. Left hand threads advance (tighten) when a fastener is turned in the counterclockwise direction. Examples of left hand threads can be found on the arbor of a radial arm saw, one side of a turnbuckle, and on the short threaded shaft used to attach a bicycle pedal to one side of the bicycle (it is the pedal stamped with an "L" on the end of the shaft).

Multiple Threads:

If you wrap a rope around a cardboard tube, the helical winding rope illustrates a single thread as illustrated in Figure THR 5A. If you take a white rope and a red rope and twist them around the tube so that there is an alternating white, red, white, red… rope pattern, it illustrates a double thread. The double thread will have two starting positions when viewed from the end of the threaded cylinder as illustrated in Figure THR 5B. A single 360° revolution of a nut on a double thread moves the nut two times the pitch or twice the distance of a single thread. In like manner a triple thread, with three starting points, moves a nut three times the pitch in one 360° rotation. A quadruple thread, with four starting points, moves a nut four times the pitch in one 360° rotation. Double, triple, and quadruple threads are used to transmit quick motion along the axis of the threaded cylinder.

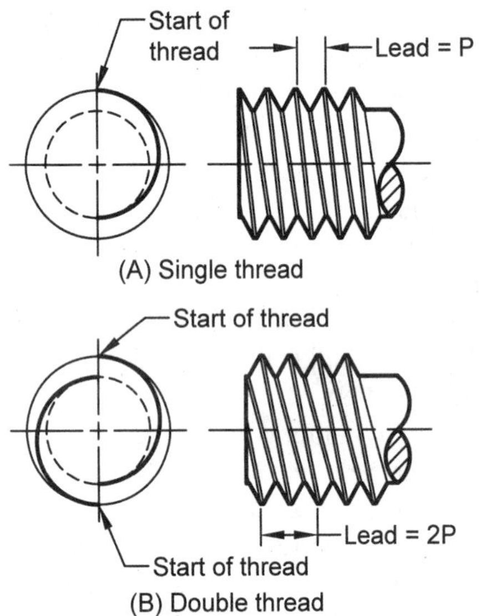

Figure THR 5: A) single thread with one starting position, B) double thread with two starting positions.

Thread Representation

There are three different ways to represent threads on a drawing: detailed, systematic, or simplified. The detailed method of representation most closely represents a given thread profile as illustrated by the detailed drawings of the unified thread profile in Figure THR 6. Note the angle of the exterior threads lean left while interior threads lean right because you are looking at the back side of the thread. However this method is seldom used in industry because it is so time consuming to draw and when modeled in a 3-D computer system it slows the computer down dramatically. However, detailed threads can be very effective when drawing a technical illustration.

The schematic and simplified methods for representing threads are the same when representing internal treads as illustrated in Figure THR 7A and THR 7B. The schematic and simplified methods differ when sectioning through internal threaded holes as illustrated by the sections in Figure THR 7A and THR 7B. External threads in elevation and section differ as illustrated in Figure THR 8A and THR 8B.

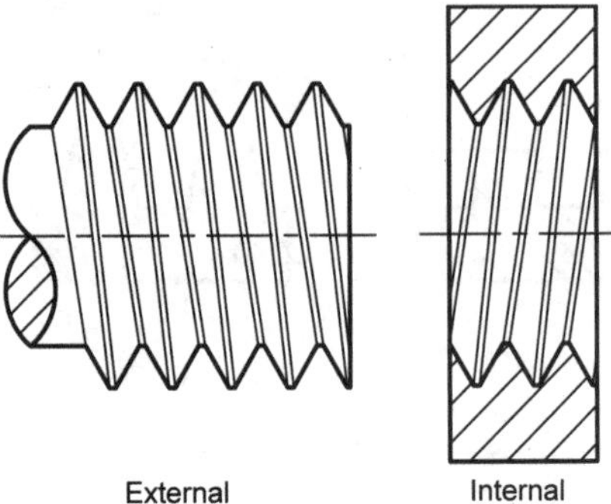

Figure THR 6: Detailed thread representation of a unified thread profile.

The schematic method of representation is not common in the industry but it works very effectively when sketching. This method is characterized by thin lines representing crests and shorter thicker lines representing roots as illustrated by the internal sectional views in Figure THR 7A and the elevation view of the external threads in Figure 8A. The sectional view in Figure 8A shows a few roots and crests but also a sharp "V" profile of the sectional threads.

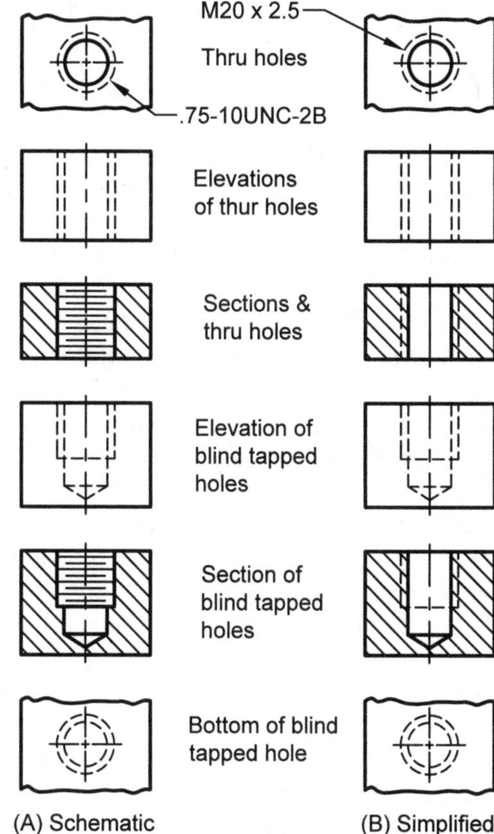

Figure THR 7: Thru hole and blind tapped holes in elevation and section for both A) schematic and B) simplified thread representation. There is a significant overlap between the two methods.

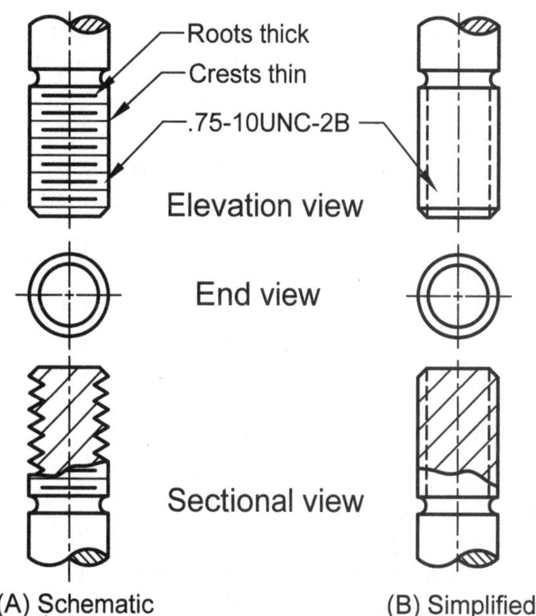

Figure THR 8: External threads in elevation and section for both A) schematic and B) simplified thread representation

The simplified method of representing threads is the most common method used on drawings because it is the easiest to draw. Figure THR 7B illustrates how internal simplified threads are drawn in different views. Figure THR 8B illustrates an external simplified thread in elevation and section. When generating 2-D working drawings from a 3-D computer model, threads are displayed using the simplified method.

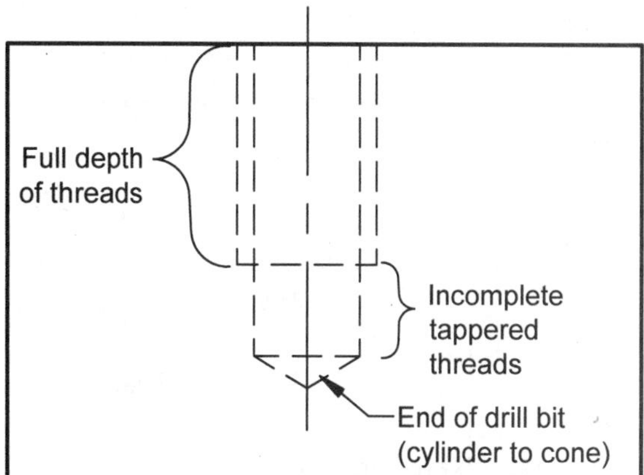

Figure THR 9: When specifying thread depth of a blind tapped hole only the full threads are included.

If a threaded hole does not pass completely through the material, it is called a blind tapped hole as illustrated in Figure THR 9. The threaded hole is first drilled with a tap drill with its size specific to the size of the thread. This accounts for the cylinder to cone shaped transition in Figure THR 9. When specifying the depth of thread on a drawing, only the full threads are included. Unless you use a special bottoming tap, which cuts threads the full length of the drilled hole, the normal thread cutting tap is tapered so incomplete threads will be cut but these incomplete threads are not included in the depth of thread call out as illustrated in Figure THR 9.

Inch Thread Specification:

Whether you use the detailed, schematic, or simplified method to represent threads on a drawing, the thread's specifications must be labeled using a note (or call-out). The minimum thread note for an inch based unified thread is illustrated in Figure THR 10. The note's components are:

- *.25* is the major (nominal) diameter of the screw thread. The major diameter is referred to as nominal because the actual diameter from crest to crest is slightly less than the major diameter because full height thread is seldom used.
- – (dash) is a separator between the major diameter and the threads per inch,
- *20* is the number of threads per inch. In this example the actual pitch is 1/20th of an inch or .05 inches.
- *UN* stands for a unified thread profile.
- *C* stands for a course thread in the unified thread profile series. Each diameter in the unified thread profile has a series made up of a C (course), F (fine) or EF (extra fine) threads per inch.

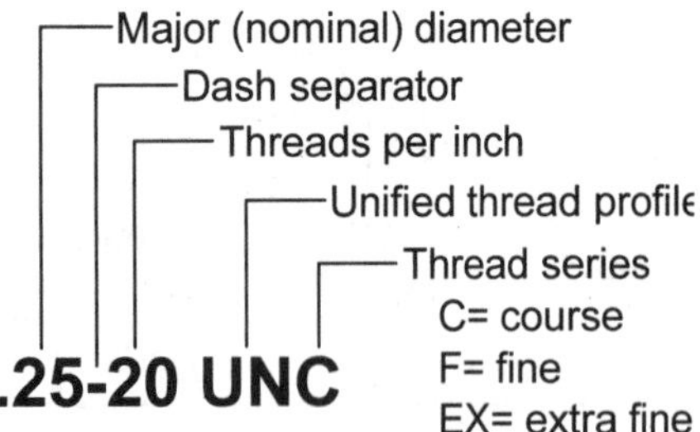

Figure THR 10: Minimum unified thread profile specification.

A more complete inch based thread specification is illustrated in Figure THR 11. The .25 – 20 UNC was explained in Figure THR 10. The additional specification components are:

- *2* describes the class of fit allowance (1 loose, 2 average, which is the default if not specified, and 3 tight).
- *A* stands for exterior (A = exterior, B = interior).
- *LH* stands for left hand thread (thread is assumed to be a right hand thread if LH is not present).
- *DOUBLE* stands for a thread with two starts. Other possibilities are triple or quadruple.
- *X* is a separator
- *2 LG* is the fasteners length in inches.
- *HEX HD* or hexagonal head is the type of head on the fastener.
- *CAP* is for cap screw which is a category of fastener.

All the information in a note may not be required. For example, in Figure THR 11, the A for exterior and 2 LG for length may be evident from the drawing and can therefore be left out of the note. However, a bill of materials or purchase order will require a complete note.

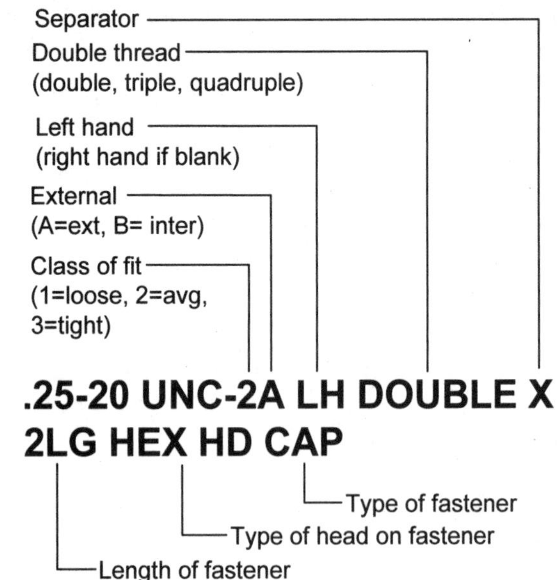

Figure THR 11: Complete unified thread specification.

Metric Thread Specification:

The minimum thread specification for the metric M profile is illustrated in Figure THR 12. The components of the specification are:

- ❑ *M* stands for the M profile series in the metric system.
- ❑ *16* is the threads major (basic) diameter in millimeters.
- ❑ *X* is a separator.
- ❑ *2* is the thread's pitch in millimeters.

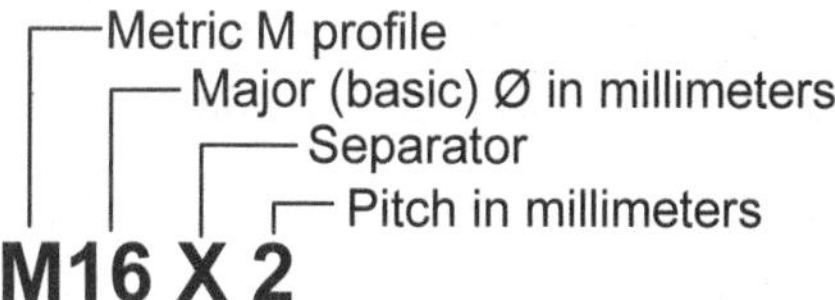

Figure THR 12: Minimum metric M profile thread specification.

The more complete M profile thread specification in Figure THR 13 adds the tolerance class. The components of the tolerance class are

- ❑ *5g* pertains to the thread's pitch diameter
  - ▪ *5* being the tolerance grade and
  - ▪ *g* the tolerance position (allowance).
- ❑ *6h* pertains to the major diameter
  - ▪ *6* being of the tolerance grade and
  - ▪ *h* being the external tolerance position (allowance).

Tolerance grade ranges from tight 3, 4, 5 medium 6 to 7, 8, and 9 loose. Tolerance position (allowance) uses small letters for exterior threads with: e = loose allowance, g = small allowance and h = no allowance. Capital letters are used with interior threads with: G = small allowance and H = no allowance.

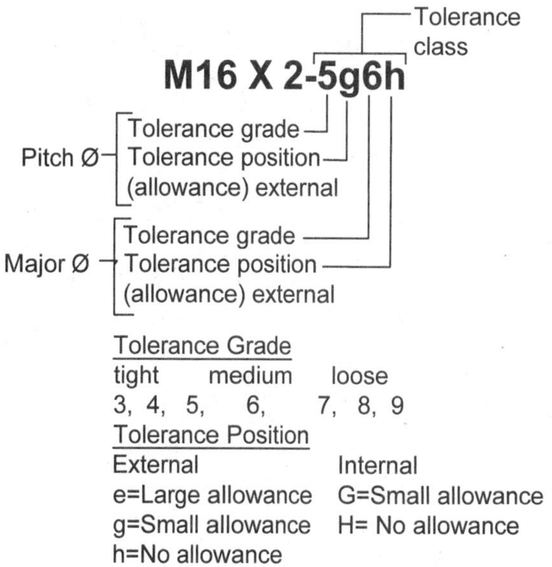

Figure THR 13: A more complete metric M profile thread specification with tolerance class added to the minimum note.

Gage Numbers in Thread Specifications:

For diameters less than .25 inches, some fasteners use a gage number to specify the major diameter. Figure THR 14 illustrates a thread note for a small fillister head machine screw. The major diameter is specified by the NO 10 gage specification. In addition, the actual diameter in inches is displayed in parentheses however this is optional. The thread gage alone, the decimal equivalent of the gage size alone, or both together are acceptable.

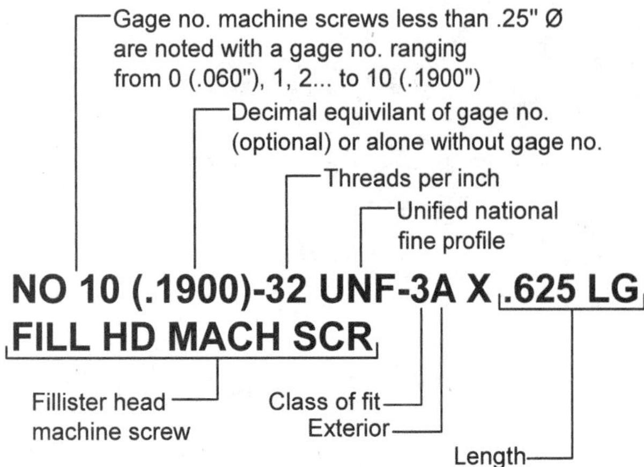

Figure THR 14: Specification using a gage number for the major diameter with its actual diameter in parentheses.

Acme Thread Specification:

An acme thread specification is illustrated in Figure THR 15.

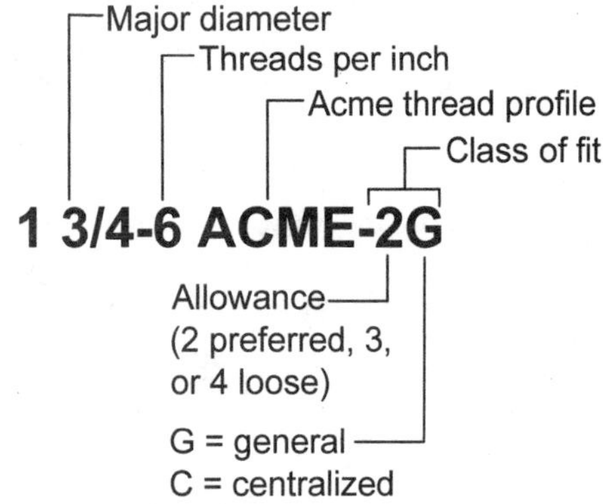

Figure THR 15: Example of an acme thread specification.

National Pipe Thread Specification:

A national pipe thread specification is illustrated in Figure THR 16. Pipe thread can be either tapered or straight. Pipe threads fall into one of four categories:

1. pressure tight joint using a sealant
2. pressure tight joint without the use of a sealant
3. loose fitting mechanical joint without being pressure tight
4. Rigid mechanical joint without being pressure tight.

There are several pipe thread notations. The following list of abbreviations provides the most common thread type but does not list all pipe thread types:

NPT = American National Standard Taper Pipe Thread
NPTR = American National Standard Taper Pipe Thread for Railing Joints
NPSC = American National Standard Straight Pipe Thread for Couplings
NPSM = American National Standard Straight Pipe Thread Free-fitting Mechanical Joints
NPSL = American National Standard Straight Pipe Thread Loose-fitting Mechanical Joints & Locknuts
NPSH = American National Standard Straight Pipe Thread for Hose Couplings
NPTF = Dry Seal USA (American) Standard Pipe Thread

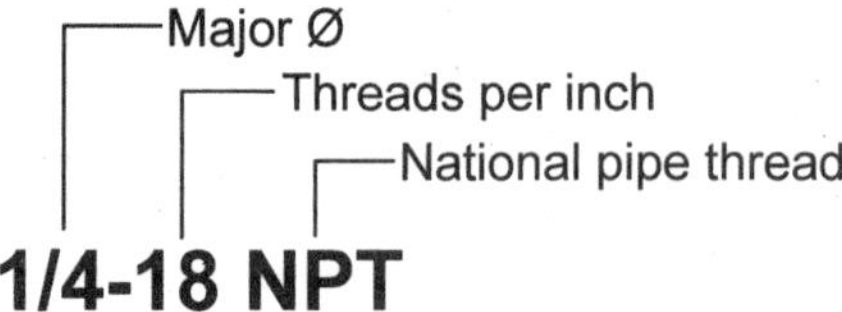

Figure THR 16: Example of a pipe thread specification.

Buttress Thread Specification:

A buttress thread specification is illustrated in Figure THR 17. The use of PULL-BUTT (labeled just BUTT) vs. PUSH-BUTT is determined by whether the external buttress thread is pushing or pulling of the load.

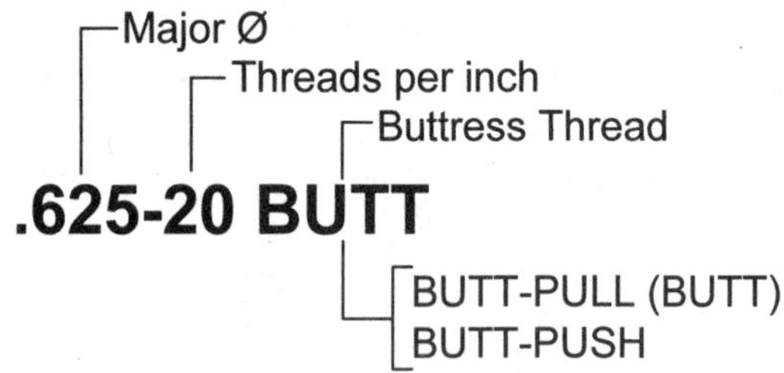

Figure THR 17: Example of a buttress thread specification.

Major Diameter:

The major diameter of a thread is commonly referred to as "nominal" with inches and "basic with metric as it can not actually be measured on the exterior (bolt) or interior (nut) thread. The major diameter is an imaginary cylinder whose diameter is slightly larger than the maximum allowable tolerance limit of the exterior thread and slightly less than the minimum tolerance limit of the interior thread. Figure THR 18 illustrates a meshing exterior and interior thread with the basic major diameter labeled. The minimum and maximum tolerance limits of both the exterior and interior threads are also labeled with the letter "T".

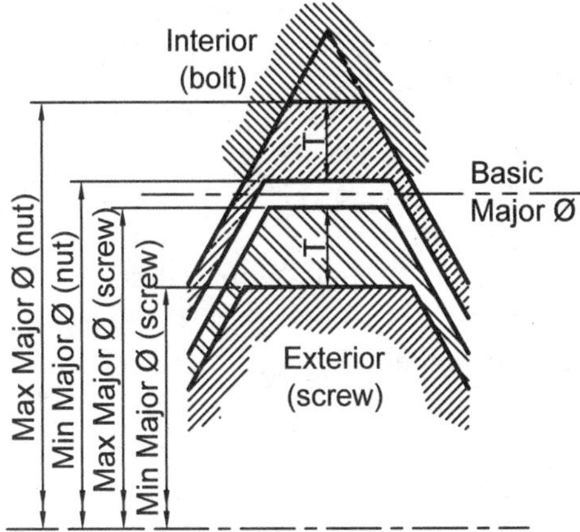

Figure THR 18: The major (basic or nominal) diameter lies between the maximum exterior diameter and the minimum interior diameter.

Forming Threads:

Threads can be formed in a variety of ways. The simplest method is to cut them manually. Figure THR 19A illustrates a thread cutting tap and tap handle used to cut interior threads. Figure THR 19B illustrates a photo of a die. The tap is inserted and rotated into a hole which has been drilled to a diameter specific to the given tap size. For example, to thread a hole for a .50-13 UNC thread, a tap drill with a diameter of .4219 inches is used. This size tap drill results in a thread with a thread height of 75% of full thread height. 75% thread height provides more than adequate strength and are much easier to cut than full height threads. Figure THR 20 illustrates a thread cutting die and die stock to hold the die for cutting exterior threads. The die is lined up with the central axis of the cylinder and turned around the cylinder.

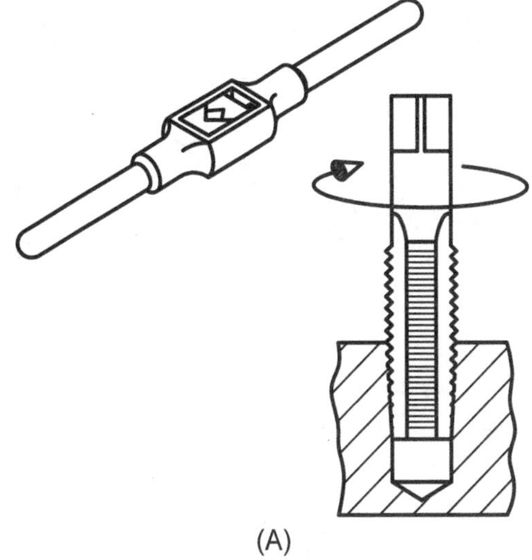

(A)

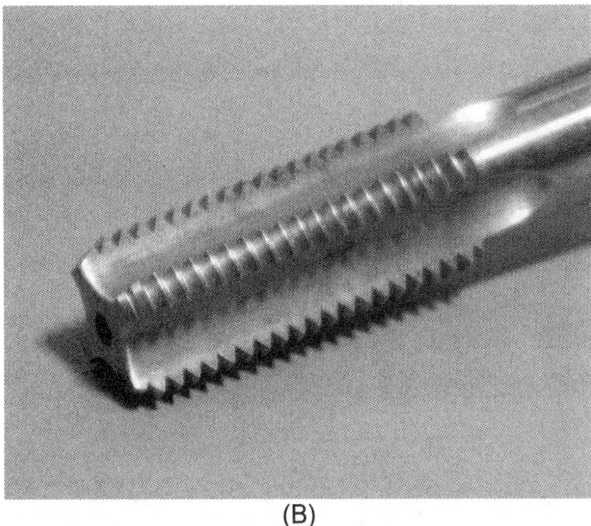

(B)

Figure THR 19: A) tap and tap handle for cutting interior threads, B) photo of a Tap.

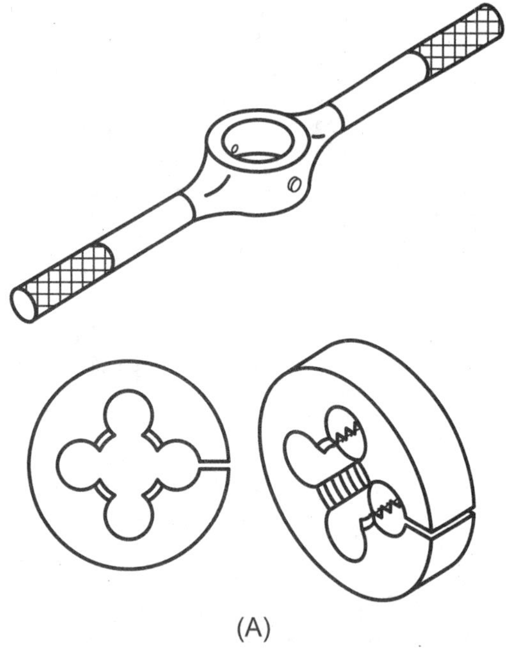

(A)

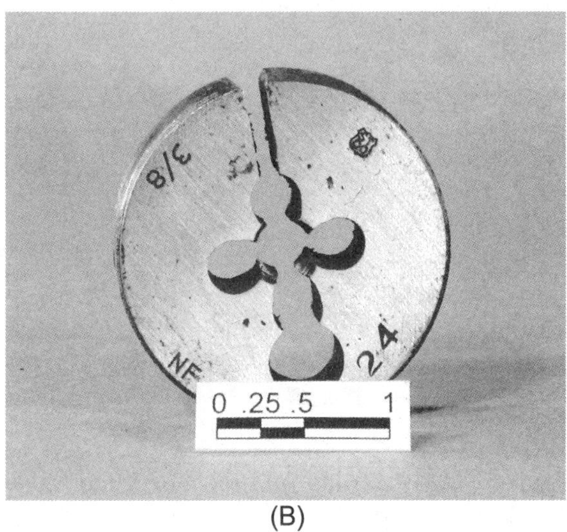

(B)

Figure THR 20: A) Illustration of a die for cutting exterior threads and a die stock for holding the die, B) photo of a 3/8-24 die.

If precision is not important, threads can be cast. Casting is a process in which a liquid material, such as metal, is poured into a mold. Plastic can also be used by softening it enough to inject it into a mold. The cap on a tub of toothpaste is an example of plastic injection molding (casting).

If precision is important, threads are cut with grinding wheels. The grinding wheel is in the shape of the thread profile being cut. Grinding is the most accurate method for cutting threads. Testing tools such as a thread plug gage used to check the accuracy of threads needs to be ground to insure accuracy.

Mass produced bolts are formed using a thread rolling process. Rolling can only be used on external threads. Rolling squeezes a cylinder under high pressure between grooved dies until threads are formed on the cylinder's surface. Figure THR 21 illustrates two rolling methods. In Figure THR 21A a cylindrical blank is positioned between two block rolling dies. One die is stationary while the other moves as indicated by the arrows in Figure THR 21A. The cylindrical blank rotates as it is being squeezed between the block dies which form the threads. In Figure THR 21B the cylindrical blank is positioned between the three cylindrical rolling dies which squeeze the threads onto the cylindrical blank.

The advantages of rolled threads over cut or ground threads are: 1) rolled threads are 10 to 20% stronger because rolling deforms the metal's grain structure rather than cutting through it, 2) the rolling process produces threads with harder surfaces which results in better wear resistance, 3) no stock is wasted.

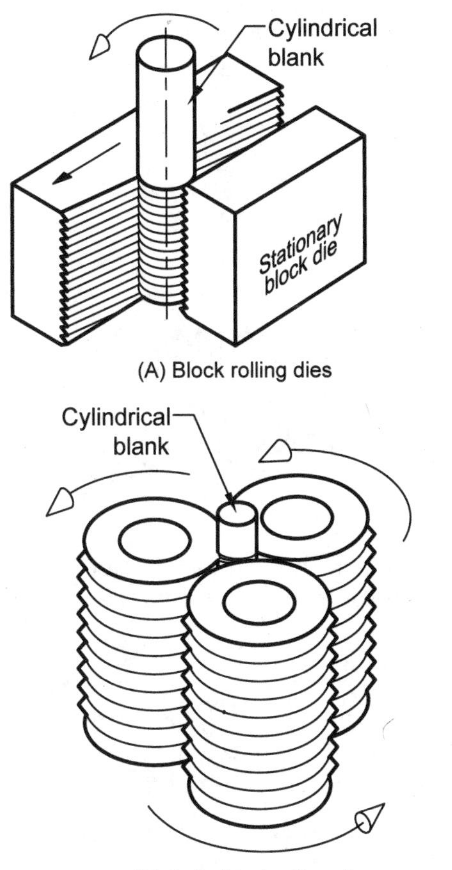

Figure THR 21: Rolling dies squeeze the thread into a cylindrical blank such as bolts.

Measuring Threads

A *screw pitch gage* is used to determine the threads per inch and pitch of a unified thread and the pitch of a metric M profile thread. The screw pitch gauge illustrated in Figure THR 22 has a series of sheet metal plates with a saw tooth pattern cut out for different threads per inch (or pitches for metric M profile). Determining the threads per inch is a two step operation. First, measure the major diameter to help narrow down the possible threads per inch. For example, if the micrometer reads slightly under .50", the possible threads per inch would be either: 13 UNC, 20 UNF, or 28UNEF, because those are the possible threads per inch for a .50" major diameter. Second, match the saw toothed edge that best fits the given threads as illustrated in Figure THR 23. Each screw pitch gage plate is labeled with the number of threads per inch and with the actual pitch, e.g., 8-.125.

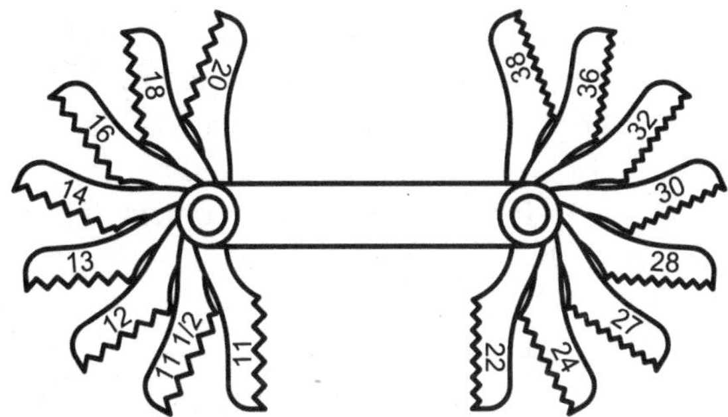

Figure THR 22: Inched based screw pitch gage

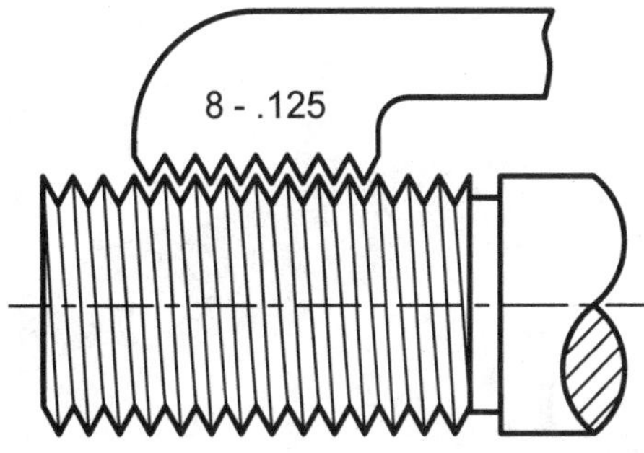

Figure THR 23: The saw toothed edge of the screw pitch gage is matched to the thread. This thread has 8 threads per inch and thus a pitch of 1/8 or .125".

A *threaded ring gage*, as illustrated in Figure THR 24, is used to check the accuracy of external threads. The "GO" end of the gage should screw on freely without excess play otherwise the pitch diameter is too large. The "NO-GO" end of the gage should start screwing on but stop after a few threads engage otherwise the pitch diameter is under specification.

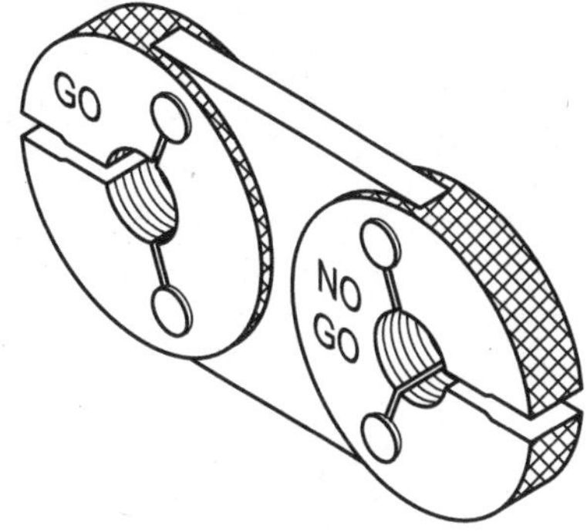

Figure THR 24: Threaded ring gage used to check the accuracy of exterior threads.

A *threaded plug gage,* as illustrated in Figure THR 25A and actual photograph in Figure 25B, is used to check the accuracy of internal threads. The "GO" end of the gage must enter the treaded hole its full length. The "NO-GO" end of the gage should not enter the hole or have a snug fit by the third thread. If the NO-GO end enters the hole farther, then the thread is oversized.

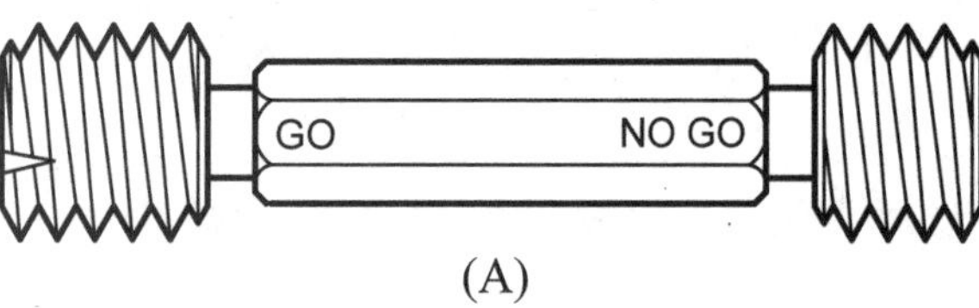

(A)

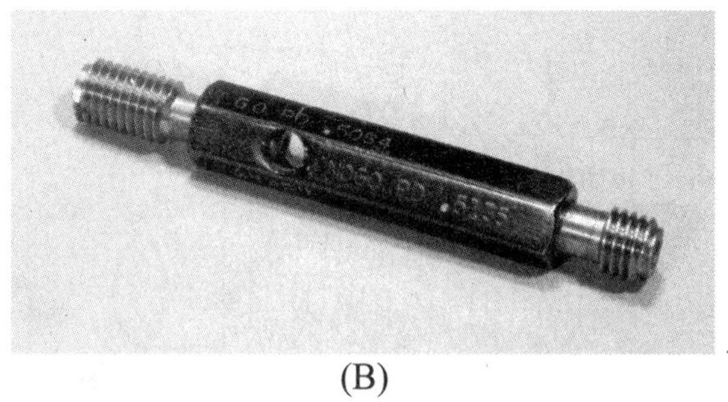

(B)

Figure THR 25: A) threaded plug gage used to check the accuracy of interior threads, B) Actual photograph of a GO-NO GO gage.

The best way to measure a thread is to use one of two micrometers. The first is a *pitch micrometer* illustrated in Figure THR 26. A pitch micrometer has a "W" shaped anvil and a cone headed spindle. The anvil fits over a crest and the spindle fits into a root opposite the anvil. The pitch diameter is then read. The second method or *three wire method* uses a normal micrometer with a flat headed anvil and spindle. As illustrated in Figure THR 27, two wires are placed into roots and a third wire is placed in the root opposite the first two. The micrometer takes the reading on the wires as illustrated in Figure THR 27. The diameter of a wire is calculated using the formula which produces a wire that is tangent at the pitch line:

W= 0.57735P

where: W = wire ∅

P = pitch of thread

The pitch diameter is determined using the following formula:

E= M + 0.86603P – 3W

where: E = pitch ∅

M = measurement over the wires

P = pitch of thread

W wire ∅

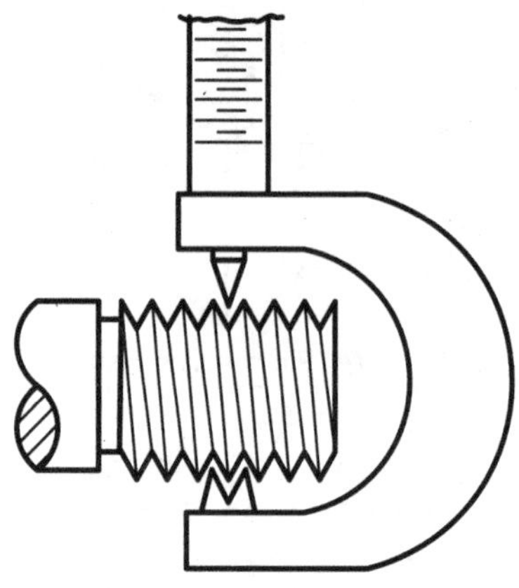

Figure THR 26: A pitch micrometer measuring the pitch diameter of an external thread.

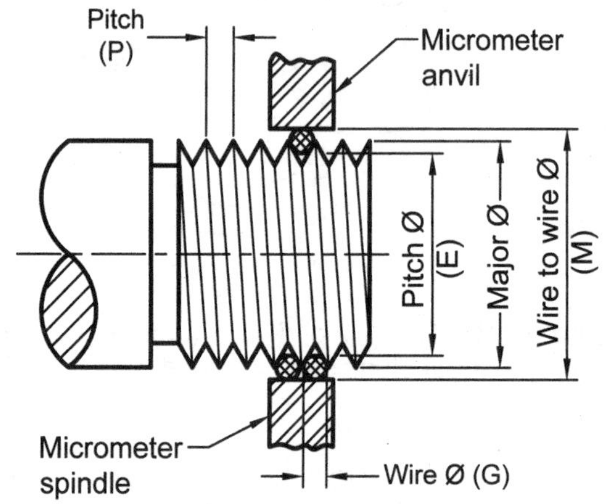

Figure THR 27: A normal micrometer measuring the pitch diameter using the three wire method.

Index:

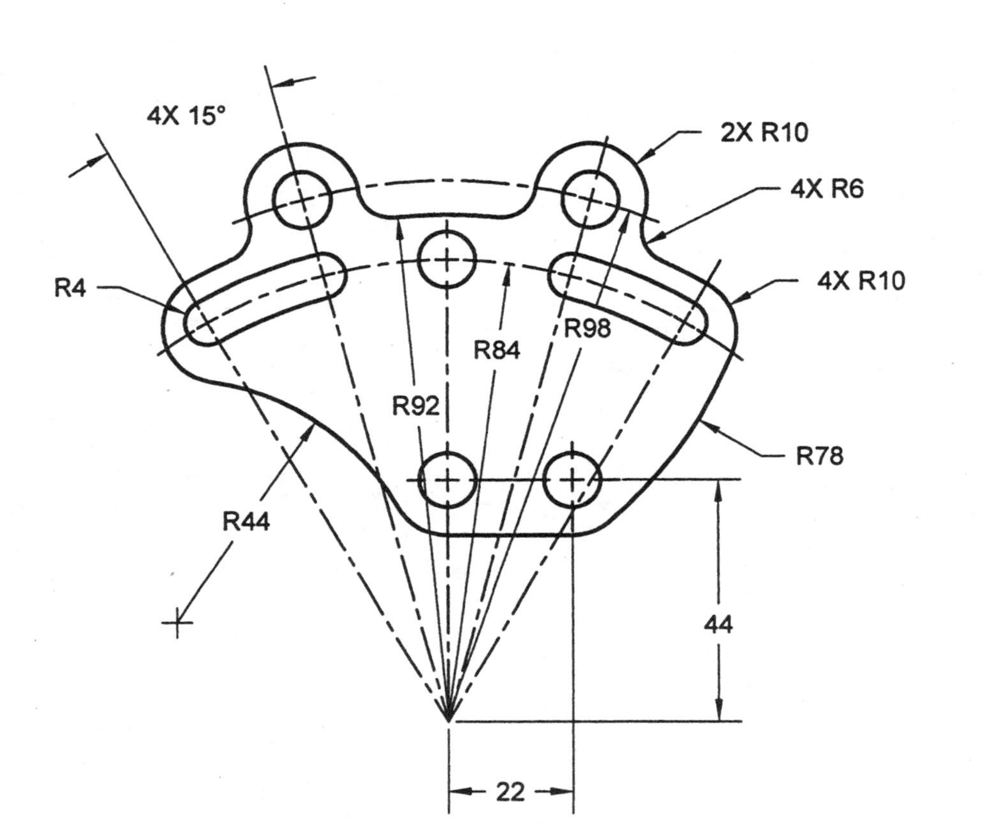

Draw Full Scale
Show All Center Lines
Do Not Show Dimension

GEOM CONSTRUCTION 2D-CAD	Name: ______ ID#: ______ Lab Hr: ______	METRIC 	GC-BA

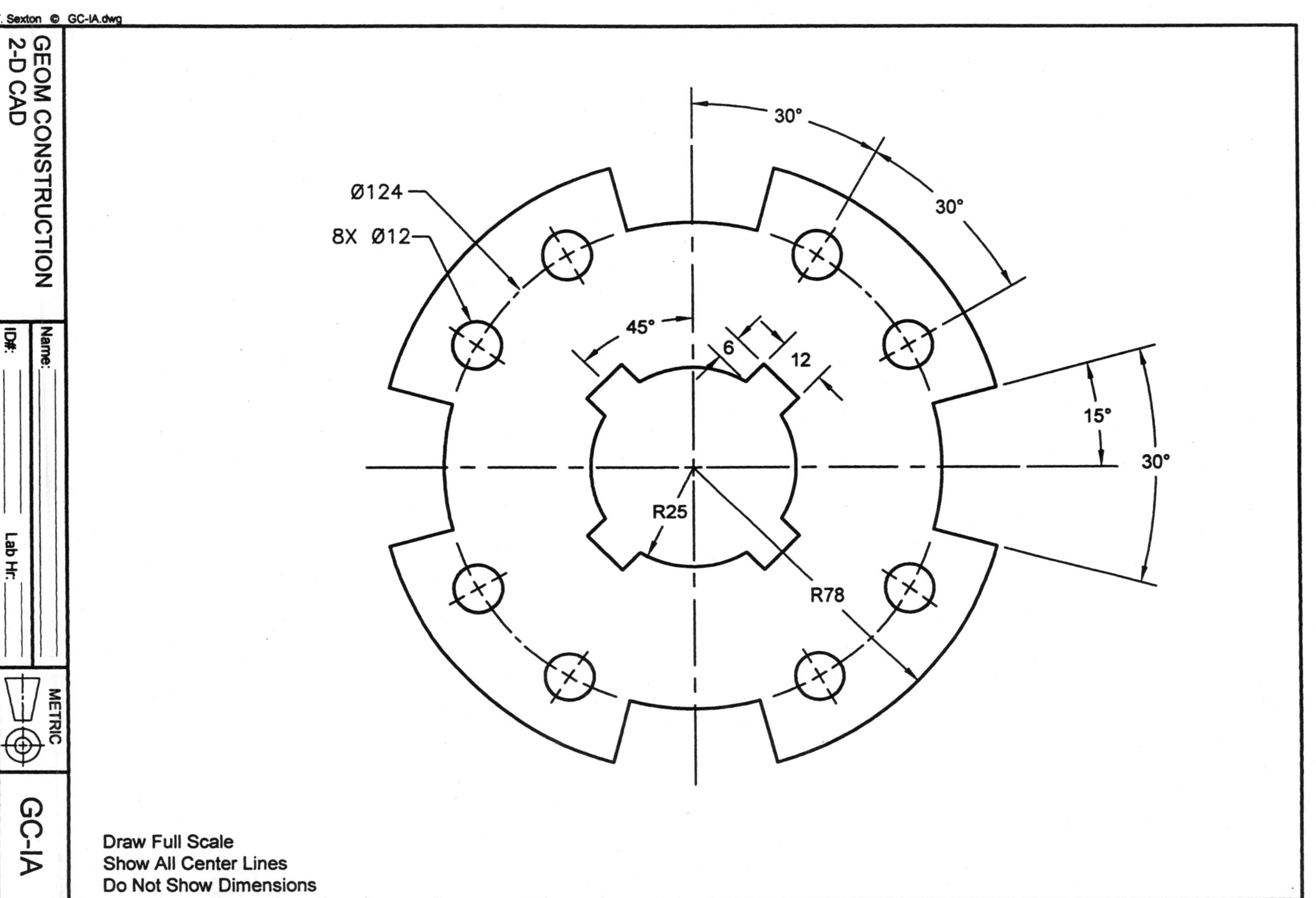
30°
Ø124
30°
8X Ø12
45°
6
12
15°
30°
R25
R78
Draw Full Scale
Show All Center Lines
Do Not Show Dimensions
T. Sexton © GC-IA.dwg
GEOM CONSTRUCTION
2-D CAD
Name:
ID#:
Lab Hr:
METRIC
GC-IA

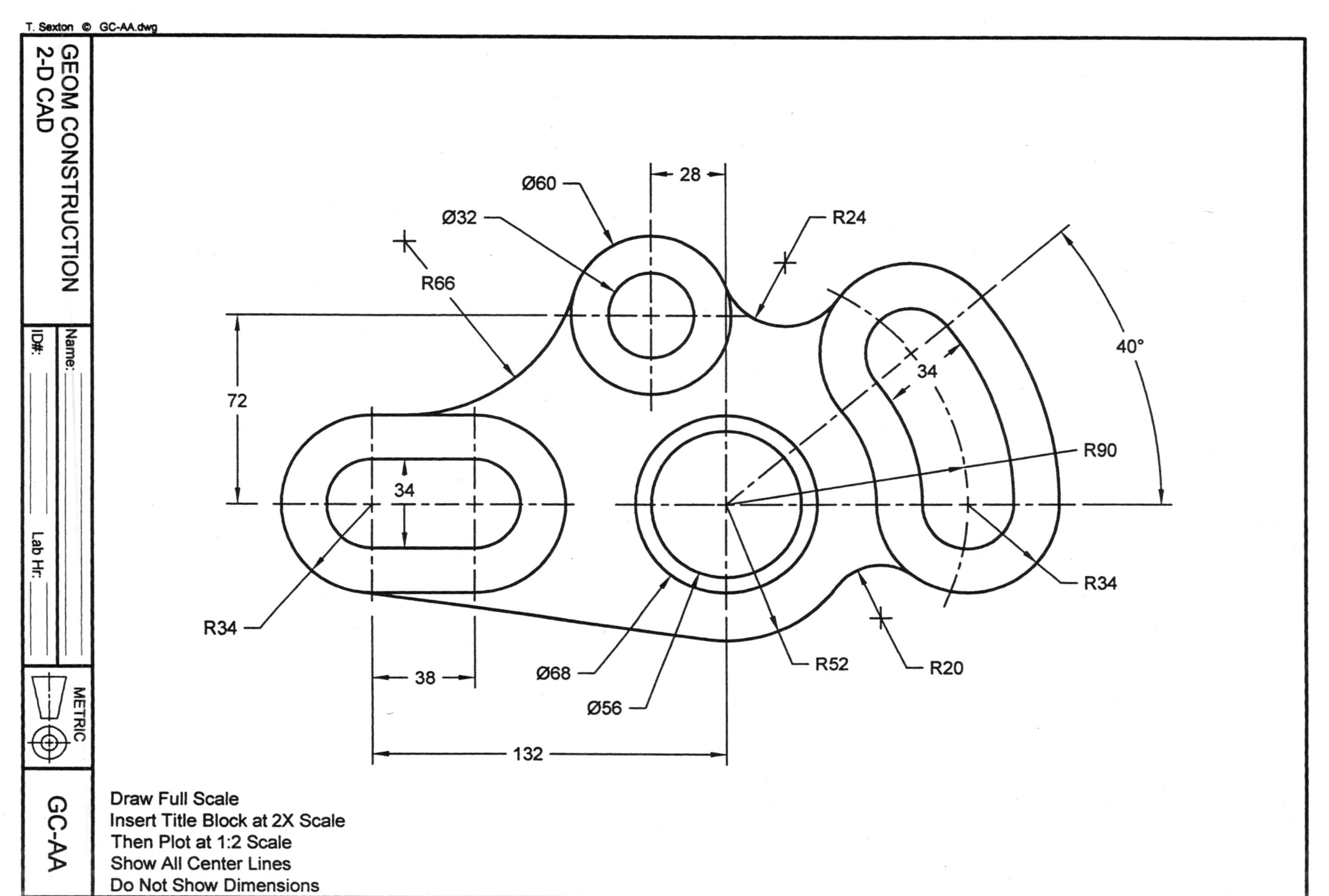

Ø60
28
Ø32
R24
R66
40°
34
72
34
R90
R34
R34
38
Ø68
Ø56
R52
R20
132
Draw Full Scale
Insert Title Block at 2X Scale
Then Plot at 1:2 Scale
Show All Center Lines
Do Not Show Dimensions
GEOM CONSTRUCTION
2-D CAD
Name:
ID#:
Lab Hr:
METRIC
GC-AA
T. Sexton © GC-AA.dwg

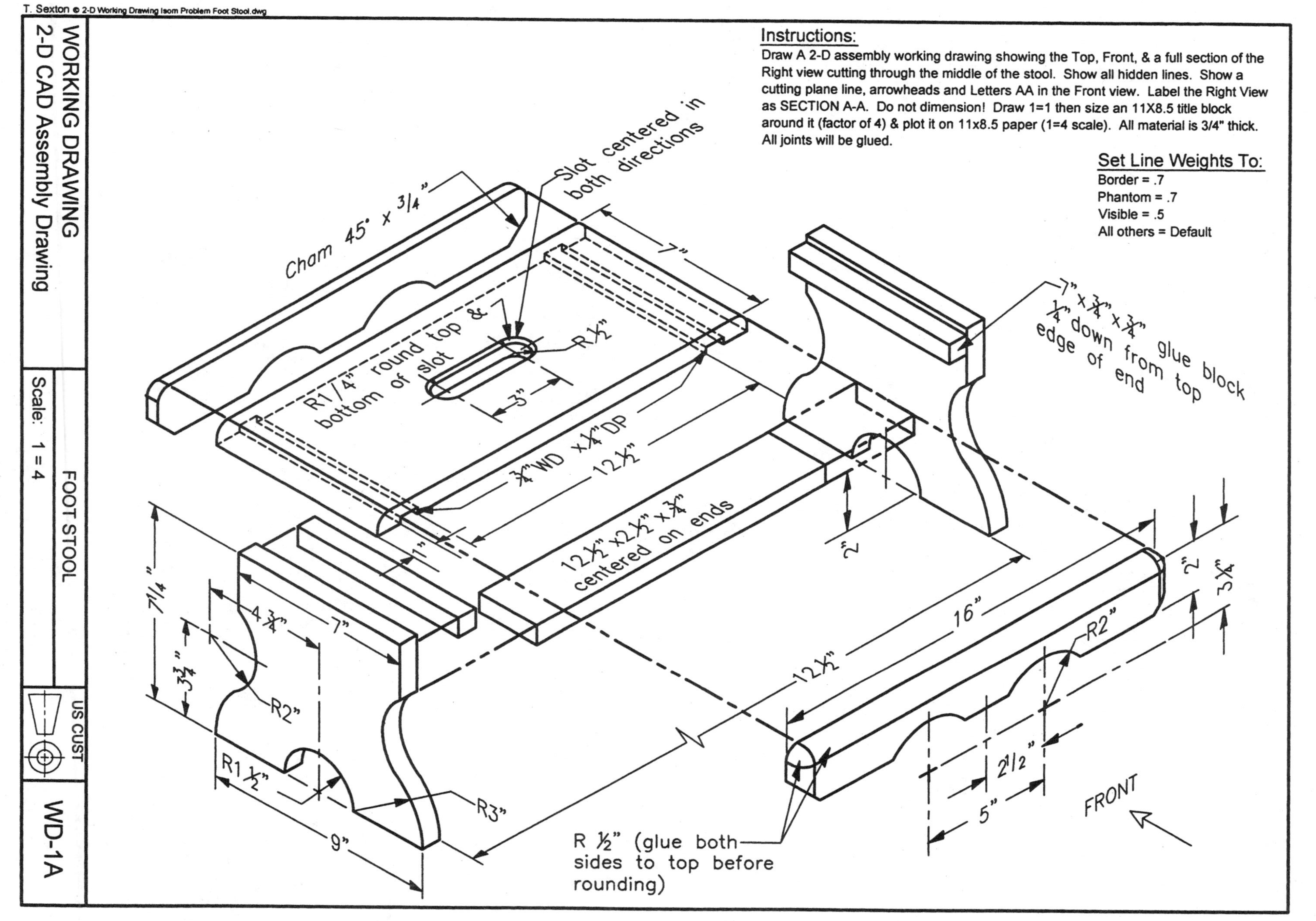
T. Sexton © 2-D Working Drawing Isom Problem Foot Stool.dwg
Instructions:
Draw A 2-D assembly working drawing showing the Top, Front, & a full section of the Right view cutting through the middle of the stool. Show all hidden lines. Show a cutting plane line, arrowheads and Letters AA in the Front view. Label the Right View as SECTION A-A. Do not dimension! Draw 1=1 then size an 11X8.5 title block around it (factor of 4) & plot it on 11x8.5 paper (1=4 scale). All material is 3/4" thick. All joints will be glued.
Set Line Weights To:
Border = .7
Phantom = .7
Visible = .5
All others = Default
Cham 45° x 3/4"
Slot centered in both directions
7"
R1/4" round top & bottom of slot
R½"
3"
¾"WD x¼"DP
12½"
1"
12½" x2½" x¾" centered on ends
2"
7" x¾" x¾" glue block ¼" down from top edge of end
7¼"
3¾"
4¾"
7"
R2"
R1½"
R3"
9"
R ½" (glue both sides to top before rounding)
16"
12½"
R2"
2½"
5"
2"
3¾"
FRONT
WORKING DRAWING
2-D CAD Assembly Drawing
FOOT STOOL
Scale: 1 = 4
US CUST
WD-1A

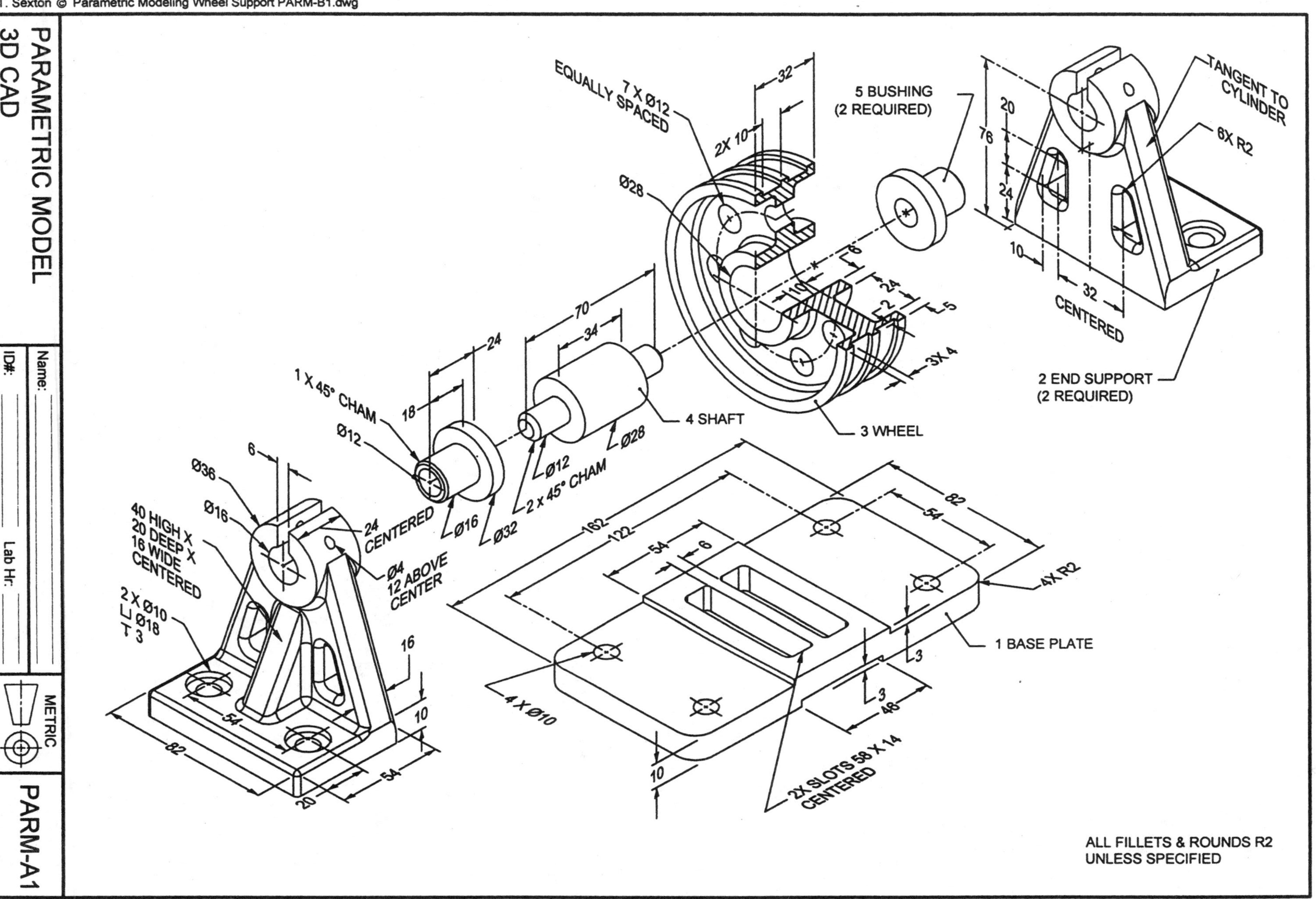

PARAMETRIC MODEL 3D CAD	Name: ______ ID#: ______ Lab Hr: ______	METRIC	PARM-A1

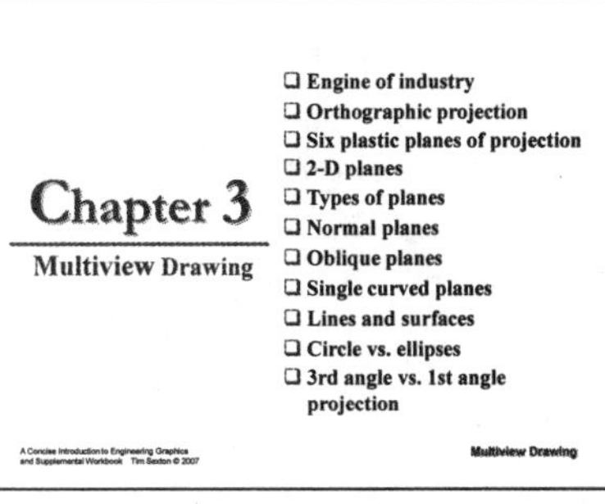
Chapter 3
Multiview Drawing
❑ Engine of industry
❑ Orthographic projection
❑ Six plastic planes of projection
❑ 2-D planes
❑ Types of planes
❑ Normal planes
❑ Oblique planes
❑ Single curved planes
❑ Lines and surfaces
❑ Circle vs. ellipses
❑ 3rd angle vs. 1st angle projection
Multiview Drawing

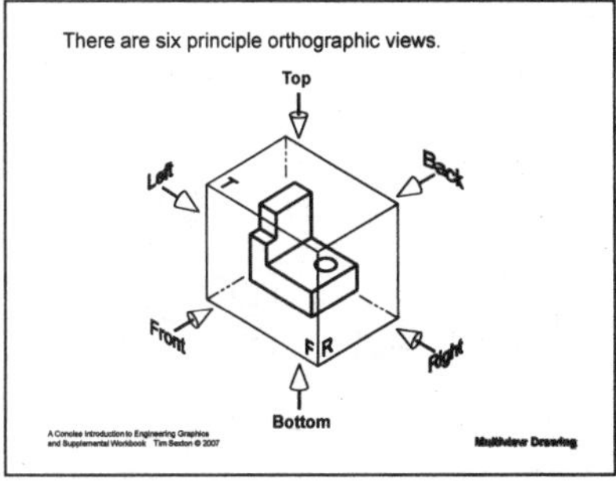
There are six principle orthographic views.
Top
Left
Back
Front
Right
Bottom
Multiview Drawing

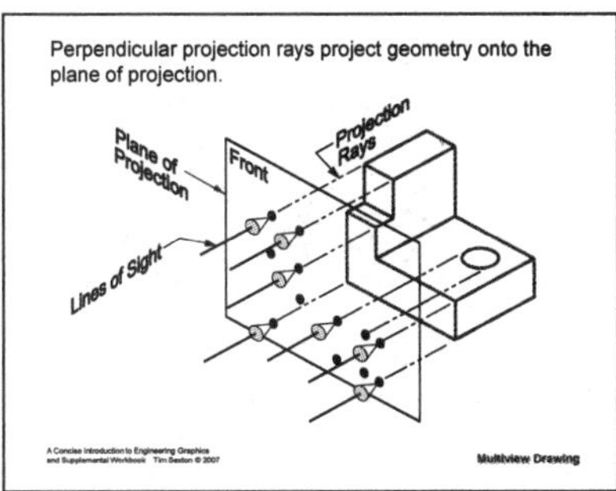
Perpendicular projection rays project geometry onto the plane of projection.
Plane of Projection
Front
Projection Rays
Lines of Sight
Multiview Drawing

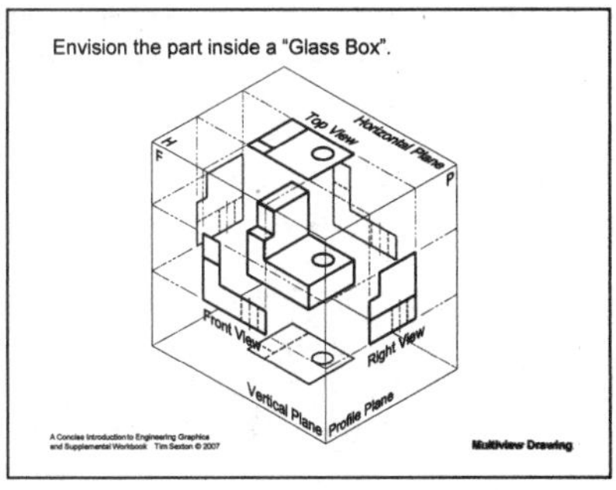
Envision the part inside a "Glass Box".
Top View
Horizontal Plane
Front View
Right View
Vertical Plane
Profile Plane
Multiview Drawing

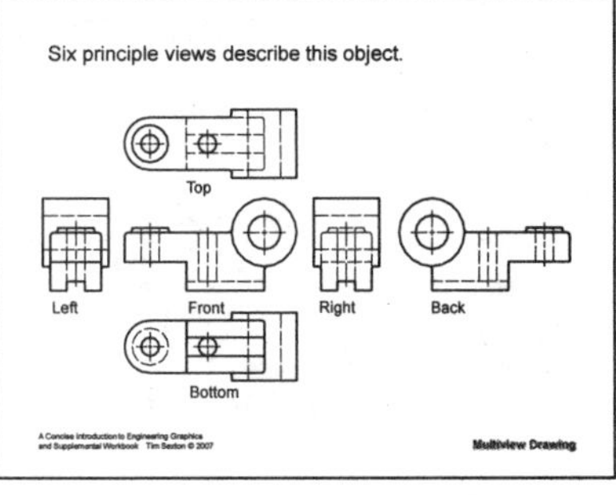
Six principle views describe this object.
Top
Left
Front
Right
Back
Bottom
Multiview Drawing

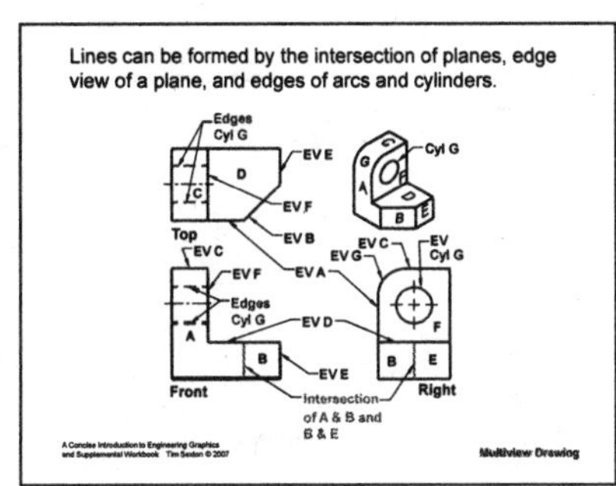
Lines can be formed by the intersection of planes, edge view of a plane, and edges of arcs and cylinders.
Edges Cyl G
EV E
Cyl G
EV F
EV B
Top
EV C
EV A
EV F
Edges Cyl G
EV D
EV E
Front
Intersection of A & B and B & E
Right
Multiview Drawing

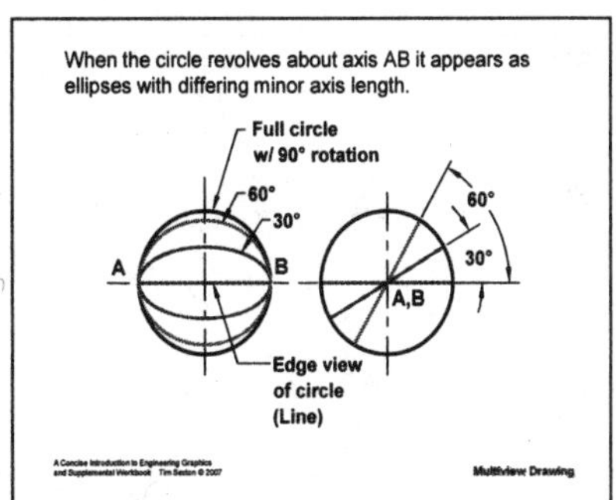
When the circle revolves about axis AB it appears as ellipses with differing minor axis length.
Full circle w/ 90° rotation
60°
30°
A
B
A,B
Edge view of circle (Line)
Multiview Drawing

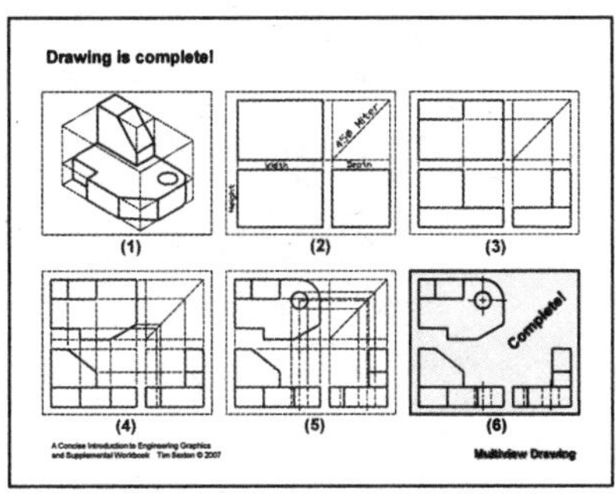
Drawing is complete!
(1)
(2)
(3)
(4)
(5)
(6)
Complete!
Multiview Drawing

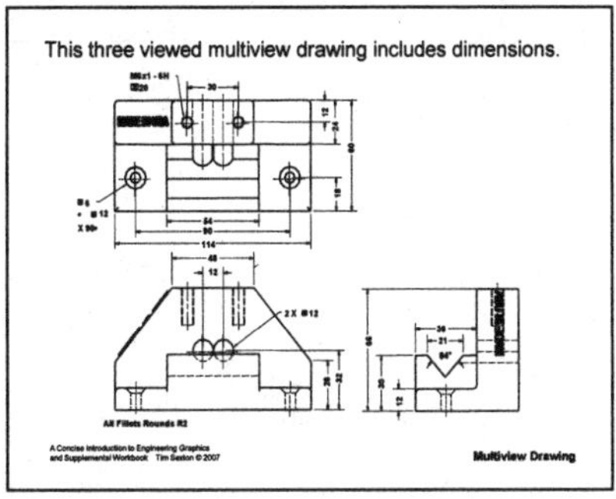
This three viewed multiview drawing includes dimensions.
All Fillets Rounds R2
Multiview Drawing

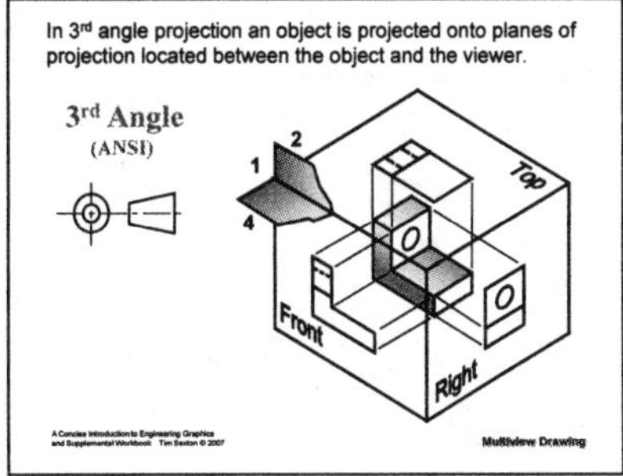
In 3rd angle projection an object is projected onto planes of projection located between the object and the viewer.
3rd Angle
(ANSI)
Top
Front
Right
Multiview Drawing

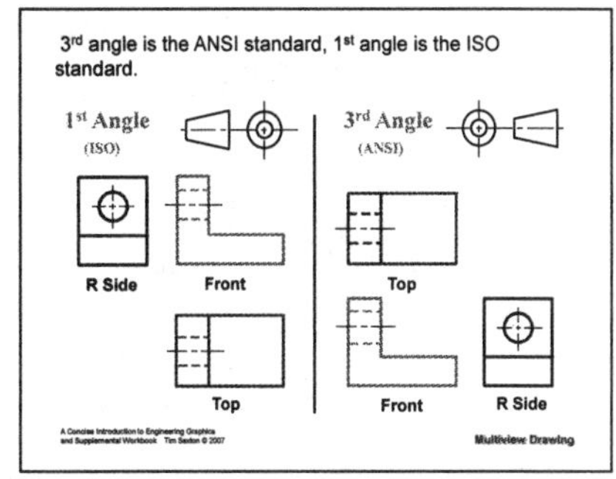
3rd angle is the ANSI standard, 1st angle is the ISO standard.
1st Angle
(ISO)
3rd Angle
(ANSI)
R Side
Front
Top
Top
Front
R Side
Multiview Drawing

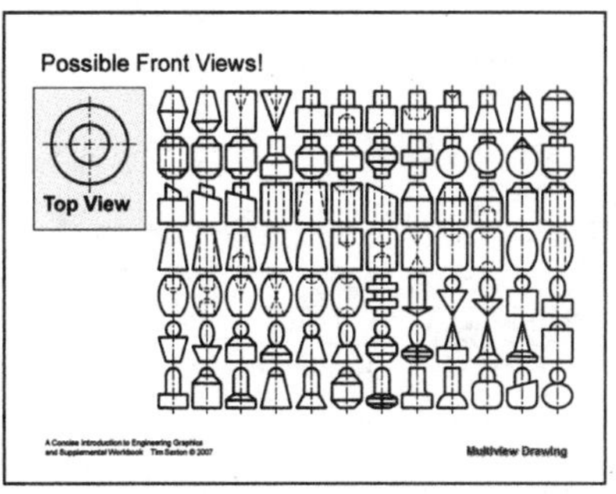
Possible Front Views!
Top View
Multiview Drawing

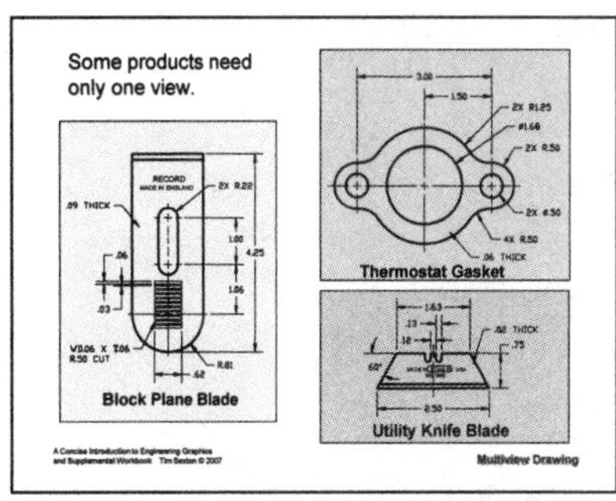
Some products need only one view.
Thermostat Gasket
Block Plane Blade
Utility Knife Blade
Multiview Drawing

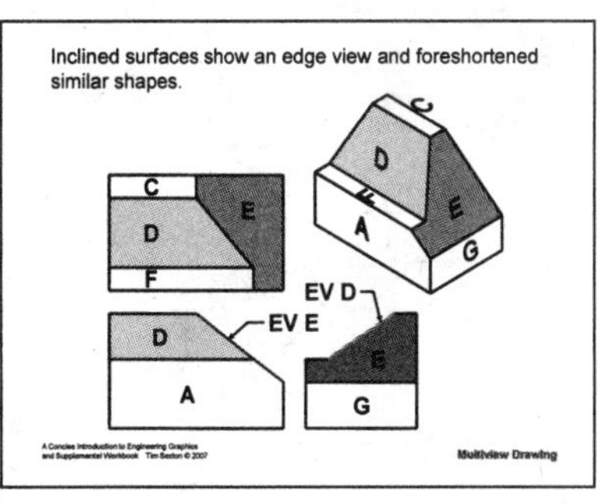
Inclined surfaces show an edge view and foreshortened similar shapes.
C
D
E
F
A
G
EV D
EV E
Multiview Drawing

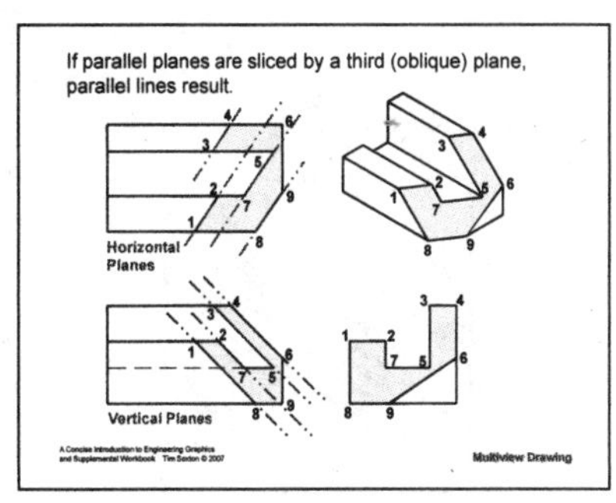
If parallel planes are sliced by a third (oblique) plane, parallel lines result.
Horizontal Planes
Vertical Planes
Multiview Drawing

Sample Slides

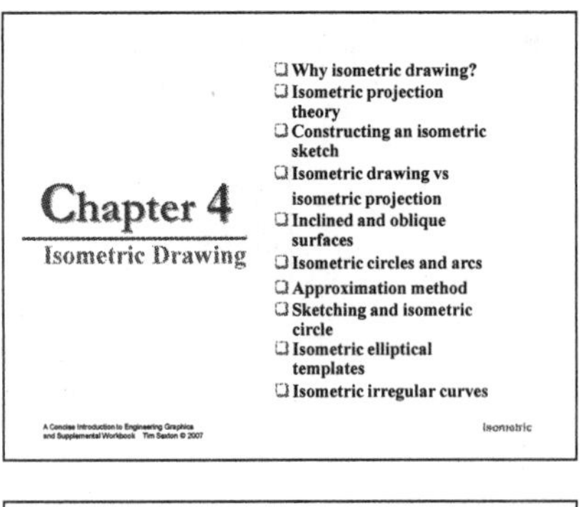

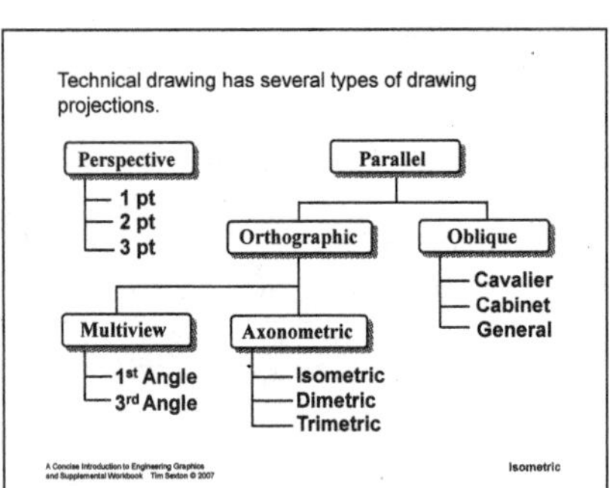

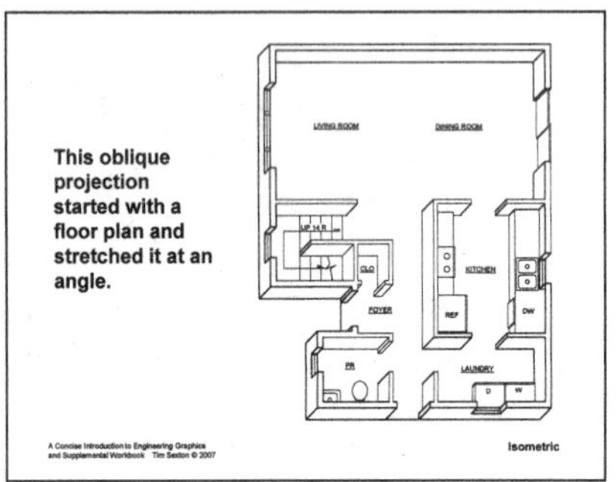

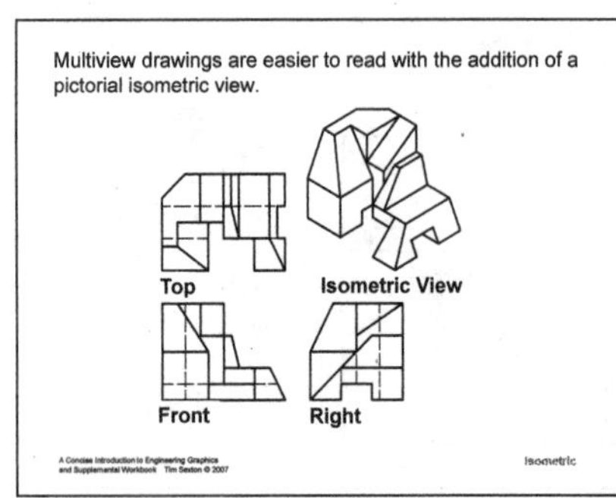

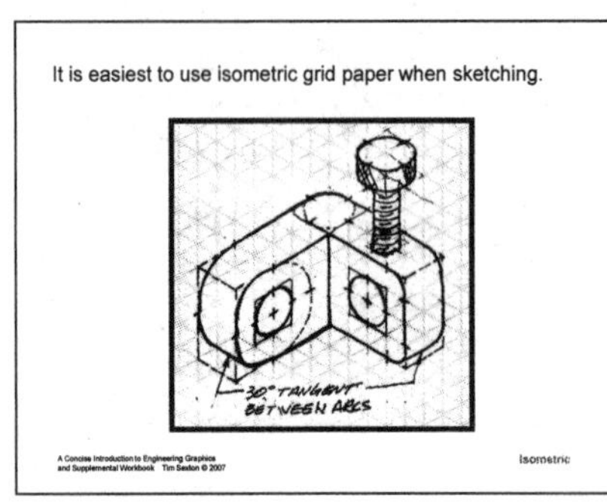

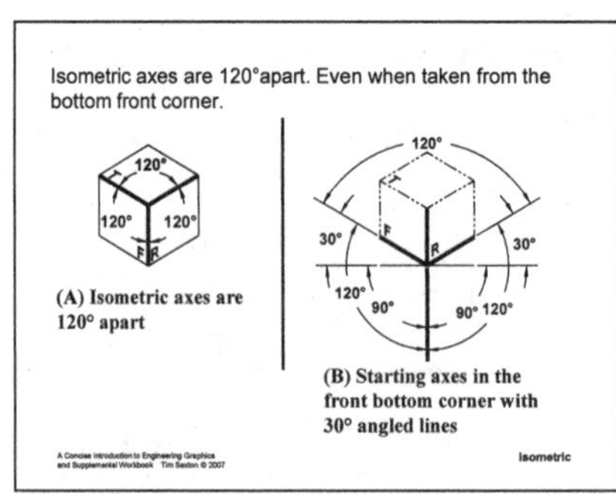

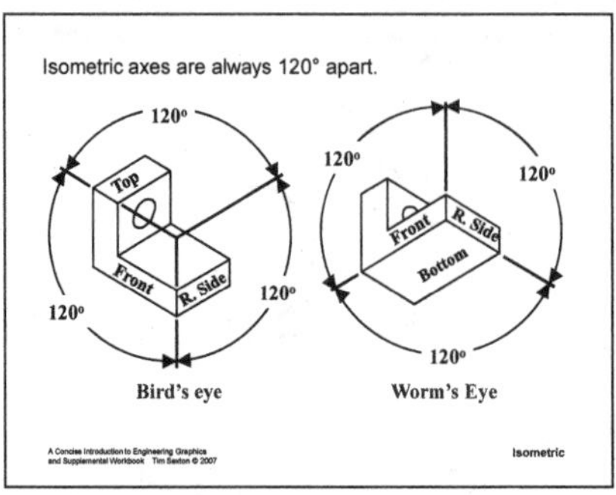

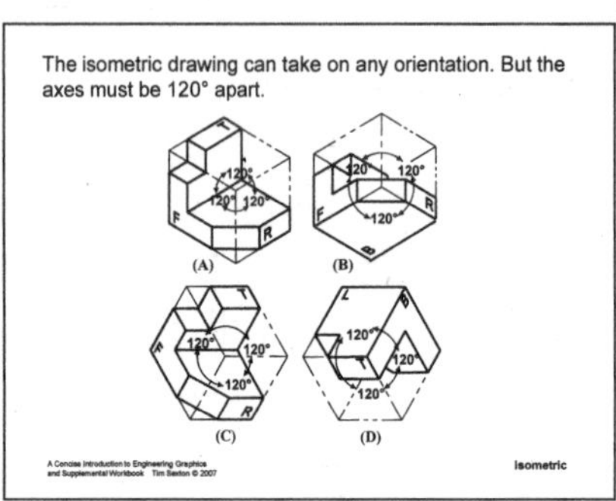

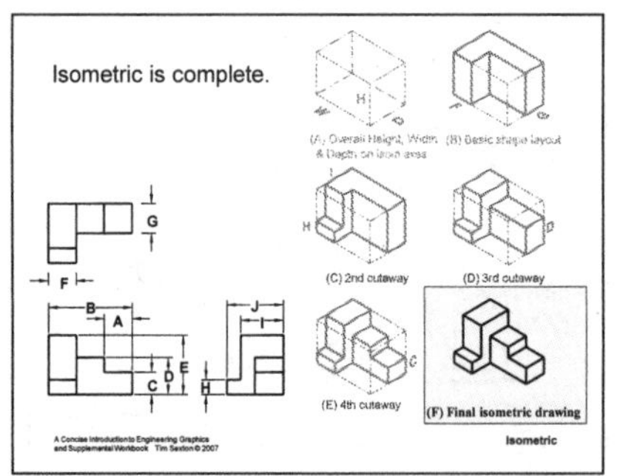

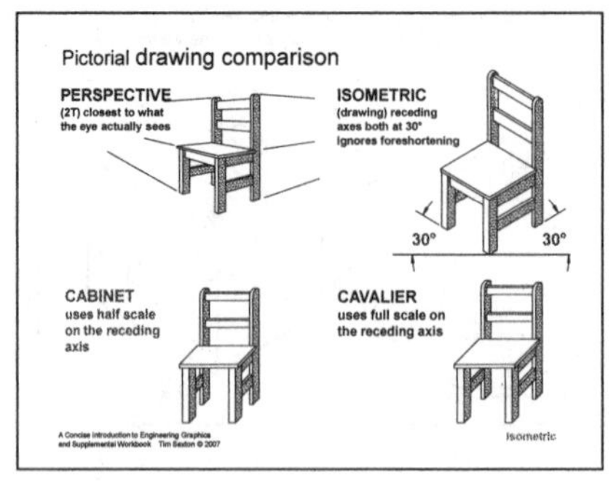

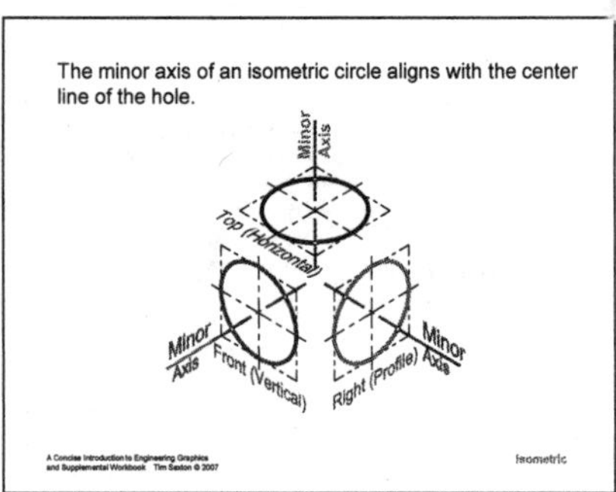

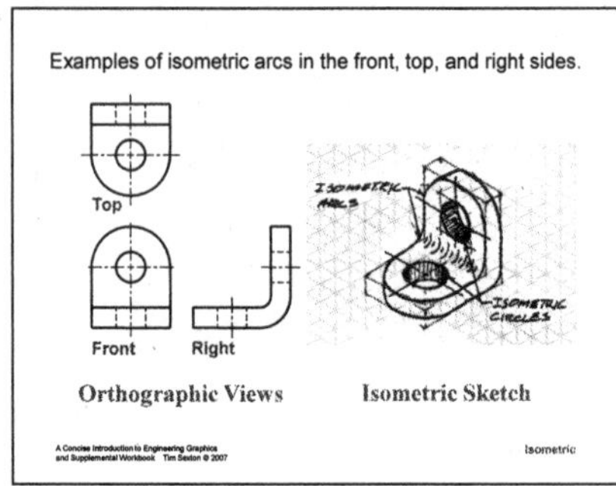

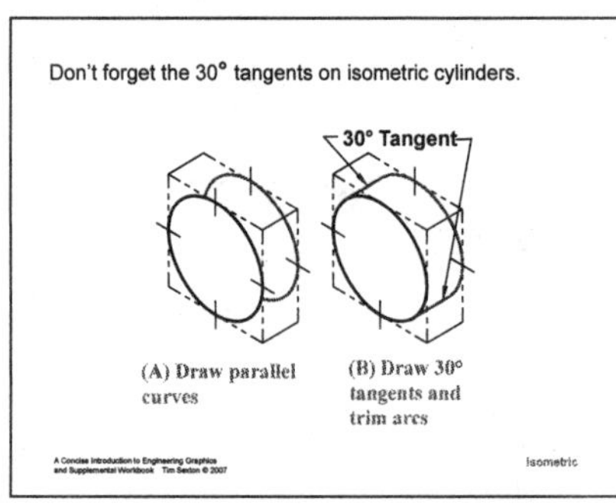

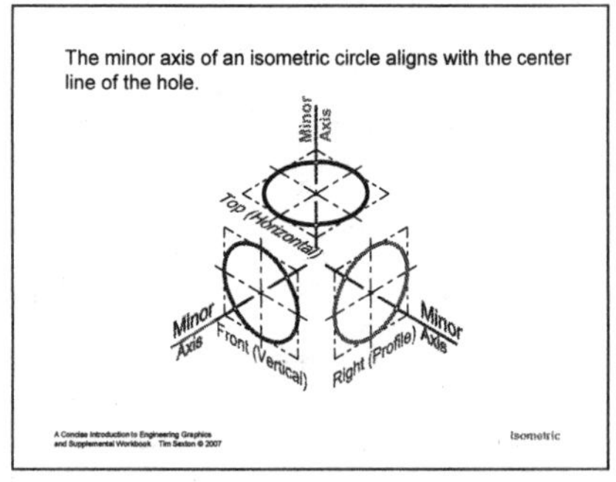

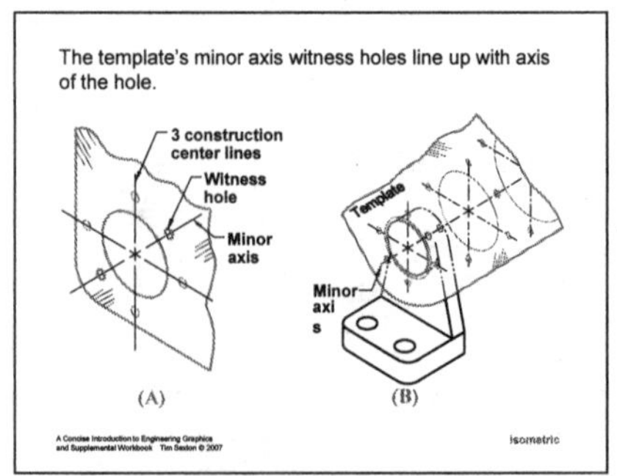